Contemporary Issues in Sustainable Development

This book analyzes different perspectives around sustainable development, risk management and managing demand across various sectors in India. Diverse theories and analytical methods from various disciplines, as well as case studies, are brought together to present an in-depth study.

The book discusses the challenges of achieving sustainability, the role of quantitative research to assess current scenarios, and the role of policy making to bring improvements in the Indian context. It examines the socioeconomic ways of pursuing sustainable development in the areas of agriculture, climate change and energy; the environment and natural resources; health and society. It also analyzes important quantitative models for sustainability policy analysis and provides case studies to understand the practical implementations of the models.

This book will be a great reference manual that covers a whole gamut of analytical techniques that are useful for students, research scholars and practitioners of economics, environmental studies, development studies, sociology, South Asian studies and public policy, among others.

Tanika Chakraborty is Associate Professor of Economics at Indian Institute of Management (IIM) Calcutta. She is also affiliated with IZA Bonn and CESifo Munich. Her research interests are in the areas of education, labor and development economics.

Deep Mukherjee is Associate Professor in Department of Economic Sciences at Indian Institute of Technology (IIT) Kanpur. His research interests are in the areas of agricultural economics and public policy.

Sarani Saha is Professor in the Department of Economic Sciences, Indian Institute of Technology Kanpur. Her research interests are in the fields of labor economics and environmental economics.

Contemporary Issues in Sustainable Development

The Case of India

Edited by Tanika Chakraborty, Deep Mukherjee and Sarani Saha

Routledge
Taylor & Francis Group
LONDON AND NEW YORK

First published 2021
by Routledge
2 Park Square, Milton Park, Abingdon, Oxon OX14 4RN

and by Routledge
52 Vanderbilt Avenue, New York, NY 10017

Routledge is an imprint of the Taylor & Francis Group, an informa business

British Library Cataloguing-in-Publication Data
A catalogue record for this book is available from the British Library

Library of Congress Cataloging-in-Publication Data
Names: Chakraborty, Tanika, editor. | Mukherjee, Deep, editor. | Saha, Sarani, editor.
Title: Contemporary issues in sustainable development : the case of India / edited by Tanika Chakraborty, Deep Mukherjee, and Sarani Saha.
Description: Abingdon, Oxon ; New York, NY : Routledge, 2020. | Includes bibliographical references and index.
Identifiers: LCCN 2020037457 (print) | LCCN 2020037458 (ebook) | ISBN 9781138227101 (hardback) | ISBN 9781003141020 (ebook)
Subjects: LCSH: Sustainable development--India.
Classification: LCC HC440.E5 C6726 2020 (print) | LCC HC440.E5 (ebook)| DDC 338.954/07--dc23
LC record available at https://lccn.loc.gov/2020037457
LC ebook record available at https://lccn.loc.gov/2020037458

ISBN: 978-1-138-22710-1 (hbk)
ISBN: 978-0-367-69231-5 (pbk)
ISBN: 978-1-003-14102-0 (ebk)

Typeset in Sabon
by SPi Global, India

Visit the eResources: www.routledge.com/9781138227101

Contents

Figures

Tables

Contributors

Ayush Agrawal (BS–MS [Dual Degree], IIT Kanpur) currently works at PricewaterhouseCoopers as a business analyst. His research interests include topics in applied microeconomics.

Bhagirath Behera (PhD, University of Bonn) is a professor of economics at the Department of Humanities and Social Sciences, IIT Kharagpur. His research interests include environmental and natural resource economics.

Anup Kumar Bhandari (PhD, Indian Statistical Institute) is an associate professor in the Department of Humanities and Social Sciences at IIT Madras. His research interests are in production economics and industrial economics.

Tanika Chakraborty (PhD, Washington University in St Louis) is an associate professor of Economics at IIM Calcutta and a research fellow at IZA Bonn. Her research interests are in the areas of education, labor and development economics.

Debalina Chakravarty (PhD, Jadavpur University) is currently an assistant professor of economics at the Department of Economics and International Business, School of Business Studies, Sharma University, Greater Noida. Her broad research areas are energy economics and urban economics.

Priyanka Chatterjee (PhD, Jawaharlal Nehru University) is a programme coordinator in Public Health Resource Network. Her broad research areas are labor economics and health economics.

Basudeb Chaudhuri (PhD, University of Paris I Panthéon-Sorbonne) is Seconded National Expert from the French government (University of Caen) to the European Commission, Directorate General of Research and Innovation. He has expertise in qualitative social research.

M. Sandeep Chowdary is an MTech. student in the Department of Industrial Management and Engineering of IIT Kanpur.

Nandini Das (PhD, Jadavpur University) is a Post-Doctoral Researcher at Global Change Programme-Jadavpur University. She works in the area of energy transition.

Shyamasree Dasgupta (PhD, Jadavpur University) is an assistant professor in the School of Humanities and Social Sciences at IIT Mandi. Her research interest encompasses energy economics and climate change.

Rohini Dutta (Ghosh) (PhD, Indian Statistical Institute) is an epidemiologist and interested in health policy issues. She is presently working as a consultant at IIT Kanpur.

Rakhshanda F. Fazli (PhD, Aligarh Muslim University) is a professor in the department of West Asian and North African Studies, Aligarh Muslim University. Her research interest lies in population geography, remote sensing and GIS.

Vikash Gautam (PhD, Indira Gandhi Institute of Development Research) is working at the Koan Advisory Group, New Delhi, as a lead economist. His research focuses on development economics and corporate finance.

Namrata Gulati (PhD, Indian Statistical Institute) is an assistant professor of economics at South Asian University, New Delhi. Her academic interest is in the area of health economics.

Aarti Gupta (PhD, IIT Kanpur) holds a PG diploma in business studies from Harvard University. Currently she is the chief investment officer at DBR Ventures.

Anjula Gurtoo (Fellow, IIM Ahmedabad) is a professor in the Department of Management Studies, Indian Institute of Science Bangalore. Her research activities are in the fields of behavioral science, informal economy, infrastructure and development.

Mohd Irfan (PhD, IIT Kanpur) is an assistant professor in the Department of Management Studies, IIT(ISM) Dhanbad. His research interests are focused on energy and environmental economics.

Karan Singh Khati (MBA, Kumaun University) is currently a senior research fellow pursuing a PhD in Department of Economic Sciences, IIT Kanpur. His research interests are in areas of banking and efficiency analysis.

Abhishek Kumar (BS-MS [Dual], IIT Kanpur) is a business analyst at JP Morgan Chase & Co. He is interested in applied econometrics.

Chitwan Lalji (PhD, IIT Kanpur) is an assistant professor of economics at IIM Kozhikode. Her research interests are in the areas of health and behavioral economics.

Rajlakshmi Mallik (PhD, Indian Statistical Institute) is the director of the Centre for Development Research, Sustainability and Technical Advancement (C-DRAṢṬĀ). She has interest in policy-oriented developmental issues.

Aparna Mishra (PhD, IIT Kanpur) is a senior manager with Ernst & Young (EY), Global Delivery Services (GDS), Bangalore.

Pulak Mishra (PhD, Vidyasagar University) is a professor of economics in the Department of Humanities and Social Sciences, IIT Kharagpur. His research interests include public policy and economics of rural development.

Deep Mukherjee (PhD, University of Connecticut) is an associate professor in the Department of Economic Sciences at IIT Kanpur. His research interests are in areas of public policy, agricultural and environmental economics.

Diganta Mukherjee (PhD, Indian Statistical Institute) is associate professor at the Sampling and Official Statistics Unit, Indian Statistical Institute, Kolkata. His research interests are in welfare economics and finance.

Debayan Pakrashi (PhD, Queensland University) is an assistant professor in the Department of Economic Sciences, IIT Kanpur. He is interested in policy issues related to health, labor and development economics.

Nitin Punit (PhD, University of Delhi) is an assistant professor in Shivaji College, University of Delhi. His research areas are sustainable regional development, remote sensing and GIS.

Dil Bahadur Rahut (PhD, University of Bonn) is a senior global program manager at the International Maize and Wheat Improvement Center. His areas of research are development and natural resource economics.

Naveen B. Ramu (PhD, Indian Institute of Science Bangalore) is a postdoctoral research fellow at the Department of Management Studies, IISc Bangalore. His research focuses on issues regarding public transport.

Thiagu Ranganathan (PhD, IIT Bombay) is an associate professor at the Centre for Development Studies, Thiruvananthapuram. He works in the areas of agricultural economics and development economics.

Joyashree Roy (PhD, Jadavpur University) is the Bangabandhu Chair Professor at the Asian Institute of Technology, Bangkok. Her subjects of expertise are energy economics, economics of climate change and environmental economics.

Sarani Saha (PhD, University of California Santa Barbara) is a professor in the Department of Economic Sciences, IIT Kanpur. Her research interests are labor economics and environmental economics.

Shreejata Samajpati (PhD, University of Central Florida) is an analytics professional at Full Life Care, Seattle, a Washington state–contracted lead agency for implementing the Medicaid Health Homes Demonstration Program. Her specialization is in health economics and policy.

Amey Sapre (PhD, IIT Kanpur) is an assistant professor at the National Institute of Public Finance and Policy, New Delhi. He works in areas of national accounts statistics and public finance.

Kaushal Kumar Saxena (PhD, University of Udaipur) is a retired professor of economics from IIT Kanpur. He was an Indo-American Fulbright Fellow and has expertise in input–output analysis.

Anjan Sen (PhD, IIT Delhi) is an assistant professor in the Department of Geography, University of Delhi. His research areas are New Economic Geography and Regional Development and Planning.

Debarun Sengupta (MA, Banaras Hindu University) is a doctoral student in Department of Economic Sciences, IIT Kanpur. His research interests are in the areas of production economics and environmental economics.

Arun Kumar Sharma (PhD, IIT Bombay) is currently serving as Emeritus Fellow in sociology in the Department of Humanities and Social Sciences at IIT Kanpur. He is interested in demography and rural sociology.

Anoop Singh (PhD, Indira Gandhi Institute of Development Research, Mumbai) is a professor in the Department of Industrial Management and Engineering at IIT Kanpur. His areas of interest include power market development, energy economics and policy.

Mahendra Kumar Singh (BS-MS [Dual Degree], IIT Kanpur) is currently pursuing PhD in Economics at Iowa State University. He is interested in agricultural economics and policy issues.

Sanjay Kumar Singh (PhD, Indira Gandhi Institute of Development Research) is the professor in the business environment area at IIM Lucknow. His research interests are focused on transportation economics and energy economics.

Mohammad Suhail (PhD, Aligarh Muslim University) is an assistant professor in the Department of West Asian and North African Studies, Aligarh Muslim University. His area of interest includes natural resource management and GIS applications.

Acknowledgements

The editors would like to thank the sponsor and all the invited speakers for the event of a Quality Improvement Programme workshop on "Sustainable Development: Policy & Management", held at IIT Kanpur during the period of 30 November 2015 to 5 December 2015. The idea of this book came up after listening to all the presentations on such diverse topics using various methodologies. Therefore, a significant number of the chapters of the book are authored by the various speakers of the event. We are grateful to the All India Council for Technical Education (AICTE) for sponsoring this event and also IIT Kanpur for facilitating the conduct of the workshop.

Introduction

Over the last two decades, *sustainability* has become a keyword in the field of development studies and economic policy making dialogues across the world. Sustainability is a broad discipline that encompasses the disciplines from natural sciences like conservation biology, ecology, hydrology, geography, engineering fields such as architecture, urban and transport planning, and various social sciences fields like demography, politics, economics, and sociology. According to the Brundtland Report for the United Nations World Commission on Environment and Development, development is sustainable if it "meets the needs of the present without compromising the ability of future generations to meet their own needs" (Brundtland et al., 1987). In this modern world, which is influenced by consumerism and dominated by an urban pattern of existence, a lot of natural resources are consumed daily. It has been estimated that the consumption of resources is 40% more than the volume that can be renewed. Hence, the concepts of sustainability or sustainable development focus on maintaining the balance between technological changes and consequent economic changes as well as the importance of protecting the environment in which we live in. To be precise, sustainability is the practice of protecting not just our natural resources but also human and ecological health through new and innovative methods and at the same time without any compromise on the way of life.

In 2005, the World Summit on Social Development identified (a) economic development, (b) social development and (c) environmental protection as the three pillars of sustainability. *Economic development* is the most complex issue as people disagree on political and economic terms as to how policies implemented will affect business, profitability, jobs and employability. Economic development is about fulfilling people's needs without any compromise in their quality of life, thereby reducing the cost on their part of protecting the environment. *Social development* is a multifaceted concept, and it is mainly about educating people on the effects of environmental protection as well as the dangers associated in case we fail to do so and therefore encouraging them to participate in environmental sustainability. *Environmental protection* refers to protecting the environment through the reuse, recycling and reduction of the consumption of power to the maximum possible extent. Production processes are regulated to reduce emission levels and incentivized for using renewable sources of energy. It is defined as to "how we should study and protect ecosystems, air quality, integrity and sustainability of our resources and focusing on the elements that place stress on the environment", such that technological advancement leads us to a greener future.

Keeping sustainable development in mind, at the Millennium Summit of the United Nations in 2000, all 191 United Nations member states and various international organizations committed to achieve the following eight Millennium Development Goals (MDGs) by 2015: (1) Eradicate extreme poverty and hunger; (2) Achieve universal primary education; (3) Promote gender equality; (4) Reduce child (age <5 years) mortality; (5) Improve maternal health; (6) Combat HIV/AIDS and malaria; (7) Ensure environmental sustainability; and (8) Build global partnership for development. Mixed progress has been observed regarding MDG target achievements. In 2012, the meeting of the United Nations Conference on Sustainable Development came up with Sustainable Development Goals (SDGs) replacing the MDGs, which were about to expire in 2015. The SDGs consists of 17 universal objectives. In addition to the MDGs, some new goals are added. For instance, a direct emphasis has been placed on tackling the effects of climate change, pollution and other environmental factors that can harm people's health, livelihoods. According to the SDG Report of 2016, the number of people living below extreme poverty has decreased by more than half, the proportion of stunted children younger than 5 reduced from 33% to 24%, and between 1990 and 2015, the global maternity ratio decreased by 44%. In 2015, 91% of the world population has been reported to use improved drinking water. Despite these achievements, efforts have not been enough to serve the purpose of sustainable development and more needs to be done on this front. In other words, sustainability still remains an area of concern.

So a sustainable economy is one that achieve both intragenerational and intergenerational equity. However, this intergenerational equity cannot be achieved if equity cannot be ensured for the present generation. In simple words, sustainability is important because the global future is dependent on it. And a sustainable future is a crucial factor owing to the three interdependent factors: (1) fossil fuel depletion, (2) climate change due to carbon dioxide emissions and (3) the increasing costs of energy and water. Since the beginning of the Industrial Revolution, the use of cheap fossil fuel energy has been prevalent in all industrialized nations of the world. Fossil fuels being a finite and nonrenewable resource, it is alarming if the demand for fossil fuels increases at a higher rate as compared to the rate at which it can be replenished. The prices of alternative sources of energy have increased steadily, thereby increasing the cost of freshwater. This has increased the burden on the economies worldwide besides shooting up the costs of an average household.

Sustainable development is truly an emerging interdisciplinary field with international appeal as many of these problems are transboundary in nature. Therefore, embracing the path of sustainable development is highly warranted to strike a synergy between the economy, energy and environment. Global climate change is a classic example. Each country's emissions of greenhouse gases contribute cumulatively to the increase of the global concentrations which affect all the countries. Thus, greenhouse gases emitted in China and India will have a global impact. Similarly, deforestation in Brazil may aggravate the problem of climate change and the long-term consequences will be felt worldwide. There are many international organizations and institutes working in this field including the United Nations, the World Resources Institute, the World Wide Fund for nature and International Institute for Sustainable Development.

Aim and scope

With a growing demand for graduate programs on sustainable development in India and other countries, there is also an immediate need for good-quality reference books for such a program or a semester-long course on the topic. Students would love to refer to interesting local developmental case studies when they are taking a course on sustainable development with a focus on quantitative analysis. However, in our humble opinion, there are very few books that strike a decent balance between theories and analytical methods from various disciplines, practice and case studies regarding sustainable developmental issues in India. This belief is the main motivation for this volume. The discussions in this book center on the various challenges of achieving sustainability, the role of quantitative research to assess the current scenarios and finally the role of policy making to bring improvements in the Indian context. This book is planned to be a reference manual that covers a whole gamut of analytical techniques that are useful for students, research scholars and practitioners alike.

This book is aimed to aid in the teaching of a semester-long postgraduate course on Quantitative Sustainable Development. The various chapters expose the readers to a variety of economic models and quantitative methods useful for research in the field of socioeconomic analysis of sustainability. The book not only covers important quantitative models for sustainability policy analysis but also provides case studies to understand the practical implementations of the models. We have requested the authors to write these chapters keeping in mind that this volume shall also be useful for graduate-level teaching purpose. Thus, undergraduate-level knowledge of mathematics, statistics and econometrics is prerequisite to consult many of the chapters presented in this volume. Besides technically demanding chapters, this book also has a couple of chapters focusing on qualitative analysis of the sustainability issues to give readers a feel of nontechnical approach to sustainability analysis. We do not claim that this book could be used as a stand-alone textbook for such a course. The current volume could be used as a complement reference along with some already-existing quality reference books such as the following:

- Kumar, Surender, and Shunsuke Managi. (2009). *The Economics of Sustainable Development: The Case of India* (Vol. 32). New York: Springer Science & Business Media (2009).
- Rogers, Peter P., Kazi F. Jalal, and John A. Boyd. *An Introduction to Sustainable Development*. London: Routledge (2012).
- Roorda, N., Corcoran, P. B., & Weakland, J. P. (2012). *Fundamentals of sustainable development*. Routledge.
- Sarkar, Runa and Sinha, Anup. *Economics of Sustainable Development*. New York: Business Expert Press (2017).

Plan of the book

The book covers various case studies regarding sustainable developmental issues in India that have happened in the last few decades. It is split into five sections: **(1) Agriculture, (2) Energy and climate change, (3) Environment and natural resources,** (4) **Health** and (5) **Society and policy.** Under these sections, there are various chapters using quantitative tools and interesting data sets to answer various questions in the context of sustainable development.

In Part I, the first chapter talks about the risk factors associated with the agricultural sector in terms of the various sources from which risk originates in that sector and the various management practices that can minimize such risks. This study has been based on the six districts of the Vidarbha region in Maharashtra. The next three chapters focus on agricultural production issues in the Indo-Gangetic plains of India. The second chapter discusses the assessment of the water footprint for wheat cultivation in Uttar Pradesh. The authors find that wheat cultivation leads to a high-water footprint due to wheat being a winter crop along with existing intensive irrigation practices. In Chapter 3, a district-level study has been conducted in the state of Uttar Pradesh to examine the productive efficiency of the agricultural sector using data envelopment analysis as the backdrop of recent policy initiatives that have been undertaken in this sector. In the fourth and last chapter of Part I, the authors analyze agricultural productivity growth at the district level in the state of Bihar. This study provides estimates of Total Factor Productivity Indices using three major crops. Based on these indices, the determining factors that explain the differences in agricultural productivity growth at the district level are identified.

Part II, titled "Energy and climate change", includes chapters that discuss multiple and varied dimensions of the concerned aspect. The first chapter in this section has tried to find out how the willingness to pay of the modern and educated youth in India for climate change–based research activity varies in accordance with their perception of the of climate issues. The analysis has been based on primary data using a discrete choice model. In Chapter 6, the authors estimate the rebound effect—a change in the end user's behavioral responses due to the change in energy efficiency. They use an almost ideal demand system model and village-level data from two districts in West Bengal for their estimation and find a high but partial rebound effect, implying that there is still a scope to save energy through energy efficiency measures. In a related study (Chapter 7), the authors have undertaken an index decomposition analysis based on Log Mean Divisia Index and its applications to show how the multidimensional facet of energy efficiency influences overall sustainability. The chapter ends with a discussion on implied policy prescriptions to achieve climate goals. The last chapter in this section has made an attempt to evaluate a new model or framework for the alternative policy measures of renewable resources under different market environments in an economically efficient way such that it minimizes cost and enhances welfare in specific states of India. The study is also relevant in the national market context for the entire country.

Four chapters appear in Part III, pertaining to the discussion of problems with environmental and natural resources. The section begins with a study (Chapter 9) that deals with the empirical analysis of district-level data from Himachal Pradesh to understand whether the levels of human development lead to environmental degradation by constructing a bivariate composite index, on the basis of indicators of human development and environmental quality. Chapter 10 sheds light on how rapid industrialization in modern times besides resulting in high economic growth has also generated high levels of environmental pollution and how this level varies with the level of affluence. This study is based on the input–output modeling framework to show the interdependence among various production sectors. Chapter 11 examines whether the Environmental Kuznet's Curve hypothesis, that is, the inverted U–shaped relationship between per capita income and emission level holds for the region known as the Indian subcontinent (India, Pakistan, Bangladesh, Nepal and Sri Lanka) using panel data regression methods. Chapter 12 identifies the factors influencing households' perceptions about changing climate in the Indian states of Odisha and West Bengal. The authors utilize primary data and information collected from 1,100 households using a structured questionnaire, focus group discussions and interactions with the stakeholders.

The relation between health and environment, being a crucial one in driving the aspect of sustainable development, finds a place in this book through a number of studies under Section IV. One such relation has been outlined in the first chapter in this section. Chapter 13 is set to analyze the socioeconomic, demographic and other latent factors that result in child mortality by using a factor analysis technique for household-level data in the suburban area of Kanpur city of Uttar Pradesh which is known for its high child mortality. Chapter 14 explores as to how the self-selection criteria for the consumption of freely available drinking water determines the occurrence of waterborne diseases thereby influencing health outcomes. Socioeconomic variables such as class structure, education level, household size and location influence have been found to influence the selection criteria on the basis of the data available for households from the 69th round of National Sample Survey. In the context of rising medical expenses in India, Chapter 15 looks at whether the awareness about health insurance exists among the lower-class poor people and what factors determine their preferences while choosing between high medical expenses and insurance premiums. This study attempts to find out the willingness to pay among individuals based on the data obtained from the primary survey in the slum areas of Mumbai. The policy implication is to fill the information gap among the poor so that they can express their desirability for health insurance schemes. The last chapter in this part (Chapter 16) focuses on the nonmarket valuation of health risks from infectious diseases like malaria. Incorporating a health production function approach and utilizing primary data from Kolkata and its suburban areas, the author explores the public–private interplay of health risks. The viability of a scaled-up public program is subsequently discussed by comparing the public treatment willingness to pay estimates with the annual estimated costs of the Kolkata Municipal Corporation incurred on account of malaria vector control.

Four essays appear in Part V concerning the policy and societal aspects of sustainable development. Chapter 17 aims to show that the Multidimensional Poverty Index, which includes deprivation in terms of education, health and standard of living, is a better measure of poverty as compared to the traditional income index approach. The observations are based on household and state-level data collected by the India Human Development Surveys, using correlation and regression analysis. In a rather descriptive approach, Chapter 18 provides insight into the development priorities in the eastern part of the Sunderbans region of West Bengal by measuring the quality of life. The conclusions are based on some descriptive analysis of policy intervention along with the empirical findings on the relationship between perceived and objective measures of well-being. Chapter 19 theoretically explores the usefulness of loan waiver schemes in India. The authors contend that in the presence of numerous arguments both in favor and against the various loan waiver schemes, this scheme has continued to be one of the main fiscal policy options of the governments, not only at the central level but also at the state level. The conclusion has been rolled out in favor of loan waiver schemes based on the study of one such prominent scheme undertaken in Uttar Pradesh. The final chapter in this book (Chapter 20) focuses on the sustainability issue in public transportation service. Using path analysis, the authors have attempted to model the cause–effect relationship of service quality with respect to three pillars of sustainability in relation to overall satisfaction of intercity bus transport based on synthetic data. The influence of technology interface on satisfaction of the service is also explored.

References

Brundtland, G. H., M. Khalid, and S. Agnelli. *Our common future*: Report of the world commission on environment and development *Med. Confl. Surviv.*, 4(1) (1987), p. 300, 10.1080/07488008808408783.

Nam, U. V. *Transforming our World: The 2030 Agenda for Sustainable Development. Division for Sustainable Development Goals*. New York: (2015).

Part I
Agriculture

1 Risk and risk management in agriculture

Thiagu Ranganathan

Risk is like love: we all know what it is, but we don't know how to define it.

– Joseph Stiglitz[1]

1.1 Introduction

Risks that affect economic activities have been a key area of study in various disciplines of social science. Agriculture is an economic activity that is fraught with varied risks. The role of these risks in agricultural production and its impact on farmers' welfare have been topics of key interest in academic and policy circles. This chapter provides an overview of various issues related to risk and risk management in agriculture, with examples from the Indian context. The prominence of risk is high in agriculture as compared to other economic activities for various reasons. First, agriculture is highly dependent on the vagaries of weather. The production cycle in agriculture is long and this means that agricultural production is vulnerable to weather-based vagaries during this long production cycle. On the marketing front, the interval between production decisions and marketing is typically higher, and the sale prices at the end of the production cycle might not be on the lines that are anticipated during the production. Sometimes, there is a greater difficulty in forming an expectation of prices and adjusting supply in agriculture than in other economic activities. Inputs like seeds, fertilizers and pesticides are also not amenable to storability and availability, and the prices of these inputs also fluctuate at crucial times. Given such prominence of risk and risk management in agriculture, it becomes imperative to discuss this while exploring aspects related to sustainable development.

The remaining chapter is structured as follows: Section 1.2 explores the nature and sources of risk in agriculture. Section 1.3 describes the models of decision-making under risk and measurement of farmers' attitudes towards risks. Section 1.4 describes the implications of risk and farmers' risk attitudes on the adoption of new technologies. Section 1.5 summarizes the typical formal and informal risk management strategies of farmers.

1.2 Nature and source of risks in agriculture

Production in agriculture is dependent on various weather variables. Rainfall, humidity, temperature and other weather variables affect yields of crops. Extreme events like hailstorms could also have a significant impact on crop production (Hardaker et al., 2015). These weather variables could also destroy the existing/stored farm stock. Production is also affected by plant diseases, pests and insects. A lack of timely availability of inputs and high input prices could also impact farmers' production, and this risk is significantly higher in recent times. Farmers also face output price risks on account of production and supply shocks. Given these various aspects of farm income risks, this section first deals with some conceptual issues related to risks.

1.2.1 Nature of risk

Uncertainty arises when the outcome of an action is not certain. While the final outcome itself may not be certain in many cases, there might be some knowledge about the possible outcomes and the degree of certainty of these outcomes. Various decisions and situations in agriculture, starting from sowing to the final sale of harvest, could be described in this manner. For instance, the farmer decides on when to sow the field based on the occurrence of first rains. When the rains will begin every year is never certain. But there is always some knowledge of it based on past experience and information from the weather department. Similarly, the price of a particular crop on a particular day may not be certain for the farmer, but there is some knowledge of it based on past experience and information. The degree of uncertainty also varies for different uncertain variables. When the likely outcomes are known, but their likelihood of occurrence is not known, such situations are described to be characterized by ***ambiguity*** (Ellsberg, 1961). Ambiguity corresponds to situations in which some events do not have a definitive probability assignment (Ghirardato and Marinacci, 2000). For instance, a farmer might know that a new variety of crop could fetch a price ranging from Rs. 3000 to Rs. 5000 per quintal but might not know the likelihood of obtaining any of the prices. But, in other situations, a farmer might be aware of both the likely outcomes and likelihood of occurrence. Such situations are termed as ***risk***, whereby there is knowledge of both the likely outcomes and the likelihood of occurrence. Most of this chapter deals with situations in which there is knowledge of likely outcomes and subjective or objective probabilities[2] of those likely outcomes.

Risk could be measured in various ways, but conceptually, two broad notions of risk are used while referring to various measures. According to the first notion, risk is seen from the point of variability in outcomes while in the second, risk is seen as the possibility of occurrence of something bad or unpleasant (such as an injury or a loss; Roumasset et al., 1979). In line with the first notion, measures of dispersion like range, standard deviation, variance and coefficient of variation of income are used to quantify risk, while according to the second notion, the probability of the income falling below some disaster level and skewness are some measures that are used to quantify risk. People might have notions of risk that are either of the two or a combination of these two, and their perceptions and attitudes towards risk are also defined by their notions. These two notions of risk also mean that whether a particular distribution is riskier than the other is sometimes dependent on the notion used to define risk. Although the variability in outcomes is an important way of characterizing risk, measures that look at the probability of loss are also important while describing risk. Some of the assumptions in conventional risk analysis undermine the importance of the distinction between these two notions. One other aspect pertaining to risk analysis is that the behavior of individuals might be more aligned to one of the notions of risk, whereas the statistical analysis might be easier while considering the other notion of risk.

Consider the following example which indicates the importance of the distinction between two risks and how certain assumptions could undermine the importance of one of the notions of risk. We consider an example from Walker and Subba Rao (1982), where they compare the distributions of net returns (Rs./ha) of farmers cultivating cotton/pigeon pea intercropping and those cultivating hybrid Sorghum in Kanzara village and Akola district in Maharashtra.

Under the assumption of normal distribution, hybrid sorghum does better than cotton/pigeon pea intercropping as the probability of earning higher than a particular return is higher for hybrid sorghum, for the most part, of the cumulative distribution function. The same is true even in the observed distribution but over a lesser range as the crossover area is considerably larger in the distribution compared to the normal distribution. Because of this assumption of normal distribution, risk, as measured by probability of loss/disaster, is considerably underestimated. For instance, the probability of net returns falling below negative Rs. 300 was double (0.08 compared to 0.04) for hybrid sorghum compared to cotton/pigeon pea under the assumption of normal distribution while it is eightfold (0.08 against 0.01) in observed distribution. This could be one of the reasons inhibiting adoption of hybrid sorghum in the region among farmers who give importance to risk as the probability of disaster. By approximating the returns distribution to normal distribution, we might be losing out less information on dispersion measures, but we see quite a significant discrepancy in measures of risk as a probability of occurrence of a loss. The measures related to probability of loss/disaster might be the ones that might be more relevant from the farmers' decision making in this case.

Over the years, risk, as measured by dispersion in outcomes, has become more popular in analysis. This popularity is also because this notion provides a convenient analytical base for various analyses. One such important application we deal with in this chapter is how this notion could be used to articulate the sources and significance of farm revenue risk. The next section details the methodology and cites some of the studies using them.

1.2.2 Sources of farm revenue risk

As seen from risk as a variability point of view, standard deviation and variance are important measures of risk. Farm revenue risk can be decomposed into those arising out of variances in prices and those arising out of variations in yield. This analysis is important as it helps in prioritizing the relevant risk management policies. The following statistical identity is used to decompose different components of revenue risk:

$$\sigma_R^2 = \bar{q}^2\sigma_p^2 + \bar{p}^2\sigma_q^2 + 2\bar{p}\bar{q}\sigma_{pq} + \varepsilon\ldots, \qquad (1.1)$$

where σ_R^2, σ_p^2 and σ_q^2 are the variances of farm revenue per acre, unit output prices and farm yield, respectively; $\bar{q}$ and $\bar{p}$ are the mean farm yield and output price, respectively; and σ_{pq} is the covariance between price and yield. ε represents the residual factors involving more than one variance term that contribute to the variance of farm revenue. The ratio of the first term on right hand side (RHS) to variance of farm revenue per acre could be seen as the contribution of price risk to farm revenue risk, the ratio of second RHS term to variance of farm revenue per acre is the contribution of yield risk to farm revenue risk and the ratio of the third term to revenue variance is the contribution of "natural hedge" (the covariance between price and yield that is generally negative) to farm revenue risk.

The statistical identity in Equation 2.1 could be modified to understand important changes over time and draw some important policy implications. Ranganathan and Gaurav (2013) use such a modified identity to understand the impact of the dismantling of Maharashtra Cotton Monopoly Procurement Scheme and the commercial release of Bt cotton on the nature of risk in six districts of Vidarbha region. They perform this analysis for the post-liberalization period from 1991 to 2011. They divide the time periods into 1991 to 2002 and 2003 to 2011 based on the dismantling of Maharashtra Cotton Monopoly Procurement Scheme and commercial release of Bt cotton in 2002–2003. To model the variances, they use the following conceptual model.

Let R_{ij} denote the gross farm revenue per hectare[3] out of cotton cultivation in the i th district during the jth period. This revenue is a product of the yield in the district, given by y_{ij} in kilograms per hectare and the price of cotton in the district, given by p_{ij} in Rs. per kilogram, and let $\Delta\bar{R}, \Delta\bar{p}$ and

$\Delta\bar{y}$ be used to denote the changes in means of the gross revenues, average price and average yield over two distinct periods, respectively. Also, let $\Delta\sigma_p^2, \Delta\sigma_y^2$ and $\Delta\sigma_{p,y}$ denote the changes in variances of prices, variance of yield and covariances between price and yield over two distinct periods, respectively. Let Δr denote the change in higher order cross-moments over the two distinct periods. Then the variance of gross revenues in period 2 is given by

$$\begin{aligned}\sigma_{R_{i2}}^2 &= \left(\bar{p}_1 + \Delta\bar{p}\right)^2\left(\sigma_{y_{i1}}^2 + \Delta\sigma_y^2\right) + \left(\bar{y}_1 + \Delta\bar{y}\right)^2\left(\sigma_{p_{i1}}^2 + \Delta\sigma_p^2\right) \\ &\quad + 2\left(\bar{p}_1 + \Delta\bar{p}\right)\left(\bar{y}_1 + \Delta\bar{y}\right)\left(\sigma_{p_{i1},y_{i1}} + \Delta\sigma_{p,y}\right) \\ &\quad - \left(\sigma_{p_{i1},y_{i1}} + \Delta\sigma_{p,y}\right)^2 + r_{i1} + \Delta r \ldots \end{aligned} \tag{1.2}$$

On simplification, we obtain the following equation:

$$\begin{aligned}\sigma_{R_{i2}}^2 - \sigma_{R1}^2 = {} & \sigma_{p_{i1}}^2 \Delta\bar{y}^2 + 2\bar{p}_1\sigma_{p_{i1},y_{i1}}\Delta\bar{y} + 2\bar{y}_1\sigma_{p_{i1}}^2\Delta\bar{y} + && \text{Changes in mean yield} \\ & \sigma_{y_{i1}}^2\Delta\bar{p}^2 + 2\bar{y}_1\sigma_{p_{i1},y_{i1}}\Delta\bar{p} + 2\bar{p}_1\sigma_{y_{i1}}^2\Delta\bar{p} + && \text{Changes in mean price} \\ & \bar{p}_{i1}{}^2\Delta\sigma_y^2 + && \text{Changes in variance of yield} \\ & \bar{y}_{i1}{}^2\Delta\sigma_p^2 + && \text{Changes in variance of price} \\ & 2\bar{p}_{i1}\bar{y}_{i1}\Delta\sigma_{p,y} - 2\sigma_{p_{i1},y_{i1}}\Delta\sigma_{p,y} - \left(\Delta\sigma_{p,y}\right)^2 + && \text{Changes in covariance} \\ & && \text{between price and yield} \\ & 2\bar{p}_{i1}\Delta\bar{p}\Delta\sigma_y^2 + \Delta\bar{p}^2\Delta\sigma_y^2 + 2\bar{y}_{i1}\Delta\bar{y}\Delta\sigma_p^2 + \\ & \Delta\bar{y}^2\Delta\sigma_p^2 + 2\Delta\bar{p}\Delta\bar{y}\sigma_{p,y} + 2\bar{p}_{i1}\Delta\bar{y}\Delta\sigma_{p,y} && \{\text{Changes in} + \\ & 2\bar{y}_{i1}\Delta\bar{p}\Delta\sigma_{p,y} + 2\Delta\bar{p}\Delta\bar{y}\Delta\sigma_{p,y} + && \text{interaction terms}\} \\ & \Delta r \ldots && \text{Changes in residuals} \end{aligned} \tag{1.3}$$

The change in the variance of farm revenue over the two periods is composed of 20 different terms.[4] These terms explain which of the various sources have contributed to the changes in revenue risk in the six districts in the study. Table 1.1 provides the findings of their study.

Table 1.1 shows that revenue variance increased in five of the districts and decreased in Nagpur. For four out of these five districts, the highest increase happened in the terms involving the covariance term. In the remaining district (Akola), the highest component increase was in yield variance terms. Interestingly, all five districts saw a decrease in price variances. These findings indicate that even though liberalization could decrease price risk, there could be an increase in revenue risk due to a decrease in the natural hedge (price–yield covariance). This could be due to higher market integration that happens because of certain policy shifts. When markets are well integrated, the prices in the region do not respond to yield vagaries in the local region.

The preceding analysis shows ways to identify the sources of changes in revenue risk in a particular region. Such an analysis would help us

Table 1.1 Variance Decomposition of Revenues in Six Districts of Vidarbha

District	*Changes in Variances of Gross Revenues*	*Changes in Mean Yield Terms*	*Changes in Mean Price Terms*	*Changes in Yield Variance Terms*	*Changes in Price Variance Terms*	*Changes in Price-Yield Covariance Terms*	*Changes in Various Interaction Terms*	*Changes in Residual Terms*
Akola 1991–2002 to 2003–11	83,32,509	8,26,261	–2,400,566	1,07,17,523	–1,192,066	48,13,033	–5,243,673	8,11,998
(% of total change in variance)		10	**-29**	**129**	-14	58	-63	10
Amravati 1991–2002 to 2003–11	84,68,578	8,99,539	–2,022,972	57,48,944	–891,649	60,49,876	–3,261,307	19,46,147
(% of total change in variance)		11	**-24**	68	-11	**71**	-39	23
Buldhana 1991–2002 to 2003–11	1,07,24,487	11,36,212	-6,070,926	–5,945,311	-716,463	61,71,818	1,542,873	14,606,284
(% of total change in variance)		11	**-57**	-55	-7	**58**	14	136
Nagpur 1991–2002 to 2003–11	–1,942,671	421,108	–2,016,661	–2,505,878	–1,464,213	2,456,927	403,181	762,866

(*Continued*)

Table 1.1 (Continued)

District	*Changes in Variances of Gross Revenues*	*Changes in Mean Yield Terms*	*Changes in Mean Price Terms*	*Changes in Yield Variance Terms*	*Changes in Price Variance Terms*	*Changes in Price-Yield Covariance Terms*	*Changes in Various Interaction Terms*	*Changes in Residual Terms*
(% of total change in variance)		–22	104	**129**	75	**–126**	–21	–39
Wardha 1991–2002 to 2003–11	84,64,530	–52,184	1,312,347	–3,213,766	–102,619	6,858,213	–70,399	3,732,937
(% of total change in variance)		–1	16	–38	–1	**81**	–1	44
Yavatmal 1991–2002 to 2003–11	1,92,92,941	882,650	–814,759	10,663,019	–342,903	5,166,473	–3,482,413	7,220,873
(% of total change in variance)		5	–4	55	–2	27	**–18**	37

Source: Author
Note:

understand the typical policy interventions that would be required to be implemented to stabilize the revenue of farmers. In the preceding analysis, we find that the price risk has reduced, so it might not be required to focus on that risk in all the years. The finding that the prices in a district do not seem to increase as much when the yield decreases indicates that the farmers might need more support in the price front in years when yield is low. We could also extend this analysis to identify the contribution of input cost risk to the income risk of the farmers. The analysis could also be extended in the context of a diversified crop portfolio and how the yield, price, and input costs of different crops contribute to the net income risk of the farmer.

1.3 Models of decision-making under risk

This section summarizes some of the major findings related to models of decision making under risk: expected utility model, cumulative prospect theory and models based on the security principle. Like there is no single definition of risk, there is no single definitive model explaining decision making under risk by the farmer. It is important to understand the various models and think of an appropriate model based on the context in which farmers make decisions.

1.3.1 Expected utility theory

Expected utility theory (von Neumann and Morgenstern, 1947) is the most often used model in economics to understand decision-making under risk. In expected utility theory, the decision-maker is assumed to make a decision based on the utility of the available alternative prospects. Given a set of assumptions, expected utility theory states that cardinal utility values could be assigned to different consequences in such a way that the expected utility of any action can be used to rank risky actions. For example, let us assume that a farmer has a logarithmic utility function. If the farmer were to choose between two risky hybrid seed varieties of which seed 1 would provide an income of Rs. 4000 and Rs. 8000 with a probability of 0.4 and 0.6, respectively, and seed 2 would provide income of Rs. 3000 and Rs. 9000 with a probability of 0.4 and 0.6, respectively, the expected utilities of the two seeds would be estimated as follows:

$$\mathrm{EU}(\mathrm{seed1}) = 0.4 \times \ln(4000) + 0.6 \times \ln(8000) = 8.71$$

$$\mathrm{EU}(\mathrm{seed2}) = 0.4 \times \ln(3000) + 0.6 \times \ln(9000) = 8.67$$

Based on this theory, the farmer will cultivate seed 1 and not seed 2. In terms of mathematical notation, if $x(x_1,p_1;x_2,p_2;\ldots;x_n,p_n)$ denotes a risky prospect where outcomes x_1, x_2 and x_n occur with probabilities p_1, p_2

and p_n, and $p_1 + p_2 + \ldots + p_n = 1$, the expected utility of the prospect is given by

$$EU(x) = p_1U(x_1) + p_2U(x_2)\ldots + p_nU(x_n) = \sum_{i=1}^{n} p_iU(x_i)\ldots \quad (1.4)$$

1.3.2 *Non-expected utility theory—rank dependent utility theory and cumulative prospect theory*

In Equation 0, we observe that the utility of the risky prospect is linear in probabilities. There is provision for the utility of the outcome being non-linear based on specification of the utility function for the farmer, but probabilities are assumed to be treated linearly by the decision-maker. But we know that farmers could also overweigh or underweigh probabilities. This aspect is also taken into account in some of the theories of decision-making under risk. For example, optimism and pessimism could be incorporated by using a probability weighting function like $\sqrt{p}$ and p^2, respectively. So, if likelihood of an event is 0.04, a farmer who weighs the probability using a function $\sqrt{p}$ will act as if the actual probability of occurrence of an event is 0.2 while the one who weighs the probability using a function p^2 will act as if the actual probability of occurrence is 0.0016.

In the ***Rank Dependent Utility Theory*** (Quiggin, 1993), the risky prospects are evaluated not just by linear/non-linear utilities but also by linear/non-linear weighing of probabilities. Denoted mathematically, the rank-dependent utility (RDU) of risky prospect $x(x_1,p_1;x_2,p_2;\ldots;x_n,p_n)$, where outcomes x_1, x_2 and x_n occur with probabilities p_1, p_2 and p_n, $x_1 < x_2\ldots < x_n$, and $p_1 + p_2 + \ldots + p_n = 1$ is given by

$$RDU(x) = \left[w(p_1 + r_1) - w(r_1)\right] \times U(x_1) + \left[w(p_2 + r_2) - w(r_2)\right] \times U(x_2)\ldots$$
$$\left[w(p_n + r_n) - w(r_n)\right] \times U(x_n) = \sum_i \left[w(p_i + r_i) - w(p_i)\right] \times U(x_i)\ldots, \quad (1.5)$$

where r_i is the rank of an outcome defined as probability of obtaining an outcome better than it and *w*() is the probability weighing function which accommodates for overweighing or underweighing of probabilities by the decision-maker. The RDU of a pessimistic farmer with a probability weighing function, $w(p) = p^2$, and a logarithmic utility function, $U(x) = \log x$, for a seed bag with the risky income prospect $x(1000,0.2;2000,0.7;4000,0.1)$ is given by.

$$\mathrm{RDU}(x) = \left[w(1) - w(0.8)\right] \times \log(1000) + \left[w(0.8) - w(0.1)\right] \times \log(2000)$$
$$+\left[w(0.1) - w(0)\right] \times \log(4000) = 0.36 \times 6.91 + 0.63 \times 7.60$$
$$+0.01 \times 8.29 = 7.36$$

Cumulative prospect theory (CPT; Tversky and Kahneman, 1992) builds on the earlier two theory of decision-making under risk and brings in some other aspects on board. This theory recognizes that people might deal with losses and gains differently, and hence, the theory treats the domain of gain and loss separately. The probability weighting function is formulated to have different parameter values in both the domains. The utility function in the loss domain is multiplied by a factor to indicate the differential treatment of gains and losses by the decision-maker. This factor is greater than 1 if losses are overweighted relative to gains by a decision-maker, and it will be less than 1 otherwise. A typical function used in defining prospect theory is

$$PT(x) = \sum_{i+}\left[w^{+}(g_i + p_i) - w^{+}(g_i)\right] \times U^{+}(x_i^{+}) + \sum_{i-}\left[w^{-}(l_i + p_i) - w^{-}(l_i)\right] x_i^{-}] \times U^{-}(x_i^{-})..., \quad (1.6)$$

where g_i means a gain ranks or ranks similar to the one defined in RDU, l_i is the loss rank which is the probability of getting an outcome worse than the same, $w^{+}()$ is the probability weighting function in the gains domain, $w^{-}()$ is the probability weighting function in the loss domain, $U^{+}()$ is the utility function in the gain domain and $U^{-}()$ is the utility function in the loss domain. Consider a farmer with the following utility function and probability weighting function in the loss and gain domains:

$$w^{+}(p) = \frac{p^{0.61}}{\left(p^{0.61} + (1-p)^{0.61}\right)^{1/0.61}} \quad (1.7)$$

$$w^{-}(p) = \frac{p^{0.69}}{\left(p^{0.69} + (1-p)^{0.69}\right)^{1/0.69}} \quad (1.8)$$

$$\begin{aligned} U(x) &= x^{0.88} && \text{if } x > 0 \\ &= 0 && \text{if } x = 0 \\ &= -2.25 \times (-x)^{0.88} && \text{if } x < 0 \end{aligned} \quad (1.9)$$

The farmer's evaluation of a seed bag with a risky income prospect $x(-1000,0.1;-500,0.1;2000,0.7;4000,0.1)$ is estimated as.

$$\begin{aligned} PT(x) &= [w+(0.1)-w+(0)] \times (4000^{0.88}) + [w+(0.8)-w+(0.1)] \times (2000^{0.88}) \\ &+ [w-(0.1)-w-(0)] \times (-2.25 \times 1000^{0.88}) + [w-(0.2)-w-(0.1)] \\ &\times (-2.25 \times 1000^{0.88}) = 0.186 \times 1478 + 0.421 \\ &\times 803 + 0.170 \times -982 + 0.087 \times -534 = 400.29 \end{aligned}$$

1.3.3 Models based on securebased on security principle – safety principle, safety first, safety fixed and maximin

This section describes a set of models that prioritize the security desires of the decision-makers. These models are particularly important for modeling the decisions of farmers who might have subsistence as their primary objective. For modeling these decisions, the objective functions of the farmers are not always derived from utility functions unlike in the previous two types of decision models. In some cases, however, it is possible to relate the implied optimal decision to equivalent decision based on utility functions with jump discontinuities. Although there are many models that could be formulated using the security principle, four have been used widely and also made empirical sense in some cases.

According to the ***Safety Principle*** (Kunreuther and Wright, 1974), the farmers minimize the probability of some attribute, usually income (say, x), falling below specified disaster level d^*, that is, min $P(x < d^*)$. For example, let us say that the disaster level is fixed as the income of Rs. 11,000 that will fulfill the household's subsistence need. If the farmer has to choose between two seed bags with expected levels of Rs. 14,000 and Rs. 20,000 but with the probability of income falling below Rs. 11,000 at 0.43 and 0.57, respectively, the farmer will choose the one with expected income of Rs. 14,000 according to this principle.

The ***Safety First Rule*** (Roy, 1952) has the disaster rule as a constraint, but the decision-maker is assumed to maximize income subject to it. If $E(x)$, the expected income, is the objective function that the farmer wants to maximize, he or she can have a constraint that probability of income falling below d^* is less than a critical probability P^*. In this case, there may be not a feasible solution either.

The ***Safety Fixed Rule*** involves the maximization of the minimum return that can be obtained with a probability of at least the critical probability P^*. If d is the disaster level that is maximized by the farmer, the constraint is expressed as $P(x \leq d) \leq P^*$. ***Maximin Rule*** is a special case of safety fixed rule where P^* is zero. In this rule, risky prospects are chosen based on the minimum value of each of the alternatives.

Over the last two sections, we have explored different definitions of risk and explored various ways in which we can model the decision-making under risk by the farmers. These two aspects related to what is risk and how farmers make decisions under risk has an influence on how technology gets adopted. We explore this in the next section.

1.4 Risk and technology adoption

Risk and technology adoption are highly related and the decision-making rules of farmers related to modern technology adoption might be related to how they view risks arising out of it. Typically, modern technology adoption might produce higher returns compared to traditional farming, but the variance while using these technologies is typically higher. This is because these techniques work well only under certain controlled conditions favorable to them. Most of the time, assured water availability is assumed while testing the effectiveness of these modern technologies. Their performance in unfavorable un-irrigated drought conditions is not generally good and might increase the risk in such regions. In this background, Ramaswami (1992) posits a hypothesis that fertilizers could be risk-increasing in absence of irrigation while it could be risk-decreasing in presence of irrigation. Empirically, Di Falco et al. (2006) find that fertilizers increase the yield variance in the context of dryland farming in Ethiopia. In another study in Ethiopia, Yesuf and Bluffstone (2009) find that although the fertilizers decrease the variance of yield, they also increase the chances of crop failure by increasing the skewness. They also find that the soil and water conservation inputs do not have any impact on yield variance but seem to reduce the skewness of crop yield, thus reducing the chances of crop failure. Kato et al. (2011) also find that soil and conservation practices act as buffers against complete crop failures and are more and more relevant in the context of climate change. The relevance of traditional practices is probably due to risk-reducing (in terms of variance reduction) and downside risk-reducing (in terms of skewness reduction) nature of these practices. In both these aspects, modern technology could sometimes have an adverse impact on the risk and welfare of farmers. It could increase not just the variance of yield but also its skewness.

Adopting low-risk, low-return technologies in the long run because of risk implications could make farmers get struck in low-state equilibrium, which is sometimes referred as "poverty traps" (Eswaran and Kotwal 1990; Barnett et al., 2008; Dercon and Christiaensen, 2011). This impact of risk on technology adoption has been widely studied in the literature. In this sub-section, we discuss in detail the risk-technology adoption scenarios which are typically pertinent to agriculture in rainfed regions as presented in Walker and Ryan (1990).

In some scenarios, the cumulative distribution of modern varieties is always lower than that of the traditional varieties. In such a scenario, there is a clear dominance, ordering is clear and the right choice of technology is easy. There could also be scenarios where the returns in the **fertilizer** case are always better than in the **no-ferilizer** case for cotton/pigeon pea inter-crop during monsoon. In such a scenario, the farmers will no doubt apply fertilizers. Farmers in the study also seemed to believe in a similar manner for cotton/mung bean intercrop as compared to a more extensive cotton/pigeon pea/local sorghum intercrop. We could also have scenarios where traditional methods were superior to "modern" method which is also a possibility in more risk-prone marginal regions. They then provide an interesting scenario which is very pertinent to agriculture in rainfed regions. Here, the "modern"/improved varieties would significantly outperform traditional varieties in abnormally good years. But these years are so less frequent in rainfed regions that they cannot make up for inferiority of modern varieties over traditional varieties in numerous years of less-than-ideal growing conditions. In such instances, farmers would be better off by choosing traditional practices most of the time. The improved varieties might perform better than traditional practices in most instances except during most unfavorable conditions. Given farmers find such unfavorable conditions more frequently in rainfed regions, their attitudes towards such technologies might not be positive, and they might still pursue traditional agricultural practices.

From the preceding discussion, we observe that technology adoption by rainfed farmers is dependent on various aspects of risks. Since most technology is typically tested in ideal conditions of agriculture, the perceived distributions by scientists and policy makers provide clear dominance. But the rainfed regions have very few seasons of ideal and abnormally good years. These distributions and choices also change over time and are influenced by other factors as well. For example, it is well known that different components of soil quality play a huge role in determining these distributions. For example, if we consider the case of fertilizer adoption in initial years, when the soils were rich in organic carbon, the decision about the application of fertilizer to soil would have clear dominance. But, over time, excessive application of chemicals and a lack of application of organic matter have reduced the organic carbon in soil, which may have now reduced the yield response to chemical fertilizers, and this might have actually changed the distributions of fertilizer and no fertilizer to scenarios with no clear dominance of distribution. There is no empirical investigation of such aspects, and it might be important for us to understand such changes in technology adoption.

The link between risk and technology was explored in this section. Farmers do understand the implications of risk and manage them using formal and informal methods. The next section provides an overview of the various risk management strategies of the farmers.

1.5 Risk management and its efficacy

Given that risk plays such a central role in agriculture, farmers have developed various *ex ante* risk management strategies and *ex post* strategies to cope with risks (Dercon, 2002). For simplicity, we refer to both risk coping and risk management strategies as risk management strategies in this chapter. These informal risk management strategies have been developed in absence of proper public policy responses to weather risks. There are a few public policy responses like providing crop insurance product, establishing futures markets and rural public works programmes with the purpose of providing revenue risk management for farmers.

1.5.1 *Informal risk management strategies of farmers in rain-fed regions*

Farmers try to smooth their income by various risk management strategies. But, in spite of these strategies, if there is a shortfall in household income due to highly unfavorable conditions in the region, households try to smooth their consumption. In this regard, they respond to such low farm income outcomes by either selling the stored produce, borrowing for consumption, liquidating assets, receiving market transfer income from friends or relatives, increasing labour market participation or migrating in search of work (Walker and Ryan, 1990; Alderman and Paxson, 1992; Rosenzweig and Binswanger, 1993; Dercon, 2002; Jodha, Singh and Bantilan, 2012). But not all farmers can adopt all of these practices. Storage of produce is difficult as farmers might have to sell the produce immediately after harvest to maintain liquidity and pay back the credit. Borrowing for consumption from money lenders is typically the most frequently used risk coping strategy in agriculture. This means is an ineffective form of managing risk, particularly if there are unfavorable weather outcomes in consecutive years. Such risk coping strategies are generally not effective in consumption smoothing without public assistance as the weather outcomes are typically covariate across households. The covariate nature of risk leads to a large private demand for such risk coping strategies in bad years. This demand typically dampens the income from these activities for individual households. Whenever unfavorable weather conditions prevail in a large region, public-sector assistance is required to help people get out of distress.

Apart from the risk coping strategies, farmers use various risk management strategies like spatial and variety diversification, staggered plantation, inter-cropping, mixed cropping and income diversification to smooth their incomes (Alderman and Paxson, 1992; Rosenzweig and Wolpin, 1993; Townsend, 1994; Dercon, 1996; Skees, 2002). These methods typically lead to the efficient usage of nutrients, moisture and light and provide greater potential for yield compensation. The usage of these methods though is also dependent on resource endowments available with the households. Households might not be able to implement these strategies because of a lack of quality land, draft power unavailability and so on. Typically, households diversify their incomes through other livelihood activities for managing the income risk. Farmers also use sharecropping tenancy to manage risks in agriculture. Table 1.2 presents the summary of risk management strategies as observed in different villages of International Crop Research Institute for Semi-Arid Tropics (ICRISAT) study.

As we observe from Table 1.2, the intensity of risk management strategies typically increases with the extent of risk in the region. These practices may not be observed with such intensity in irrigated regions. Various factors determine the effectiveness of these risk management strategies. For example, a crop diversification strategy will be effective only if the crops used in diversification suffer from different sources of risk; that is, the sources of yield risk of the crops are less covariate. Else, both the crops will suffer from damage when the same source of risk is observed in the region. Walker, Jodha and Singh (1983) study this aspect in their study of four ICRISAT villages. In the villages of Akola and Sholapur districts, they find that a 10% increase in diversification index results in 4.8% and 13% reduction

Table 1.2 Summary of Risk Management Strategies in ICRISAT Villages

Indicators of Risk Management Strategies				
Details of Degree of Risk and Adaptation	*Akola*	*Mahbub Nagar*	*Sholapur*	*Jodhpur*
A. Characteristic of Weather Risk				
Annual Average Rainfall (mm)	820	786	690	382
Probability of favorable soil moisture conditions for rainy season cropping	0.66	0.56	0.33	0.21
Length of growing season (days)	200	176	155	60
Incidence of crop failure (Average of 3 yr)				
Plots with complete crop failure (%)	4	15	17	33
Plots with partial crop failure (%)	7	32	24	–
B. Indicators of Spatial Diversification				
Scattered land fragments (no.) per farm	2.8	3.4	5.8	7.3
Split plots (no.) per farm	5	3.5	11.2	–
Fragments (no.) per farm by distance from village				
Upto 0.8 km	1.5	1.8	1.4	1.2
0.8–1.6 km	1.1	1.1	3.4	4.3
More than 1.6 km	0.2	0.32	1	1.8
C. Indicators of Crop-Based Diversification				
Extent of inter-cropping (%)	83	76	35	100
Total sole crops planted (no.)	20	32	34	12
Total combinations of mixed crops planted (no.)	43	–	56	30

(Continued)

Table 1.2 (Continued)

Indicators of Risk Management Strategies				
Details of Degree of Risk and Adaptation	*Akola*	*Mahbub Nagar*	*Sholapur*	*Jodhpur*
D. *Crop–Livestock Based Mixed Farming*				
Ratio of land and livestock values (1976–1977 prices)	93.7	95.3	91.9	63.37
Ratio of crop and livestock incomes (1976–1977 prices)	80.2	82.08	71.29	69.31
E. *Occupational and Institutional Adjustments*				
No. of occupations (sources of income) per household	1.5	1.6	2.3	2.9
Households with more than 2 occupations (%)	14	22	18	39
Households with seasonal out-migration (%)	2	19.6	24	33
Cases of land tenancy induced by risk sharing/ management consideration (no.)	9	–	66	–

Source: Jodha et al. (2012)

in coefficient of variation of net crop incomes, respectively. But crop diversification does not protect crop incomes in villages of Mahbubnagar and Sholapur villages as the crops suffered from similar sources of yield risk. This issue of needing a negative covariation is the problem with most risk management strategies used by farmers. If the yield of the crop used in intercropping suffers from similar vulnerability to pests as the main crop and suffer from similar sources of risk, the incomes of the intercropped farms might not stabilize much because of these practices. For example, Walker and Jodha (1986) find that the correlation between the yield of sorghum and pearl millets within the same field was significantly and positively correlated at 0.63. Although this a problem, the counterfactual of having a single crop might have much larger correlation (most probably 1). So lesser covariability is desirable for effectiveness of these risk management practices. Livestock and other income-generating activities might also suffer from the same sources of risk that affect the income from crops. So having livestock that would be drought resistant and having other income-generation options that are not dependent on rainfall is crucial for them being effective in times of drought.

1.5.2 Formal risk management strategies of farmers in rainfed regions

Crop insurance seems to be the most important formal risk management policy that would stabilize farm incomes. Apart from this, the establishment of futures markets could help farmers in stabilizing prices. The third response is a rural public works scheme which could 'self-target' distressed households in times of bad weather. But all these responses work only under certain conditions. The issue with crop insurance is that it is generally costly to provide an individual farmer-based crop insurance product. Apart from direct transaction costs involved, indirect costs of moral hazard and adverse selection make providing individual insurance product typically economically unsustainable. So area-based insurance is offered in most countries which compensates the farmers based on average losses occurring in an administrative division like a taluka or district. Also, the following conditions are important for the insurance product to be beneficial to the farmers (Walker and Ryan, 1990):

- Crop income should form a high component of household income. This would help households derive significant risk benefits from crop insurance product. But, typically in rainfed regions, crop income could contribute less to the total household incomes.
- Farm households should specialize in a few crops. This would ease the design of insurance product and have a product that has less of basis risk (risk of having an individual farm not being correlated with the average area yield). But, as we observed before, rainfed regions are characterized by high diversification with more crops than in irrigated regions.

- Output prices should be relatively stable.
- Crop supply should not depend heavily on agroclimatic conditions so that the link between weather-induced fluctuations in area and crop income should not be broken. Here again, we might find that area variability is higher in rainfed areas as farmers respond to low rainfall by cultivating a different set of crops or not cultivating anything in some area of their land.
- Yield from the insured crops should be exposed to only a few sources of risk so that indemnity assessment based on an area approach is efficient in stabilizing incomes for most farmers in the region. This is in conflict with the previous condition which says that area cultivated should remain somewhat constant. In that case, multiple risks will affect the crop yield.

All these contradictions between the ideal conditions and conditions on the ground have meant that the National Agricultural Insurance Scheme implemented by government has not been greatly effective in most parts of the country. The government has tried to implement Weather Based Crop Insurance Scheme and Modified National Agricultural Insurance Scheme recently. The most recent scheme of the Indian government called Pradhan Mantri Fasal Bhima Yojana combines aspects of previous schemes and its effects need to be evaluated in the next few years.

Futures markets also have been established to enable price risk management. But, as we observed before, yield risk is a larger risk than price risk for the farmers, particularly for those in rainfed regions. The risk benefits of price stabilization are minimal as compared to yield stabilization for the farmers. Also, high transaction costs involved in trading in these markets and low awareness among farmers have meant that the markets are presently quite irrelevant to the farmers (Ranganathan and Ananthakumar, 2014).

1.6 Conclusion

Risk is an integral part of any economic activity. With reference to farming, the dependence of weather variables for production, the long production cycle and the consequent uncertainty of market prices at the time of decision-making related to production make agriculture a significantly risky activity. It is in this context that an understanding of risk and risk management is crucial for understanding agriculture

This chapter has provided some basic principles of risk and risk management applicable to agriculture. To begin with, we provided the difference between risk, ambiguity, and uncertainty. The various measures of risk seen from the notion of variability and the notion of the likelihood of occurrence of a disaster were discussed. Then we discussed the ways in which the sources of risk could be identified.

The chapter also explored various theories that explain the decision-making of farmers under risk. Theories that involved linear and nonlinear utility functions and linear and nonlinear probability weighting functions were explored. Also, the cumulative probability theory which treated the gain and loss domains separately was explored. Then we looked at various theories that consider the farmer to be motivated by the security for subsistence.

Given this background, the chapter then explored the implication of risk attitudes on technology adoption. We then looked at both the formal and informal risk management strategies used by farmers in the Indian context.

The centrality of risk to agriculture demands a sustained study of various aspects related to risk, its measurement, risk attitudes, implications of technology adoption on risk and risk management. It is also important to note that the farmers face a multiplicity of risks, and there are also various studies that analyze the decision-making by farmers in presence of multiple background risk. The scope of this chapter was not to look into those aspects, but they do need important attention in the future course of study of risk and risk management in agriculture.

Notes

1 Quoted in https://www.miragenews.com/tools-being-developed-for-price-risk-management/ and other places

2 The distinction between objective probability and subjective probability is also crucial. Objective probability is defined based on the frequency of occurrence while subjective probability is defined based on a person's degree of belief. Both the versions would be used in the chapter at different places.

3 From this point, when we refer to gross farm revenue, we mean gross farm revenue per hectare. The area under cotton cultivation is assumed to be 1 ha throughout the analysis.

4 These include five types of "pure" effect terms (the first 11 terms in the equation (7)) which capture the changes in mean price, mean yield, variance of price, variance of yield and covariance between price and yield. Apart from these, there are eight different interaction effects between some of these terms. There is one remaining term that indicates the changes in higher order cross-moments between price and yield. The changes in higher order terms are typically small.

References

Alderman, Harold, and Christina H. Paxson. "Do the Poor Insure?: A Synthesis of the Literature on Risk and Consumption in Developing Countries." World Bank Policy Research Working Paper No. 1008 (1992).

Barnett, Barry J., Christopher B. Barrett, and Jerry R. Skees. "Poverty Traps and Index-Based risk Transfer Products." *World Development* 36, no. 10 (2008): 1766–1785.

Dercon, Stefan. "Risk, Crop Choice, and Savings: Evidence from Tanzania." *Economic Development and Cultural Change* 44, no. 3 (1996): 485–513.

Dercon, Stefan. "Income Risk, Coping Strategies, and Safety Nets." *The World Bank Research Observer* 17, no. 2 (2002): 141–166.

Dercon, Stefan, and Luc Christiaensen. "Consumption Risk, Technology Adoption and Poverty Traps: Evidence from Ethiopia." *Journal of Development Economics* 96, no. 2 (2011): 159–173.

Di Falco, Salvatore, and Jean-Paul Chavas. "Crop Genetic Diversity, Farm Productivity and the Management of Environmental Risk in Rainfed Agriculture" *European Review of Agricultural Economics* 33, no. 3 (2006): 289–314.

Ellsberg, D. "Risk, Ambiguity, and the Savage Axioms." *The Quarterly Journal of Economics* 75, no. 4 (1961): 643–669.

Eswaran, Mukesh, and Ashok Kotwal. "Implications of Credit Constraints for Risk Behaviour in Less Developed Economies." *Oxford Economic Papers* 42, no. 2 (1990): 473–482.

Ghirardato, P., and M. Marinacci. "Ambiguity Made Precise: A Comparative Foundation." Social Science Working Paper 1026 (2000), California Institute of Technology.

Hardaker, J. B., R. B. M. Huirne, J. R. Anderson, and G. Lien. "Coping with Risk in Agriculture: Applied Decision Analysis." CABI Publishing, 2015. Boston, USA

Jodha, Narpat S., Naveen P. Singh, and Maria Cynthia Serquiña Bantilan. "Enhancing Farmers' Adaptation to Climate Change in Arid and Semi-arid Agriculture of India: Evidences from Indigenous Practices: Developing International Public Goods from Development-Oriented Projects." Working Paper Series no. 32 (2012).

Kato, Edward, Claudia Ringler, Mahmud Yesuf, and Elizabeth Bryan. "Soil and Water Conservation Technologies: A Buffer against Production Risk in the Face of Climate Change? Insights from the Nile Basin in Ethiopia." *Agricultural Economics* 42, no. 5 (2011): 593–604.

Kunreuther, H., and G. Wright. "Safety-First, Gambling, and the Subsistence Farmer." Fels Discussion Paper No. 59 (1974). University of Pennsylvania/The Fels Centre of Government/Institute of Public Policy Analysis/Government Study Center. Available at http://ageconsearch.umn.edu/record/257756/files/magr-penn-state-100.pdf. Last accessed on 9 September 2018.

Quiggin, J. "Generalized Expected Utility Theory: The Rank-Dependent Model." Boston: Kluwer Academic Publishers (1993).

Ramaswami, Bharat. "Production Risk and Optimal Input Decisions." *American Journal of Agricultural Economics* 74, no. 4 (1992): 860–869.

Ranganathan, Thiagu, and Usha Ananthakumar. "Does Hedging in Futures Market Benefit Indian Farmers?" *Studies in Economics and Finance* 31, no. 3 (2014): 291–308.

Ranganathan, Thiagu, and Sarthak Gaurav. "An Inquiry into the Composition of Farm Revenue Risk – The Case of Six Vidarbha Districts." *Economic and Political Weekly* 48, no. 26–27 (2013): 125–128.

Rosenzweig, M. R., & Binswanger, H. P. "Wealth, Weather Risk and the Composition and Profitability of Agricultural Investments." *The Economic Journal*, 103, no. 416 (1993): 56–78.

Rosenzweig, Mark R., and Kenneth I. Wolpin. "Credit Market Constraints, Consumption Smoothing, and the Accumulation of Durable Production Assets in Low-income Countries: Investments in Bullocks in India." *Journal of Political Economy*, 101, no. 2 (1993): 223–244.

Roumasset J. A., J.-M. Boussard, and I. Singh (eds.). "Risk, Uncertainty, and Agricultural Development." South Asian Regional Center for Graduate Study and Research in Agriculture College, Laguna, Philippines and Agricultural Development Council, New York (1979).

Roy, A. D. "Safety First and the Holding of Assets." *Econometrica*, 20, no. 3, (1952): 431–449.

Skees, J. R. "New Approaches to Agricultural Insurance: What Have We Learned?" Presentation at Brown Bag Lunch sponsored by the Sustainable Development Department of the InterAmerican Development Bank and the Agriculture and Rural Development Division of the World Bank, Washington, DC, December 10, 2002.

Townsend, Robert M. "Risk and Insurance in Village India." Econometrica: Journal of the Econometric Society, 62, no. 3, (1994): 539–591.

Tversky, A., and D. Kahneman. "Advances in Prospect Theory: Cumulative Representation of Uncertainty." *Journal of Risk and Uncertainty*, 5, no. 4, (1992): 297–323.

von Neumann, J., and O. Morgenstern. "Theory of Games and Economic Behaviour." 2nd ed. Princeton, NJ: Princeton University Press (1947).

Walker, T., and N. Jodha. "How small farmers adapt to risk." *in* Valdés, Alberto, P. B. R. Hazell, and Carlos Pomareda. *Crop insurance for agricultural development: Issues and experience* (1986). IICA Biblioteca, Venezuela, pp. 17–34.

Walker, T. S., R. P. Singh, and N. S. Jodha. "Dimensions of Farm-Level Diversification in the Semi-Arid Tropics of Rural South India." *Economics Program Progress Report-International Crops Research Institute for the Semi-Arid Tropics* (1983). FAO.

Walker, Thomas S., and James G. Ryan. *Village and Household Economics in India's Semi-arid tropics*. Baltimore, MD: Johns Hopkins University Press, 1990.

Walker, T. S., and K. V. Subba Rao. "Risk and the Choice of Cropping Systems: Hybrid Sorghum and Cotton in the Akola Region of Central Peninsular India." *Economics Program Progress Report-International Crops Research Institute for the Semi-Arid Tropics* (1982). FAO.

Yesuf, Mahmud, and Randall A. Bluffstone. "Poverty, Risk Aversion, and Path Dependence in Low-income Countries: Experimental Evidence from Ethiopia." *American Journal of Agricultural Economics* 91, no. 4 (2009): 1022–1037.

2 Assessment of water footprint under wheat cultivation in Uttar Pradesh

Mohammad Suhail and Rakhshanda F. Fazli

2.1 Introduction

Among all resources, water is the most significant and strategic resource. It provides life on earth and governs all atmospheric activities. The global economy also depends on it as most industries require water for various purposes. It is a renewable resource through the hydrological cycle wherein water evaporates from the world's oceans into the atmosphere and precipitates on the land to infiltrate beneath and run off on the ground. Thus, it is completed a cycle of return after passing from the land to the ocean. Most water on earth is saline and present in the oceans, whereas freshwater is limited and stored in aquifers, glaciers, rivers and available on the ground as runoff. Moreover, the distribution of water on the earth is uneven, both in the temporal and spatial domains (Shiklomanov, 1996). The deterioration and depletion of water resources are causing scarcity in different parts of the world. The blending of both natural and anthropogenic substances into water are further triggering water quality issues. Among others, contamination from industry, mining, chemicals used in agriculture, and intensive livestock husbandry are primarily contributing to the contamination of drinking-water supplies. A number of serious disasters caused by water contamination are reported in the literature, some of which includes Minamata – 1968, Woburn – 1970 and 1979–80, Love Canal – 1977, Bhopal – 1984, Red Rhine – 1986, Chernobyl – 1986, Cancer Alley – 1987, Lanzhou – 1987 and 2002, Camelford – 1988, Valdez – 1989, Walkerton – 2000, Ghana – 2001, Sydney tar pounds – 2005, Chennai – 2014, Elk River – 2014 and Cape Town – 2018 (Smith and Smith, 1975; Andersson, 1990; Baxter, 1990; Okada and Peterson, 2000). The problem of water contamination is becoming a global agenda. A study revealed that between 2001 and 2014, there were nearly 183 typical water pollution incidents took place in China alone (Gaojie et al., 2018). Despite the contamination of most precious water resources, availability at a judicious level is also an important issue, which is of increasing relevance in current literature. It is of concern that water contamination is adversely increased since the 1990s in most rivers in Africa, Asia and Latin America (UNEP, 2016). The wastage and contamination of water can be controlled, to a greater extent, by proper measures of wastewater treatment, leakage prevention, efficiency enhancement, improvement of quality and conservation. In the Asia region, several countries are facing serious water shortage, which has resulted in a decline in the availability of freshwater resources, illhealth and even death (Gleick, 1993). Falkenmark (1989) has suggested that the water scarcity is the lack of sufficient amount of unpolluted water to meet the demand in a region, which can be divided into water stress, water shortage and water crisis in a region. However, human interventions in the form of dams and pipelines are widely in practice to cope with water scarcity by diverting the natural flow of water resources to meet local demand. As the need for water grows, tensions over water resources are likely to increase.

In addition to serious health considerations, the scarcity of water is more or less affecting every place in the world. One estimate shows that about 2.8 billion people around the world are severely suffering from water scarcity at an average rate of at least one month per year (WWAP, 2012). The population of the world has tripled since the beginning of 19th century (United Nations, 2015).

From 1973 to 2010, the rate of the world's energy consumption has increased by 186 percent while water usage has risen up to 157 percent for industrial purposes in the same period (UN Water, 2014). Furthermore, agriculture is one of the biggest water-consuming sectors, and it accounts, with irrigation, for 70 percent of total water use globally (FAO, 2016). The remaining 30 percent water is consumed by the domestic (12 percent) and industrial (18 percent) sectors. This is an average global water consumption pattern that varies across the countries. Excessive use and misuse of this vital resource by industrial, agriculture and residential activities have threatened the survival of all living organisms. It is reported that almost 30 to 40 percent, or sometimes more, of water, remains unaccounted for due to water leakage and illegal tapping in several countries (UNDP, 2006). The encroachment of water bodies has been identified as a "major cause" of flash floods in Mumbai (2005), Uttarakhand (2013), Jammu and Kashmir (2014) and Chennai (2015) in the past one-and-half decades. Their revival and optimum management seem even more significant, if the NITI (National Institute of Transforming India) Aayog's warning is to be taken seriously. It is stated that groundwater levels in 21 major cities will dry up completely by 2020, affecting more than 100 million people. In many places, the distribution and availability of water resources continue to decline, and insufficient efforts are being made to monitor, manage and ensure remedial measures by developing institutional and social setups. However, water managers admit that better management should include efficient and sustainable use rather than availability of water in the region. The equal distribution of a water resource can be addressed by conservation measures and smarter technologies around the world. There are also robust efforts, reported in the literature, to correct the situation and maintain a continuous flow throughout the years. Positive examples exist in many regions that must be shared, learned from and replicated (Alexandra et al., 2012). However, pertinent literature on water resources and on its management is developed in the beginning of 20th century (Wang et al., 2011), specifically, after the Unites Nations' ministerial declaration of The Hague 2000 that was aimed at the progress of water resource management and construction of a framework for global research agenda on water resources. However, it is observed that the aim of water resource management is diverse and multifaceted as a number of applications are associated with it (Suhail, 2015). Although tactics of water management and conservation are not universally applicable to administer water issues, indeed, previous high-quality literature and engineering designs have helped a lot in determining effective ways for water resource management and conservation.

To cope with water resource scarcity and management, a lot of notions and formulation of theories came into existence. Recently, there are also some emerging concepts in environmental concern, for example, carbon footprint, ecological footprint, biological footprint, and so forth. Similarly, water itself has its footprint regarding consumption or productivity. Humans use a lot of water for various purposes, such as cooking, drinking and washing but even more for producing commodities. Chapagain and Hoekstra (2004) have developed a water scarcity index based on water footprint (WFP), which explained that Jordan tops the list at 713 percent followed by Egypt (119 percent), Spain (84 percent) and Germany (82 percent) on the scarcity of water resources. On this index, India (52 percent) is on par with average water-scarce countries, accompanied with France (54 percent) and the United Kingdom (50 percent), while, on the other hand, Canada, Brazil and Australia stand as having the fewest water-scarce groups. The authors considered that the measurement of the water deficit of a region is the level of its food imports, because water is required in the production of foodstuffs, while other variables affecting the demand for freshwater are the growth of population and increasing per capita use of it (Chapagain and Hoekstra, 2004, 2008). From a water resource perspective, the transfer of goods and services from one region to another region is also pertinent. By doing this, a nation can save its scarce water resources through the import of water-intensive goods, while the export of less water-consuming goods will contribute in water balance positively (J. Allan, 2003; Chapagain and Hoekstra, 2003). Conversely, water-abundant nations can be benefitted by the export of water-intensive goods and vice-versa (Hoekstra, 2003). The concept of a 'water footprint' has much relevance in this regard. Considerable development has been conducted to enrich and establish a WFP assessment by many workers including (Hoekstra and Hung, 2002; Chapagain and Hoekstra, 2003; Chapagain and Hoekstra, 2004; Chapagain, et al. 2006; Hoekstra and Chapagain, 2007; Kampman et al., 2008; Chapagain and Hoekstra, 2008; Aldaya and Llamas, 2008; Hoekstra et al., 2009; Chapagain and Hoekstra, 2010; Aldaya et al., 2010a; Aldaya et al., 2010b; Mekonnen and Hoekstra, 2010; Romaguera, et al., 2010 and others).

2.2 Defining Water Footprint

The conceptual development of WFP took place after the introduction of the term *virtual water*, which was coined in 1993 by Allan (1997) at the School of Oriental and African Studies, wherein Allan referred that virtual water is the volume of water required to produce a commodity. Furthermore, he stated that "more water flows into the Middle East each year as 'virtual water' than flows down the Nile into Egypt for agriculture" (T. Allan, 2001, p. 29. Earlier to virtual water, Malin Falkenmark (1989) has introduced the concept of blue and green water, whereby blue and green water is evident and non-evident based on its use, respectively.

Until the recent past, very few researchers have studied the science and practice of water management with regard to water consumption and pollution. In addition to this, the science of water management rarely observed production and supply chains. Consequently, little research on volume of water used in a particular commodity, along with production and supply chain, has been conducted so far, which not only influences water volumes but also affects the temporal and spatial distribution of water resources. Thus, the mechanism of WFP is also considered as the distribution of water consumption and pollution in final estimates. Every commodity or service, when transferred from one region to another region, significantly contributes to the transfer of water, too, as water consumed during the processing or production is a component of WFP. For example, a pair of bovine leather shoes requires about 8000 litres of water to produce, if processed and prepared in an efficient way, while 200 litres of water is necessary to produce 200 mL of milk. Similarly, one mug of tea (250 mL) and coffee (125 mL) requires an amount of 27 and 140 litres of water when processed and distilled in a skilful way, respectively (UNEP, 2008; The Guardian, 2013).

There is a minor distinction between virtual water and WFP as former relates to direct use only while WFP includes direct use as well as supply chain use. In 2002, Hoekstra proposed the concept of "WFP", which is subsequently elaborated (Chapagain and Hoekstra, 2008). This elaboration is developed into a comprehensive framework to analyze the linkage between human consumption and efficient use of the world's freshwater. WFP has been considered an indicator of freshwater use, which is considered not only as the direct consumption by a consumer or producer but also as the indirect consumption along the whole supply chain (Hoekstra et al., 2011). Meanwhile, the component analysis, that is, blue, green and grey WFP components, provides a rational and holistic approach to understanding direct and indirect water consumption in an efficient way. It has been proved by many researchers that WFP is a multidimensional indicator, which can inculcate a reasonable estimate of the volume of water consumed by source, polluted volumes by type of pollution and specified spatial and temporal coverage of all components. The manual of WFP refers that blue water is the volume of both surface and underground water consumed by the commodity during production, whereas the green WFP is the soil moisture that stored in the soil during a rain event. Moreover, a grey WFP indicates the required volume of freshwater for understanding the load of contamination based on country-specific ambient water-quality standards. A well-developed methodology and assessment framework are available in *The Water Footprint Assessment Manual – 2011* (Hoekstra et al. 2011; Mekonnen and Hoekstra, 2011).

In India, the concept of a WFP has vivid and lucrative implications for achieving water-use efficiency and security (Kumar and Jain, 2007). Conflict over water resources has not been considered due to the mere scarcity of physical absence, but it should be considered since it includes malpractice and the mismanagement of resources. Since India has a vast agricultural expanse, which spreads from one of most fertile Indo-Gangetic plain in the north to long stretches of coastal areas in the south. Despite a considerable amount of water resources, both underground and surface, the economy of India depends on agriculture-based outputs. Predominantly, wheat and rice are important cereal crop cultivated in India. Although a sizeable section of the population is still suffering from hunger and malnutrition in India, it is because of the sub-optimum use of agricultural and water resources, which consequently is a major cause of non-viable agriculture and low productivity in India. Therefore, the management of water resources requires special provisions in India, especially cultivation per se. However, India witnesses a seemingly paradoxical situation every year in some parts wherein some areas are affected by floods and others experience droughts. It could be addressed cautiously by adopting an appropriate institutional support system, fair and robust implementation of the programs and a revision of plans based on the changing situations. The WFP approach can be taken a as ready-made solution to understand water and agriculture regime. Therefore, the present research is an attempt to assess WFPs for wheat cultivation in Uttar Pradesh at the district level. It has also tried to fill in the gaps of the water regime's understanding with the help of sub-components of the WFP concept. Thus, with the following objectives, this study has been carried out:

- To quantify overall WFP and its fluxes at the district level
- To quantify the spatial pattern of blue, green and grey WFP in the study area
- To trace the ways of water sustainability through the concept of a WFP

Uttar Pradesh has been a major producer of wheat crops, which administratively cultivated in 75 districts. This state is the most populated and fourth-largest in India. Data pertaining to the study have been collected and compiled from *Statistical Abstract – Uttar Pradesh* (a yearly statistical bulletin) for four years (2010–2013). Other ancillary data such as soil map (NBSS and LUP, 2003), rainfall and temperature (CLIMWAT, 2.0) and others were also used in the present study.

2.3 Methods and methodology

We have adopted a well-developed assessment approach for the estimation of WFP for wheat cultivation in all 75 districts of Uttar Pradesh. It also included the consideration of blue, green and grey components of WFP. A comprehensive methodology for the assessment of WFP can be accessed from *The Water Footprint Manual – 2011* (Hoekstra et al., 2011). It is, therefore, considered that the total WFP is the sum of blue, green and grey WFP as given in equation 2.1:

$$WFP_{Total} = WFP_{Blue} + WFP_{Green} + WFP_{Grey}\left[\text{Volume / Mass}\right] \qquad (2.1)$$

Here, the blue and green WFPs for wheat crop ($m^3\ t^{-1}$) were estimated by using equations 2.2 and 2.3, wherein the blue and green WFPs refers to the actual seasonal evapotranspiration (ET, $m^3\ ha^{-1}$) used during the entire period of crop growth from planting to harvest and then divided by the crop yield (Y, t ha^{-1}), respectively. Present notion of WFP is fundamentally referring to the primary crop, that is, wheat crops at the farming field, and, therefore, excluded entire secondary or tertiary processing from the present estimates:

$$WFP_{blue} = \frac{10^{*}\sum_{d=1}^{lgp} ET_{blue}}{y}, \ \left[\text{Volume / Mass}\right] \qquad (2.2)$$

$$WFP_{green} = \frac{10^{*}\sum_{d=1}^{lgp} ET_{green}}{y}. \ \left[\text{Volume / Mass}\right] \qquad (2.3)$$

Here, *lgp* refer to the length of the growing period from the first day and ending at the time of harvesting. The accumulated water consumption was, then, calculated by summation of daily ET for the blue and green components by Food and Agriculture Organization's (FAO, 2003) CROPWAT Ver. 8.0 model (M. Smith, 1992).[1] The model has two alternative options, that is, (1) 'crop water requirement'(CWR) and (2) 'irrigation schedule'. Moreover, the irrigation schedule option is more accurate as compared to the crop water requirement model. The CWR model calculates ET under optimal condition, which will be equivalent to CWR. Therefore, it refers to optimum conditions wherein well-fertilized soil, disease-free crops, large growing area and soil suitability are essential for achieving the full growth of plants (Allan et al., 1998). This model also requires effective rainfall that is, however, difficult to determine and, thus, calculated by using United States Geological Survey's soil conservation services (USGS-SCS) curve method.

Thus, the irrigation schedule option has been chosen as fit for the present study due to non-optimal conditions of the wheat grown region. It incorporates the soil–water balance approach, which tracks the content of soil moisture over the growing period using a daily time step. Thus, the equation 2.4 under an optimum condition with the soil–water regime is

$$ET_{adjusted} = K_{soil} + ET_{crop}\left[\text{length / time}\right] \qquad (2.4)$$

Here, K_{soil} denotes the effect of water stress on crop transpiration. It would be K_{soil}=1 and K_{soil}= < 1 under a no-soil-stress and water-limiting condition, respectively.

Input files for the model have been prepared as per given guidelines for each representative climatic station located in wheat-producing regions for all 75 districts. Soil and precipitation variables were calculated from Figure 2.1a and b.

Both planting and harvesting were obtained from the crop calendar prepared by the Ministry of Agriculture, Government of India, and verified with the farmers in the field through the scheduled interview method. In India, wheat is cultivated as a winter crop, which is planted in mid-November to early December and harvested in late March to early April of each year. It requires irrigation at a certain interval, usually 35 to 45 days after planting. Subsequently, the second and third irrigations are applied with intervals of 30 and 35 days, respectively. Thus, the crops need three to four irrigations throughout the whole growing season. The supply of water is usually ensured by farmers from the well or canal irrigation method. The contribution of rainfall is less in wheat cultivation as the rainfall during winter season remains less frequent. In Uttar Pradesh, wheat yield is reported to be about 31 tons per hectare. Other required inputs, including crop coefficient, critical depletion level, yield response factor, crop length, rooting depth, and soil data, were compiled from published reports, government publications and expert opinions.

Moreover, grey WFP is the pollutant factor loading that explains the pollution assimilation capacity of water for a particular crop. It is calculated (WFP_{Grey}, m^3 ton^{-1}) by following global water footprint standard manual – 2011. The grey WFP is the rate of chemical application per hectare (*CAR*,

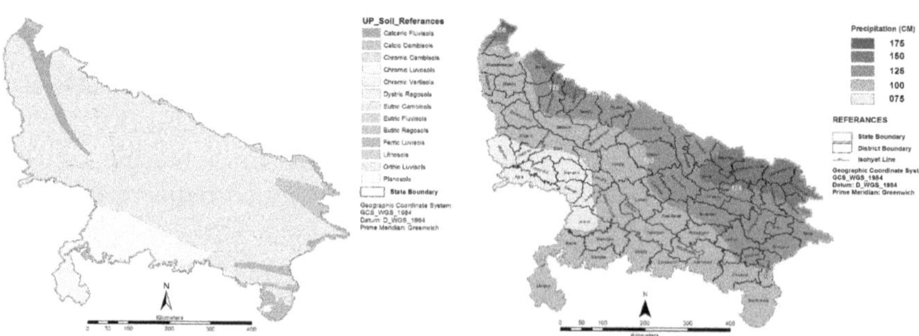

Figure 2.1 (a) Pattern of Precipitation in Uttar Pradesh; (b) Distribution of Soils in Uttar Pradesh

kg ha^{-1}), times the leaching factor (α) divided by the maximum acceptable concentration (C_{max}, kg m^{-3}) minus the natural concentration for pollutant considered (C_{nat}, kg m^{-3}) and then finally divided by the crop yield (Y, ton ha^{-1}). We calculated the grey WFP for nitrogen, phosphorus and potassium compounds using equation 2.5:

$$WFP_{Grey} = \frac{(\propto \times AR)(C_{\max} - C_{nat})}{Y}.[\text{Volume / Mass}] \qquad (2.5)$$

The required data for the grey WFP were compiled from *Statistical Abstract – Uttar Pradesh – 2013* along with fertilizer application rate of nitrogen, phosphorus and potassium applied on the net area sown. As per Hoekstra and Chapagain (2007), it was assumed that a total 10 percent of total fertilizer application rate (kg ha^{-1} yr^{-1}) will be induced by the quantity of nitrogen, which is responsible to pollute free-flowing freshwater. Precise estimates of fertilizer-induced water pollution were not available so far, while the maximum concentration of nitrogen was adopted as 10 mg/litre due to unavailability of ambient water-quality standards. Thus, this limit was assumed to calculate the required volume of freshwater to wear off the load of pollutants. However, the effect of herbicides, nutrients and pesticides on the environment has not been taken into consideration due to the unavailability of appropriate data.

2.4 Results and discussion

Total WFP, along with blue, green, and grey components, has been estimated for the wheat crop at the district level for all 75 districts of Uttar Pradesh. Comparative statements have also been prepared regarding average WFP consumption, variability, and areal extent of the wheat-cultivated area. The comprehensive share of each component, that is, blue, green and grey WFP components, of wheat cultivation has been estimated at about 91, 6 and 3 percent, respectively, in Uttar Pradesh.

Wheat cultivation largely depends on freshwater supply under irrigation conditions as the region receives scanty rainfall due to western disturbance of the monsoon. In the winter season, the region experiences high-density fogwith clouds, which helps in the yield-response factor. Moreover, the soil condition varies significantly from district to district whereas most of the soil group is dominated by citric fluvisols and cambisols, which means that the soil has a dominance of new fluvial and nutrient-rich soil composition. It is susceptible to soil degradation, salinity, acidification and alkalinity in wheat cultivated areas. Thus, it requires a much larger amount of fertilizers; that is why the share of grey WFP is found to be high as compared to the green component. The grey WFP, accounting for 6 percent of total WFP, has been extended beyond the given concept regarding the nitrogen-, phosphorus- and potassium-induced water pollutants' contribution.

However, a certain assumption has been applied in sub-categorization, that is, with regard to maximum concentration (ambient water-quality standards) and leaching fraction to free-flowing water bodies, with the help of a nitrogen-based leaching fraction. Both maximum concentration and leaching fraction of phosphorus and potassium are assumed equivalent to the nitrogen application rate (already given earlier). It is also estimated that the nitrogen shares are near about two-thirds of the total grey WFP, followed by phosphorous and potassium, with 25 and 4 percent, respectively.

Moreover, a geographic information system (GIS) environment was used to prepare figures for total WFP as well as its sub-components, which were estimated in cubic meter per ton of water consumption at the district level. Table 2.1 presents the top 10 WFP-producing districts for each category. Sonbhadra and Banda districts were found to have the highest total WFP in production, followed by Mahoba, Allahabad, Varanasi, Faizabad, Lucknow, Ghazipur, Rae Bareilly and Amethi districts. Similarly, the blue WFPs were also calculated for each district. Varanasi replaced by Ghazipur district, moving from eighth to fifth place after competing Varanasi, Faizabad and Lucknow, while others remain in the same position. The amount of total WFP also varied from 2619 to 1,655 m^3 per ton for Sonbhadra and Amethi districts in the top 10 list, respectively. Furthermore, the blue WFP amount ranges from 2,522 to 1,545 m^3 per ton for Sonbhadra and Lalitpur districts, respectively. The scenario has changed regarding green WFP due to soil properties and the available moisture balance in the soil capillaries. Thus, the top positions (highest green WFP) were estimated for Shravasti and Bahraich districts, followed by Lakhimpur Kheri, Fatehpur, Kasganj, Mainpuri, Saharanpur, Mahoba, Basti and Siddharth Nagar districts. While the available green WFP amount ranges from 217 to 74 m^3 per ton for Shravasti and Siddharth Nagar among top 10 districts, respectively, subsequently, the grey WFP amount was as high as 286 m^3 per ton in Ghaziabad district due to a high fertilizer application rate per unit area. It was 217 m^3 per ton for Varanasi, followed by Kanpur Nagar, Moradabad, Allahabad, Lucknow, Deoria, Farrukhabad, Faizabad and Maharajgani (119 m^3 per ton) districts.

Moreover, Figure 2.2 and Table 2.2 represent the average, range and variability of WFP for wheat cultivation in Uttar Pradesh at the district level for each component along with the total WFP at production. All 75 districts were categorized into five categories based on statistical range. They are categorized into 840–1,000, 1,000–1,500, 1,500–2,000, 2,000–2,500 and above 2,500 m^3 per ton groups for the total and blue WFP whereas the green and grey WFPs were classified as 0–50, 50–100, 100–150, 150–200 and above 200 m^3 per ton categories, respectively. The average total WFP for all 75 districts was estimated at 1,342 m^3/ton, while the district-wise distribution has variations as suggested by the standard deviation value of 333 m^3/ton. Similarly, total blue WFP is estimated as 1,215 m^3/ton, followed by 43 and 87 m^3/ton for green and grey WFPs, respectively.

Table 2.1 Top 10 Water Footprint (WFP) Producer Districts under Subcategories of Water Footprint

Top 10 Blue WFP	*Quantity m³ per ton*	*Top 10 Green WFP*	*Quantity m³ per tonne*	*Top 10 Grey WFP*	*Quantity m³ per tonne*
Sonbhadra	2522	Shravasti	217	Ghaziabad	286
Banda	2488	Bahraich	209	Varanasi	217
Mahoba	2159	Lakhimpur Kheri	186	Kanpur Nagar	181
Allahabad	1793	Fatehpur	151	Moradabad	178
Ghazipur	1634	Kasganj	139	Allahabad	154
Faizabad	1611	Mainpuri	119	Lucknow	134
Amethi	1577	Saharanpur	103	Deoria	132
Varanasi	1575	Mahoba	89	Farrukhabad	128
Lucknow	1572	Basti	81	Faizabad	123
Lalitpur	1545	Siddharth Nagar	74	Maharajganj	119

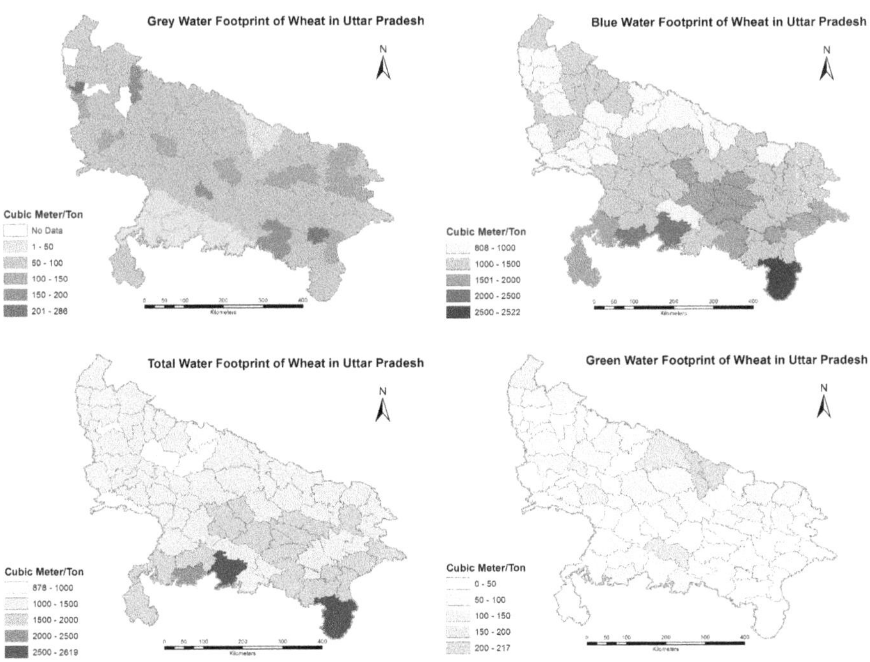

Figure 2.2 (a) Share of Total Water Footprint; (b) Share of Blue Water Footprint; (c) Share of Green Water Footprint; (d) Share of Grey Water Footprint

Table 2.2 Average and Variability of Water Footprint (WFP) and Its Components among a Group of Districts in Uttar Pradesh

WFP Range (m³/ton)		*840–1000*	*1000–1500*	*1500–2000*	*2000–2500*	*Above 2500*	*Total*
Total	No of Distt	05	48	19	01	2	75
	Average (SD*)	948	1193	1642	2280	2577	1342
		(44)	(106)	(117)	(0)	(42)	(333)
Blue	No of Distt	21	39	12	02	1	75
	Average (SD*)	907	1179	1576	2324	2522	1215
		(58)	(136)	(76)	(165)	(0)	(340)
WFP Range(m³/ton)		**0–50**	**50–100**	**100–150**	**150–200**	**200 Above**	**Total**
Green	No of Distt	51	17	03	02	02	75
	Average (SD*)	21	61	120	168	213	43
		(13)	(11)	(15)	(18)	(4)	(44)
WFP Range(m³/ton)		**1–50**	**50–100**	**100–150**	**150–200**	**200 Above**	**Total**
Grey	No of Distt	08	49	10	03	02	72
	Average (SD*)	37	78	119	171	252	87
		(9)	(14)	(10)	(12)	(34)	(42)

Source: Calculated by the author.
* Standard deviation.

Moreover, the green component has much variability due to soil regime change and considerable variation in incoming solar radiation during winter months, which resulted in the availability of soil moisture at the time of planting. Thus, the role of irrigation in wheat cultivation, specifically in Uttar Pradesh, cannot be underestimated for full growth and yield-response factor of the plant. Subsequently, the grey component of WFP exceeded the green WFP due to the high application of fertilizers, specifically nitrogen. The soil of Uttar Pradesh has a salinity and alkalinity problem at a large level. Thus, the deficiency of soil, during planting of wheat, is endorsed by nitrogen, phosphorus and di-ammonium phosphate. However, sufficient precaution was taken during the estimates of the grey component, but the leaching coefficient of nitrogen and ambient water-quality standards may alter the result of present findings, although not by a large level.

Out of 75 districts in Uttar Pradesh, two-third of the districts (48 districts) fall into the category of 1,000–1,500 m^3/ton, whereas average total WFP for wheat cultivation was estimated at 1,193 m^3/ton, with the standard deviation of 106 m^3/ton. Figure 2.2a presents the spatial variability of total WFP, which is categorized according to their respective WFP at the production level. The spatial distribution of the previous category is accompanied by the soil type and irrigation infrastructure available in each district. Findings could be associated with Figure 2.1 as boththe soil type and isohyet are the major factors to change in WFP estimate of wheat cultivation. Moreover, the highest total WFP was estimated for two districts with an average WFP 2,577 m^3/ton under the above 2,500 m^3/ton category. In the 2,000–2,500 and 1,500–2,000 categories, the total average WFP of 2,280 and 1642 m^3/ton was found for Districts 1 and 19, respectively, while the latter category has high variability as compared to other range categories. The range of 840–1,000 m^3/ton category is found in five districts, that is, Badaun, Shahjahanpur, Pilibhit, Agra and Hapur, which are also the districts with the lower estimated WFP among all total WFP districts. The average WFP was estimated at 948 m^3/ton, with the standard deviation of about 44 m^3/ton, means less variability, in the previous category.

Subsequently, the amount of blue WFP at the producer level was also high in all 75 districts, which means that districts have a maximum share of blue WFP (Figure 2.2b). It was estimated with an average value of 1,179, 907 and 1576 m^3/ton in 39, 21 and 12 districts in the categories of 1,000–1,500, 840–1,000 and 1,500–2000, respectively. Similarly, the green WFP was estimated with an average 21 m^3/ton in 51 districts, followed by 61 and 120 m^3/ton in districts 17 and 3, respectively (Figure 2.2c). The highest green WFP was recorded with an average of 168 and 213 m^3/ton for two districts in each category. However, the spatial distribution of green WFP has dominance in a lowersoil–moisture regime in the entire state. Most of the districts fall into the below 50 m^3/ton category, which means that the condition of the soil needs much improvement for wheat cultivation to minimize water loss.

Furthermore, the grey WFP was highest in 2 districts, with an average of 252 m^3/ton, followed by 171 m^3/ton in 3 districts out of 72 districts. There was insufficient data available for three districts, that is, Hapur, Sambhal and Shamli, due to the delimitation of the district boundary. Thus, it has been included in the account of the present study. Out of 72 districts, most of the districts (49) were estimated with an average grey WFP of 78 m^3/ton, while another 10 districts were estimated with an average of 119 m^3/ton, followed by 37 m^3/ton in 8 districts. However, these estimates show pertinent variability when comparing the individual-level WFP to the production-level WFP.

The present study also takes an advanced approach towards sub-categorization of grey WFP as an extension of WFP concept and possibly not reported so far in the literature of WFP, into nitrogen, phosphorus and potassium components of WFP in Uttar Pradesh. The components of grey WFP were classified into each constituent of fertilizer, that is, nitrogen, phosphorus, and potassium, while the nitrogen component was classified into five classes starting from 1–50 m^3/ton category to the above 200 m^3/ton category. Similarly, phosphorus has also been classified into three classes, that is, 1–25, 25–50 and above 50 m^3/ton categories, while potassium-related WFP ranges between the 1–5, 5–10 and above 10 m^3/ton categories. The collective results are presented in Table 2.3, along with the spatial distribution of each component at the district level in Figure 2.3a–c, respectively.

Nitrogen-related grey WFP has an almost two-third share (71 percent) with an average district-level WFP of about 62 m^3/ton at the production of all 72 districts in Uttar Pradesh. Between the 50–100 and 1–50 m^3/ton categories, there were 41 and 26 districts estimated with an average WFP of about 68 and 37 m^3/ton, respectively. It shows a commendable fertilizer application rate per unit area. The remaining five districts were estimated with the highest WFP due to excess of fertilizer application rate of about 129 and 208 m^3/ton between the 100–150 and above 200 m^3/ton categories, respectively.

However, none of the districts reported in the under 150–200 m^3/ton category, while the maximum nitrogen-related WFP was found in Ghaziabad only. It could be a data discrepancy due to delimitation of district boundary of Ghaziabad and shift of some area in Hapur district. Thus, Ghaziabad district includes estimates of the sum of the total fertilizer application rate applied in both the districts collectively, also mentioned in the 2013 statistical abstract of Uttar Pradesh. Subsequently, the other four districts, that is, Allahabad, Kanpur Nagar, Moradabad and Varanasi, which were calculated in the 100–150 m^3/ton category, were found to be above the normal distribution. It could be affected by small landholding size due to a commendable share of land utilization belonging to urban use. Hence, the rate of fertilizer application might be more to ensure maximum yield in small size of landholdings.

Therefore, the normal distribution of nitrogen-generated grey WFP could be considered below 100 m^3/ton or an average of about 68 m^3/ton per unit

Table 2.3 Grey Water Footprint (WFP) and its Subcomponents with Average and Variability

WFP Range (m^3/ton)		*1–50*	*50–100*	*100–150*	*> 150*	*Total*
Nitrogen (N)	No of Distt	26	41	4	1	72
	Average (SD*)	37	68	129	208	62
		(11)	(12)	(13)	(0)	(30)
WFP Range (m^3/ton)		**1–25**	**25–50**	**> 50**	**Total**	
Phosphorus (P)	No of Distt	55	14	3	72	
	Average (SD*)	17	30	61	23	
		(4)	(3)	(7)	(11)	
WFP Range (m^3/ton)		**1–5**	**5–10**	**>10**	**Total**	
Potassium (K)	No of Distt	57	14	1	72	
	Average (SD*)	2.6	7	14	04	
		(1.2)	(1.2)	(0)	(2)	

Source: Calculated by the author.
* Standard deviation.

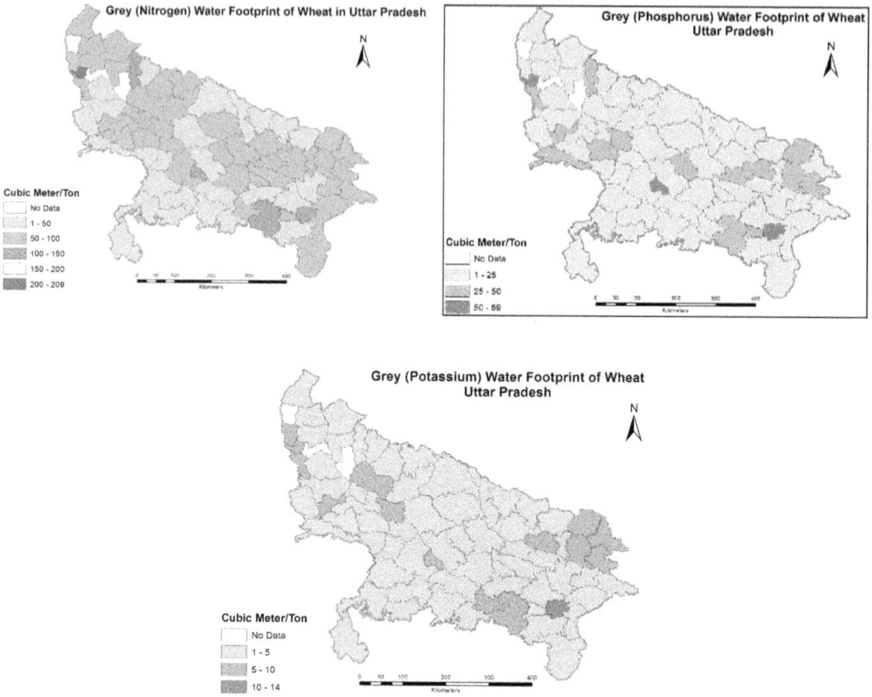

Figure 2.3 (a) Distribution of Grey (N) Component; (b) Distribution of Grey (P) Component; (c) Distribution of Grey (K) Component

area in most of the district. Similarly, phosphorus-generated grey WFP was estimated with an average 23 m^3/ton for all 72 districts, while this average was 17 m^3/ton in 55 districts, followed by 30 m^3/ton in 14 districts. Only three districts, namely Ghaziabad, Kanpur Nagar and Varanasi, were found above 50 m^3/ton categories, with an average 61 m^3/ton for each. It is again due to the delimitation of Ghaziabad and excess of the urban land use along with small-size landholding as in the case of nitrogen WFPs. The data of Hapur, Shamli and Sambhal districts were unavailable. Furthermore, potassium-related contribution in grey WFP shows that there were 57 districts with an average WFP of about 2.3 m^3/ton, followed by 7 m^3/ton in 14 districts. Only Ghaziabad again reported in the high-WFP category with 14 m^3/ton while the overall average was estimated 4 m^3/ton for all 72 districts. Here, also, this phosphorus-related grey WFP was highest in Ghaziabad for the same reason discussed earlier.

2.5 Conclusion

The present study presents findings of WFP for wheat cultivation. The estimate was obtained as per the guidelines of the international WFP network at the production level (on farm) for the period from 2010 to 2013.

Separate estimates were prepared for blue, green and grey WFP components. Additionally, an attempt has also been made to extend the concept of WFP regarding the grey WFP's sub-component-wise contribution, previously not reported so far. Findings have been presented for entire Uttar Pradesh except for grey WFP in three districts, Hapur, Shamli and Sambhal, due to unavailability of appropriate data. Subsequently, the study discloses the importance of blue WFP, which contributes nearly 91 percent of total WFP. While the share of green and grey WFPs was estimated nearly 6 and 3 percent, respectively, which is consumed by the wheat crop during the sown-to-harvest period. Moreover, the spatial variability was also reported in blue, green and grey WFP estimates at the district level, which is assimilated by the soil regime and climate barriers. Also, the wheat crop is grown as a winter crop and hence, it largely depends on the irrigation practice. The share of blue WFP was high in semi-arid and arid areas. Consequently, Sonbhadra district, along with the Bundelkhand region's districts,has a hefty share of blue WFP. The south-western region of Uttar Pradesh has been recognized to be a high-WFP region as compared to north-eastern, central and eastern parts of the state. Additionally, the contribution of grey WFP is estimated to be more than the green WFP due to the high rate of fertilizer application per unit area. Subsequently, the share of grey WFP imparted by nitrogen, phosphorus and potassium, as a separate estimate for each component, is calculated as an addition to the concept of WFP. Grey WFP was dominated by nitrogen-related WFP by 71 percent, followed by phosphorus and potassium as 25 and 4 percent, respectively. Therefore, a balanced approach among blue, green and grey WFP has been recommended to address the increasing freshwater demand in wheat cultivation. It can also be recommended regarding grey WFP's sub-components, as the residence time and the leaching fraction of fertilizer contributed much to water pollution. The present study also provided a space for future research regarding the economic value of water. The study offers some expert suggestions that could be implemented to improve future water efficiency and, thus, the WFP estimates. Pertinent suggestions are as follows:

- The WFP approach is bound with planting and harvesting dates, which affect estimates of WFP and variation among districts.
- The length of the growing season (present case: 130 days) is adversely stimulus to water consumption. However, the study exploits a few expert guesses, which were collected through scheduled interviews and from literature regarding the growing period of the crop and crop phenology and thus do not demonstrate possible variation.
- A novel aspect can be to enhance crop productivity and decrease the length of the growing period by adopting technical assistance and improvement in various seeds hybridization.
- The absence of a sufficient database is also critical as it limits the scope of the study and future estimates.

Furthermore, a pilot study has been recommended to improve databases, determine the reliability of the estimates and prepare a prediction strategy. The present study lacks the validation of its results as appropriate guidelines and measurement tools were not available. However, wheat cultivation largely depends on irrigation in absence of sufficient rainfall during the growing season. However, soil regime, a cause of rooting depth of the crops, is not found to be an influential factor due to easy access to the water table. It is significant in rain-fed agricultural land where water-holding capacity defines the growth of plants. Subsequently, the consistency and validity of data were not tested due to the data being compiled from the government agencies; however, it may affect present results. It is evident that south-western districts,situated in rain-fed agriculture zone,thus, have a low rate of fertilizer application, whereas the north-eastern, central and eastern parts of Uttar Pradesh state, particularly Badaun, Philibhit, Shahjahanpur, Saharanpur, Muzaffarnagar, Meerut, Bijnor, Jyotiba Fule Nagar, Lakhimpur Kheri, Sitapur, Hardoi, Gonda, Basti, Bara Banki, Azamgarh Jaunpur and others, are the reverse.

Note

1 For further guidelines, one can consult with available information at http://www.fao.org/nr/water/infores_databases_cropwat.html.

References

Aldaya, M. M., J. A. Allan, and A. Y. Hoekstra. "Strategic Importance of Green Water in International Crop Trade." *Ecological Economics* 69, no.4 (2010a):887–894.

Aldaya, M. M., A. Garrido, M. R. Llamas, C. Varelo-Ortega, P. Novo, and R. R. Casado. "Water Footprint and Virtual Water Trade in Spain." In A. Garrido and M. R. Llamas (eds.), *Water Policy in Spain*. Leiden, Netherlands: CRC Press (2010c), 49–59.

Aldaya, M. M., and M. R. Llamas. (2008). "*Water Footprint Analysis for the Guadiana River Basin*." Value of Water Research Report Series No 35, UNESCO-IHE. Delft, Netherlands. www.waterfootprint.org/Reports/Report35-WaterFootprint-Guadiana.pdf

Aldaya, M. M., G. Muñoz, and A. Y. Hoekstra. (2010b). "Water Footprint of Cotton, Wheat and Rice Production in Central Asia." *Value of Water Research Report Series No 41, UNESCO-IHE*. Delft, Netherlands.

Alexandra, E., V. Evans, A. Munir, Hanjra Yunlu Jiang, Manzoor Qadir and Pay Drechsel. (2012). "Water Quality: Assessment of the Current Situation in Asia." *International Journal of Water Resources Development* 28, no. 2 (2010): 195–216. doi:10.1080/07900627.2012.669520

Allan, J. A. (1997). ""Virtual Water': A Long-Term Solution for Water Short Middle Eastern Economies?" Water Issues Group. London: School of Oriental and African Studies (1997). Retrieved 31 December 2014 from http://www.soas.ac.uk/faculties/lawsocialsciences.cfm?navid=2811

Allan, J. A. "Virtual Water – the Water, Food, and Trade Nexus: Useful Concept or Misleading Metaphor?" *Water International* 28, no. 1 (2003): 106–113.

Allen, R. G., L. S. Pereira, R. Dirk, and M. Smith. (1998). "Crop Evapotranspiration: Guidelines for Computing Crop Water Requirements." *FAO Irrigation and Drainage Paper 56*, Food and Agriculture Organization, Rome.

Allan, T. *The Middle East Water Question: Hydropolitics and the Global Economy*. London: I B Tauris and Co. (2001).

Andersson, N. "Disaster Epidemiology: Lessons From Bhopal." In V. Murray (ed.) *Major Chemical Disasters – Medical Aspects of Management*. London: Royal Society of Medicine Services Ltd. (1990), 177–189.

Baxter, P. "Review of Major Chemical Incidents and Their Medical Management." In V. Murray (ed.) *Major Chemical Disasters – Medical Aspects of Treatment*. London: Royal Society of Medicine (1990), 7–20.

Chapagain, A. K., and A. Y. Hoekstra. "Virtual Water Flows between Nations in Relation to Trade in Livestock and Livestock Products." *Value of Water Research Report Series No.13*, UNESCO-IHE. Delft, Netherlands (2003).

Chapagain, A. K., and A. Y. Hoekstra. "Water Footprints of Nations." Value of Water Research Report Series No.16, UNESCO-IHE. Delft, Netherlands (2004). www.waterfootprint.org/Reports/Report16Vol1.pdf

Chapagain, A. K., and A. Y. Hoekstra. "The Global Component of Freshwater Demand and Supply: An Assessment of Virtual Water Flows between Nations as a Result of Trade in Agricultural and Industrial Products." *Water International* 33, no. 1 (2008), 19–32.

Chapagain, A.K., and A. Y. Hoekstra. "The Green, Blue and Grey Water Footprint of Rice from Both a Production and Consumption Perspective." *Value of Water Research Report Series No.40*, UNESCO-IHE, Delft, Netherlands (2010).

Chapagain, A. K., Hoekstra, A. Y., and Savenije, H. H. G. "Water Saving through International Trade of Agricultural Products" *Hydrology and Earth System Sciences Discussions* 10, no. 3 (2006):455–468.

Falkenmark, M. (1989). "The Massive Water Scarcity Now Threatening Africa: Why Isn't It Being Addressed?" *Ambio*, 18, 112–118.

Falkenmark, M., and C. Widstrand. "Population and Water Resources: A Delicate Balance." *Population Bulletin* 4 (1992):1–5.

FAO. (2003). CROPWAT model. (Software). Rome: Food and Agriculture Organization.

FAO. (2010). CLIMWAT 2.0 database, FAO, Rome. www.fao.org/nr/water/infores_databases_climwat.html

FAO. (2016). AQUASTATE website. Retrieved on December, 15, 2015 from http://www.fao.org/nr/water/aquastat/water_use/index.stm

Gaojie, W., C. Wenzhi, L. Lusan, and W. Feifei (2018). "Water Pollution Management in China: Recent Incidents and Proposed Improvements." *Water Supply* 18, no. 2: 603–611. doi:10.2166/ws.2017.139

The Guardian. "How Much Water Is Needed to Produce Food and How Much Do We Waste." 13 January 2013. Retrieved online from http://www.gaurdian.com/news/datablog/2013/jan/10/how-much-water-food-production-waste.

Gleick, P. H. *Water in Crisis*. London: Oxford University Press (1993).

Hoekstra, A. Y. "Virtual Water Trade: Proceedings of the International Expert Meeting on Virtual Water Trade." Value of Water Research Report Series No. 12. UNESCO-IHE. Delft, Netherlands (2003). www.waterfootprint.org/Reports/Report12.pdf

Hoekstra, A. Y., and A. K. Chapagain. "Water Footprints of Nations: Water Use by People as a Function of Their Consumption Pattern." *Water Resources Management* 21, no. 1 (2007):35–48.

Hoekstra, A. Y., Chapagain, A. K., Aldaya, M. M., and Mekonnen, M. M. "*Water Footprint Manual: State of the Art 2009.*" Water Footprint Network, Enschede, Netherlands (2009). www.waterfootprint.org/downloads/WaterFootprintManual2009.pdf

Hoekstra, A. Y., A. K. Chapagain, M. M. Aldaya, and M. Mekonnen (2011). *The Water Footprint Assessment Manual: Setting the Global Standard*, Earthscan, London.

Hoekstra, A. Y., and P. Q. Hung. "Virtual Water Trade: A Quantification of Virtual Water Flows between Nations in Relation to International Crop Trade" Value of Water Research Report Series No 11, UNESCO-IHE. Delft, Netherlands (2002). www.waterfootprint.org/Reports/Report

Kampman, D. A., A. Y. Hoekstra, and M. S. Krol. "The Water Footprint of India." *Value of Water Research Report Series No. 32*. UNESCO-IHE. Delft, Netherlands (2008).

Kumar, V., andS. K. Jain. "Status of Virtual Water Trade from India." *Current Science* 93 (2007): 1093–1099.

Mekonnen, M. M., and A. Y. Hoekstra. "A Global and High-Resolution Assessment of the Green, Blue and Grey Water Footprint of Wheat." *Hydrology and Earth System Sciences* 14, no. 7(2010): 1259–1276.

Mekonnen, M. M., and A. Y. Hoekstra. (2011). "The Green, Blue and Grey Water Footprint of Crops and Derived Crop Products." *Hydrology and Earth System Sciences* 15, no.5 (2011): 1577–1600.

NBSSandLUP. "Soils of Uttar Pradesh for Optimising Land Use." *NBSS Publ. 68 (Soil of India Series at Scale of 1:500000)*. Nagpur, India, National Bureau of Soil Survey and Land Use Planning (2003).

Okada, M., and S. A. Peterson. "*Water Pollution Control Policy and Management: The Japanese Experience.*" Ginza, Japan: Gyosei (2000).

Romaguera, M., A. Y. Hoekstra, S. Zhongbo, M. S. Krol, and M. D. Salama. "Potential of Using Remote Sensing Techniques for Global Assessment of Water Footprint of Crops." *Remote Sensing* 2, no. 4 (2010): 1177–1196.

Shiklomanov, I. A. "Assessment of Water Resources and Water Availability in the World." In Shiklomanov, I. A. (ed.), *Background report to the Comprehensive Freshwater Assessment*, St. Petersburg: State Hydrological Institute (1996).

Smith, M. "CROPWAT – A Computer Program for Irrigation Planning and Management." *Irrigation and Drainage Paper 46*, FAO: Rome (1992).

Smith, W. E., and A. M. Smith. *Minamata: An Alskog-Sensorium Book*. New York: Holt, Rinehart and Winston Publishers (1975).

Suhail, M. (2015). "Water Resource Management in Saudi Arabia." Unpublished PhD Thesis. Aligarh Muslim University, India.

UN Water. "Statistics: Statistics Detail." (2014). Retrieved from UN-WATER: http://www.unwater.org/statistics/statistics-detail/en/c/246692/

UNDP. *Human Development Report 2006: Beyond Scarcity – Power, Poverty and the Global Water Crisis*. United Nations Organisation, London, UK: Macmillan Palgrave (2006).

UNEP. *Vital Water Graphics: An Overview of the State of the World's Fresh and Marine Waters*. Nairobi, Kenya: UNEP (2008).

UNEP. "*A Snapshot of the World's Water Quality: Towards a Global Assessment.*" United Nations Environment Programme, Nairobi, Kenya (2016).

United Nations, Department of Economic and Social Affairs, Population Division. "World Population Prospects: The 2015 Revision, Data Booklet." ST/ESA/SER.A/377 (2015).

Wang, M.-H., J. Li, and Y.-S. Ho. "Research Articles Published in Water Resources Journals: A Bibliometric Analysis." *Desalination and Water Treatment* 28, no. 1 (2011): 353–365.

WWAP. *Water in a Changing World: World Water Development Report 3*. Paris: UNESCO/Earthscan (2009).

WWAP. *The United Nations World Water Development Report 4: Managing Water Under Uncertainty and Risk*. Paris: UNESCO (2012).

3 Productive efficiency of agricultural sector in Uttar Pradesh

Karan Singh Khati, Anup Kumar Bhandari and Deep Mukherjee

3.1 Introduction

The *World Development Report 2008* (Byerlee et al., 2008) shows that agricultural growth is at least twice as effective in reducing poverty in developing countries, compared to growth originating in nonagricultural sectors. Agricultural growth is a prerequisite to meet Sustainable Development Goals, as stated in the 2030 Agenda for Sustainable Development that succeeded the Millennium Development Goals from 2016. According to the Food and Agriculture Organization of the United Nations, the 2030 Agenda sets a challenge for planners and policy makers to make agriculture and food systems more efficient, that is, improving efficiency in the use of resources. There are several approaches to analyzing agricultural sustainability, mostly through some index, technical efficiency and environmental efficiency being two of them. De Koeijer et al. (2002) assume that agricultural sustainability is a 'mix' of environmental and economic performances and present a modeling framework for quantifying sustainability on the basis of productive efficiency theory from economics.

The sustainable development of agriculture is one of the major thrust areas in the mind of the government of India as a potential for the agriculture-based rural economy in India is juxtaposed by the low levels of profitability and entrepreneurship. For most of the important crops grown in India, productivity is still lower than the international averages. Thus, a key issue in agricultural development is the need to improve productivity given limited resources. This warrants the necessity for in-depth analysis of current productive efficiency and renewed initiatives for productivity enhancement. Factors like business environment, infrastructure, and social institutions influence the efficiency of production units. Some of the basic issues behind the productive efficiency of Indian agriculture are the revitalization of cooperative institutions, enhancing the availability of agricultural credits, rural infrastructure (irrigation, roads, electricity, and communications) development, and post-harvest management. Thus, agricultural development in India requires supportive investment in rural infrastructure and inclusive markets. In this context, this chapter studies the case of Uttar

Pradesh (UP)—the largest agrarian state of India. Because UP has a large presence in the country's agricultural economy, any improvement in agricultural productive efficiency will have a strong influence on the sustainable development indicators of the country.

Recently, Khan and Ansari (2018) provide econometric evidence of the importance of the agricultural sector on the economic growth of UP. They have applied Granger causality technique on annual data for the period between 1990 and 2015 and show that agriculture is a driver of the unregistered manufacturing sector, the transport and communication sectors, and the overall UP economy as a whole. Thus, the need for agricultural development for the overall development of the state, in particular, and the country, in general, cannot be undermined. However, agricultural productivity in UP and particularly at the district level has not been rigorously studied in the recent past. To bridge the research gap and as a fact-finding exercise for a future road map toward higher agricultural productivity growth, we analyze the productive efficiency of the agricultural sector at the district level in the state of UP.

The chapter is arranged as follows: Section 3.2 presents a summary picture of UP's agriculture; Section 3.3 describes the theoretical models for computing the productive efficiency measures; Section 3.4 describes the data and sources, outlines the empirical models, and explains results; and Section 3.5 concludes the chapter with a discussion on policy perspectives.

3.2 Agriculture in Uttar Pradesh: Past and present

Population-wise, UP is the largest state of India, and its economy is fifth largest in terms of gross state domestic product (GSDP) in 2011–2012. A report by the National Sample Survey Organization reveals that UP hosts about 20% of total agricultural households in the country in the 2012–2013 crop year. About 68.5% of rural households in UP are such that at least one member is self-employed (either in principal or subsidiary status) in farming during the last 365 days, and about 69% of rural household income comes from agricultural activities. As per a study conducted by the Associated Chambers of Commerce and Industry of India, agriculture and allied sectors contribute about 22% to UP's GSDP as of 2013–2014, and the state has emerged as the leading contributor to India's agriculture and allied sector with a share of about 13% for 2013–2014.

Historically, the state of UP is a major producer of food-grains in the country. The state accounted for the largest share by area (16.05%), as well as production (18.9%) in food grains in 2013–2014, and ranked first in wheat production and second in rice and pulses production for 2012–2013. Other important crops of UP are sugarcane and potato, and the state ranked first in the production of both the crops in 2012–2013. However, published government statistics also reveal that the productivity of rice and wheat has stagnated over the past decade. It is a matter of concern as wheat and rice are the staple crops and UP is a major contributor in the food-grain pool of the country. Potato has also performed dismally in productivity. In the last decade, the productivity has not increased but has shown a decrease from 2002–2003 to 2008–2009, showing a fluctuating increase after that. Sugarcane productivity has increased over the years but has shown only a small improvement in the later years. In addition, there has been a lot of fluctuation. It is a major commercial crop for the state and is important for its economic growth.

The performance of UP has been dismal in agricultural growth. In the 10th Plan period, the state lagged the national average. During the 10th Plan period, while the overall economy of UP grew by 5%, agriculture and allied sectors grew only by 1.4%. The 11th Plan period (2007–2012) strived to restructure the state policies for faster, broad-based inclusive growth, targeting an economic growth of 10% by the end of the plan period. Keeping in mind the importance of agriculture in the economy, the highest priority was given to the development of agricultural infrastructure and increasing productivity so as to double the income of farmers. Consequently, the share of agriculture and allied sectors in planned outlay was increased from 7.75% in the 10th Plan to 10.57%, targeting a growth of 5.7%. The intervention was planned at various levels. It was realized that, although fertilizer use is low as compared to the neighboring states, the indiscriminate use of it has deteriorated the soil quality. Soil testing and restoration were necessary for long-term sustainable growth of productivity. Irrigation is a major factor in deciding agricultural productivity. Over the years, erratic monsoons and inability of storage have led to extensive use of groundwater, thus depleting it. Also, the indiscriminate use of fertilizers has degraded the groundwater quality. Plans were formulated targeting water harvesting and groundwater recharging. Improved technology is a must for increasing productivity. Technology dissemination and extension through agricultural universities and other agencies was envisaged. Agriculture Technology Management Agency was initially set up in 32 districts of UP with plans to expand to all the districts. Another important aspect to be targeted was availability of institutionalized credit. The focus was on public–private partnerships and cooperatives to enhance the short-term cooperative credit structure through credit societies and cooperative banks. Apart from the State Agricultural Plan funding, there was provision of additional central assistance schemes through the Rashtriya Krishi Vikas Yojna. Another area of concern is the logistics of getting the farm produce to the final consumer. The non-availability of markets and dependency on middlemen for disposal of harvested crops have been a major reason for farmers being exploited. Thus, the need to provide mandis as close to the farmers as possible to avoid wastage of crops and exploitation was also a thrust area. All these provisions were directed to enhance the productivity and growth of the agricultural sector of the state.

The large size, along with the geographic, climatic, and sociocultural variations, has created a disparity in the agricultural performance of the districts in the state. Raman and Kumari (2012) conducted a district-level analysis of agricultural development for 1990–1991 and 2008–2009 using 13 agricultural development indicators. They found that the overall agricultural growth is dismal, but there are wide inter-region and inter-district differences in the agricultural performances. Districts in the western and central regions have performed better compared to the Bundelkhand and eastern regions. Districts in the western region not only have shown the best performance in terms of agricultural development indicators but also have low inter-district variation. Irrigation facility is the major differentiator in agricultural performance and is the key reason for the lacking performance of districts in the Bundelkhand region. Another noteworthy point is that the disparity among districts, specifically on technology-based indicators, has increased over the period. The technologically advanced districts have improved while the backward districts have lagged further. This emphasizes the need for technological advancements for agricultural development.

3.3 Methodology

Benchmarking is the analytical process of comparing the performance outcome of a decision-making unit (DMU) to what is reckoned to be the best achievable outcome by its peers. First, the analyst has to define the 'best' outcome, and that can then be used as a benchmark for evaluation of the actual performance outcome of the DMUs. The method has potential to improve the DMU's performance by identifying lapses in the production process or operations and suggesting best demonstrated practices followed by its peers. The technical efficiency (TE) of a DMU is its performance relative to the benchmark, the best possible performance, with the given technology. Best possible performance is an abstract and is not known and hence needs to be estimated empirically. The data envelopment analysis (DEA) methodology helps us construct this benchmark, given the performances of all the DMUs under consideration. DEA is a mathematical programming-based technique that uses linear programming (LP) for finding the combination of DMUs to construct the benchmark. It is particularly useful when the production data consist of multiple inputs and outputs, making comparisons difficult.

A formula for relative efficiency incorporating multiple inputs and outputs follows, and the development of the DEA models that measure relative efficiency is explained subsequently. The intellectual origin of the DEA in economics can be traced back to the early 1950s. The measurement of relative efficiency in the multi-inputs and multi-outputs setup was first addressed by Farrell (1957), who himself mentioned Debreu (1951) as an inspiration for developing his measure of TE. In fact, this has led some scholars to call it the "Debreu–Farrell measures of efficiency." In his pioneering work, Farrell assumed constant returns-to-scale (CRS) technology in production. In simple terms, a common measure for relative efficiency is given by

$$\text{Efficiency} = \frac{\text{weighted sum of outputs}}{\text{weighted sum of inputs}}.$$

Initially, it is assumed that this measure requires a common set of weights to be applied across all DMUs. However, Charnes et al. (1978) recognized the difficulty of obtaining a common set of weights to determine the relative efficiencies. They introduced DEA – a new mathematical programming-based method. In simple words, they proposed that each DMU should be allowed to adopt a set of weights which represents it in the most favorable way in comparison to the other DMUs. We present the fundamentals of DEA in a simple manner that is appropriate for economics and management students of the graduate level, with a minimum of technical language and readers' prior knowledge in mathematics. Needless to say, this piece is a minor addition to already prevailing excellent reviews of the field (e.g., Førsund and Sarafoglou, 2002; Ray, 2019) aimed mostly at advanced researchers. The theoretical discussion presented here draws heavily on the work of Coelli et al. (2005) and Ray (2004).

The very nature of decision-making or the objective of a DMU determines the orientation of DEA models and associated TEs. If the stakeholder is interested in knowing whether it is possible to produce more from the same input bundle and, if so, how much more, realizing the full output potential is of primary importance and output-oriented TE measures are required. On the other hand, if the stakeholder is interested in knowing whether it is possible to economize on inputs, and if so, by how much, conserving inputs has priority and input-oriented TE measures are needed.

Let us, first of all, discuss those previously mentioned TE concepts with the help of a diagram for the simplest one-input and one-output case. Under variable returns-to-scale (VRS) technology, let ABB_1 (in Figure 3.1) be the production frontier. An output-oriented measure of TE of a firm is defined as the ratio of actual output to the frontier level of output for the given level of input used by the firm. Thus, the output-oriented TE of firm E is given by $\frac{EX_1}{B_1X_1}$. In general, it measures the extent to which one can proportionately expand a DMU's output vector without changing the input scale. The same concept can be defined for the CRS technology as well. For example, in Figure 3.1, ODD_1 depicts the production frontier for CRS technology, and the output-oriented TE of firm E is $\frac{EX_1}{D_1X_1}$. Notably, irrespective to the technology specification (CRS or VRS), TE is equal to unity at all points on the frontier. Similarly, we can define TE for input orientation as the proportion to which input use can be reduced without changing the output quantity. For instance, the firm's input-oriented TE measures would be the

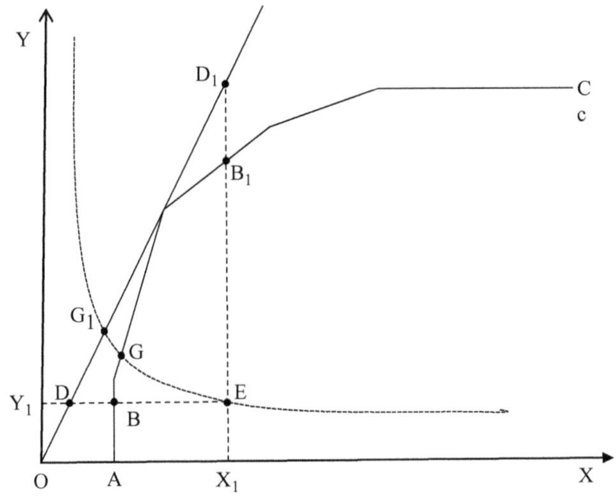

Figure 3.1 Single-Input Single-Output Production Frontier

ratio of Y_1B to Y_1E and that of Y_1D to Y_1E, respectively, for VRS and CRS technologies.

Although we have explained the TE concepts through a production function diagram, note that in DEA, there is no prior necessity of defining the production function. Some salient features of DEA are (1) no parametric specification of the frontier and (2) reliance on some fairly general assumptions about the nature of the underlying production technology, which would be consistent with many production functions. The CRS assumption for the production technology implies that DMUs operate on optimal scales. Such a presumption, in reality, may not be always reasonable as different firms operate under different types of market power, financial constraints, and externalities. Banker et al. (1984) relaxed the CRS assumption and proposed the DEA formulation for VRS technology. This model, popularly known as the BCC model, distinguishes between *technical inefficiency* and *scale inefficiency* by defining and estimating the former at a given scale of operation under the assumption of a unique optimum. We intentionally present the BCC model first to help the readers understand the CCR model better, although the latter was published earlier.

The model, in addition to a VRS structure, assumes the following fairly general axioms[1] for the production technology of firms: (a) all the observed input–output bundles are *feasible*; (b) the production possibility set is *convex*, implying that given a set of N feasible input–output bundles, *any* weighted average of these N input bundles can produce the same weighted average of the corresponding N output bundles, where N is the number of firms within the industry; and (c) any input or output is *freely disposable*. Enabled by these assumptions, utilizing the DEA method, we can construct a production possibility frontier on the basis of the observed inputs-outputs bundles of a given set of DMUs. The *frontier* is constructed by solving a sequence of LP problems—one for each DMU in the sample. The frontier, thus obtained, is essentially a piecewise linear surface that envelops the data points. It then yields, as a by-product, the extent of the technical inefficiency of a unit in terms of the distance between the observed data point corresponding to the unit and the frontier so constructed. We briefly describe the method next.

Let the firm i be observed to produce Y_i, an r-component (column) vector[2] of quantities of outputs, by using the input bundle X_i, a k-component (column) vector of quantities of inputs, the jth element of X_i (Y_i) is taken to be zero, if the ith firm does not use (produce) the jth input (j^{th} output). Using the axioms (a) through (c) mentioned earlier, we can easily construct the corresponding (VRS) production possibility set (PPS) as

$$T^V = \left\{ (X,Y) : \sum_{i=1}^{N} \lambda_i Y_i \geq Y, \sum_{i=1}^{N} \lambda_i X_i \leq X, \sum_{i=1}^{N} \lambda_i = 1, \lambda_i \geq 0 \ \forall i \right\}$$

The DEA method intends to construct a frontier on the basis of the observations on inputs and outputs of the N firms, by solving a set of N LP problems, one for each firm. The problem for the firm s, it can be easily understood keeping the notion of proportional measure of TE and the PPS defined earlier in the backdrop, is to find a scalar φ and an N-component vector $\lambda_s = (\lambda_{si})$, which solve the following LP:

$$\left(P_s^M\right) \qquad \textit{Maximize} \quad \varphi$$

subject to (1) $\sum_{i=1}^{N} \lambda_{si} Y_i \geq \phi Y_s$, (2) $\sum_{i=1}^{N} \lambda_{si} X_i \leq X_s$, (3) $\sum_{i=1}^{N} \lambda_{si} = 1$ and (4) $\lambda_{si} \geq 0 \ \forall i$.

Let (P_s^M) have an optimal solution, say, $[\phi_s^{M*}, \lambda_s^M = \left(\lambda_{si}^M\right)]$. The optimal value, ϕ_s^{M*}, then indicates the *maximum* possible *proportional expansion*[3] *of* the output vector that could be achieved by the sth firm while keeping the input quantities fixed at X_s. This *proportion* may then be used to get a measure of the *output-oriented* TE[4] of the sth firm relative to the frontier ($TE_{y,s}^M$) as defined

$$TE_{y,s}^M = 1 / \phi_s^{M*}.$$

Let us now focus on the interpretation of the outcome obtained from the preceding DEA-LP problem. The DMUs which have optimal objective function value equal to one are located on the frontier and are called "efficient" or "best practice" units. In a DEA-LP problem, at least one of the studied DMUs is located on the frontier. Those away from the frontier are deemed to be inefficient. Those "efficient" DMUs are considered benchmarks for inefficient units. When different outputs can be expanded at different rates, φ^* is the lowest of these expansion factors. For example, in a two-output case, if one output can be expanded by a factor of 2 and the other by a factor of 1.5, φ^* equals the lower of the two values. In this case, it is possible to expand the output bundle itself by at least 50%. In this case, the output-oriented technical efficiency is 0.67, indicating that it has realized only two-third of the potential output that can be produced from its input bundle.

One may also define and measure an *input-oriented* TE of a firm. The problem for the firm s is then to find a scalar θ and an N-component vector $\alpha_s = (\alpha_{si})$, which optimizes the following LP:

$$\left(P_s^I\right) Minimize \quad \theta$$

subject to (1) $\sum_{i=1}^{N} \alpha_{si} Y_i \geq Y_s$, (2) $\sum_{i=1}^{N} \alpha_{si} X_i \leq \theta X_s$, (3) $\sum_{i=1}^{N} \alpha_{si} = 1$ and (4) $\alpha_{si} \geq 0 \; \forall \, i$.

Let the problem (P_s^I) have an optimal solution, say, $[\theta_s^{I*}, \alpha_s^I = \left(\alpha_{si}^I\right)]$. The optimal value, θ_s^{I*}, then gives the maximum possible *proportional contraction*[5] of the input vector that could be achieved by the sth firm, while retaining its level of output unchanged at Y_s. An (input-oriented) *TE* of the firm s is then given by θ_s^{I*} itself.

It can be easily understood that we proportionately change either the input or the output vector in the two previous LP problems, while maintaining the status quo for the other. To be specific, we expand (contract) the entire output (input) vector in output-oriented (input-oriented) LP, while input (output) vector is held fixed at the observed level. One can also think of a simultaneously expanding output vector and a contracting input vector at the same proportion to reach another measure of technical efficiency, known as the *graph hyperbolic measure*. The problem for the firm s in that case would be to find a scalar δ and an N-component vector $\gamma_s = (\gamma_{si})$, which solve the following LP:

$$\left(P_s^{GH}\right) \qquad Maximize \quad \delta$$

subject to (1) $\sum_{i=1}^{N} \gamma_{si} Y_i \geq \delta Y_s$, (2) $\sum_{i=1}^{N} \gamma_{si} X_i \leq \frac{1}{\delta} X_s$, (3) $\sum_{i=1}^{N} \gamma_{si} = 1$ and (4) $\gamma_{si} \geq 0 \; \forall \, i$.

If (P_s^{GH}) have an optimal solution, say $[\delta_s^{GH*}, \gamma_s^{GH} = \left(\gamma_{si}^{GH}\right)]$, the graph hyperbolic measure of technical efficiency of the sth firm would then be $(1/\delta_s^{GH*})$. To visualize the concept of graph hyperbolic efficiency, we once again refer to Figure 3.1. Under the VRS assumption, the efficient projection of DMU E on PPF is G, where the output is increased to δY_1 while, simultaneously, the input is decreased to X_1/δ and the efficiency score is $1/\delta$. In a similar fashion, for CRS assumption, we obtain the efficient input–output combination at point G_1. Note that E, G, and G_1 lie on a rectangular hyperbola thus explaining the name.

Note that we can obtain the output-oriented and input-oriented CCR TE scores for the sth firm if we solve the previously stated LP problems without constraint (3). To see why it is so, let us first construct the PPS that follows CRS, in addition to the axioms (a) through (c) mentioned earlier. Let us define $\hat{Y} = \sum_{i=1}^{N} \lambda_i Y_i$ and $\hat{X} = \sum_{i=1}^{N} \lambda_i X_i$ with $\sum_{i=1}^{N} \lambda_i = 1$ and $\lambda_i \geq 0\ \forall i$. If we assume that the production technology follows CRS, it implies that if $\left(\hat{X}, \hat{Y}\right)$ is a feasible bundle under the given technology, then $\left(k\hat{X}, k\hat{Y}\right)$ is also feasible under the same technology for any $k \geq 0$. Define $\tilde{X} = k\hat{X}$ and $\tilde{Y} = k\hat{Y}$ for some $k \geq 0$. Also define $\mu_i = k\lambda_i\ \forall i$. Hence, $\lambda_i \geq 0$ implies that $\mu_i \geq 0\ \forall_i$ since $k \geq 0$. Again, $\sum_{i=1}^{N} \lambda_i = 1$ implies that $\sum_{i=1}^{N} \mu_i = k \sum_{i=1}^{N} \lambda_i = k \geq 0$. But when $\mu_i \geq 0\ \forall_i$, it is trivial that $\sum_{i=1}^{N} \mu_i \geq 0$. Thus, if we impose that the system is constrained by the restriction $\mu_i \geq 0\ \forall i$, $\sum_{i=1}^{N} \mu_i \geq 0$ is simply a redundant constraint for the system. Therefore, we can construct the PPS that follows CRS as follows:

$$T^C = \left\{ \left(\tilde{X}, \tilde{Y}\right) : \sum_{i=1}^{N} \mu_i Y_i \geq \tilde{Y}, \sum_{i=1}^{N} \mu_i X_i \leq \tilde{X},\ \mu_i \geq 0\ \forall i \right\}.$$

The output-oriented CRS TE[6] is $TE_{y,s}^{CRS} = 1/\phi_s^{CRS*}$, where $\phi_s^{CRS*} = \max \phi_s^{CRS}$ subject to (1) $\sum_{i=1}^{N} \mu_{si} Y_i \geq \phi Y_s$, (2) $\sum_{i=1}^{N} \mu_{si} X_i \leq X_s$, and (3) $\mu_{si} \geq 0\ \forall i$. Similarly, the input-oriented CRS technical efficiency $TE_s^{CRS} = \min \theta_s^{CRS}$ subject to (1) $\sum_{i=1}^{N} \alpha_{si} Y_i \geq Y_s$, (2) $\sum_{i=1}^{N} \alpha_{si} X_i \leq \theta X_s$, and (3) $\alpha_{si} \geq 0\ \forall i$, and the graph hyperbolic technical efficiency score would be $1/\delta_s^{CGH*}$, where $\delta_s^{CGH*} = \max \delta_s^{CRS}$ subject to (1) $\sum_{i=1}^{N} \gamma_{si} Y_i \geq \delta Y_s$, (2) $\sum_{i=1}^{N} \gamma_{si} X_i \leq \frac{1}{\delta} X_s$, and (3) $\gamma_{si} \geq 0\ \forall i$. It is to be noted that the input- and output-oriented technical efficiency scores are exactly the same and graph hyperbolic technical efficiency score is its square root under the CRS assumption (see Ray, 2004, for further details).

TE scores help us ranking the DMUs on the basis of their productive performance. However, as the efficiency score depends on the model used, it may alter the ranking of the firm as we switch from one model to another for our analyses. To ascertain how the model choice affects the ranking we use Spearman's rank correlation coefficient (ρ). It is a nonparametric measure of statistical dependence among the rankings of two variables. If there are no tied ranks, then the rank correlation coefficient is given by

$$\rho = 1 - \frac{6\sum_{i}^{n} d_i^2}{n\left(n^2 - 1\right)},$$

where d_i = difference in paired ranks and n = number of cases. However, if there are tied ranks then the preceding formula doesn't yield good results, and one can use the alternative formula as follows:

$$\rho = \frac{\sum_{i}^{n}(x_i - \bar{x})(y_i - \bar{y})}{\sqrt{\sum_{i}^{n}(x_i - \bar{x})^2 \sum_{i}^{n}(y_i - \bar{y})^2}},$$

where x and y are the ranks and $\bar{x}$ and $\bar{y}$ are the mean ranks. The value of Spearman's rank correlation coefficient varies between +1 and −1, where the magnitude of the value indicates the strength of correspondence while the sign indicates the direction of movement.

The actual output produced by a DMU from a given bundle of inputs depends on a number of exogenous factors that influence its ability of efficient utilization of inputs. Some of these factors may be favorable and enhance efficiency while others may be detrimental to productive performance. There are two different ways the production process is modeled to include these attributes (say, a). One approach would specify the production frontier as $y^* = f(x, a)$, where y^* is the maximum possible output, x is the input bundle, and a is the set of attributes affecting output. In the alternative approach, the frontier production function does not depend on the attributes and they either facilitate or hinder resource utilization affecting TE only. In this approach, actual output relates to the frontier as $y = f(x) \times \mathrm{TE}(a)$. This formulation leaves the attributes out of the DEA specification, and once the TE scores are obtained, a second-stage k-variable regression model is estimated to determine how any individual attribute a_j affects the TE scores. However, there is no consensus about what type of regression model is to be used for the second stage. Earlier researchers have used the ordinary least squares (OLS) regression model. Hoff (2007) compares within-sample prediction performance of two-limit tobit, OLS, quasi-maximum likelihood estimation, and the unit-inflated beta model for second-stage DEA analysis. He concludes that, first, the tobit approach will, in most cases, be sufficient in representing second-stage DEA models and, second, the OLS may actually in many cases replace tobit for that purpose.

3.4 Empirical analyses and findings

The DMUs of our analysis are districts of UP. We only consider an output-oriented DEA model. This is a reasonable choice when the objective of the farmers is to produce as much as they can, given the resources at their disposal. While the agricultural output in UP is very diverse, the study considers only the major crops produced in the state. So we have opted for four outputs and three inputs technology set. The outputs are *total food-grains* (Y_1), *sugarcane* (Y_2), *oilseeds* (Y_3), and *potato* (Y_4), and the inputs are *gross cropped area* (X_1), *total NPK fertilizer* (X_2) and *number of tube-wells* (X_3). All the outputs and Nitrogen-Phosphorus-Potassium (NPK) fertilizer are measured in metric tons while the gross cropped area is measured in hectares. The number of agricultural laborers is an important input, but due to the unavailability of the data, we have not considered it in the present study. Data on agricultural inputs and outputs for 2012–2013 and 2013–2014 are obtained from the *Statistical Abstract UP* 2014 and 2015 editions. Although there are 75 districts, two districts (Hapur and Shamli) are excluded due to the unavailability of data.

In the present study, different attributes portraying agricultural systems and infrastructural facilities of the districts are taken as explanatory variables for TE scores in a fixed-effect panel regression model. In general, the model can be written as

$$y_{it} = \beta_1 x_{it1} + \beta_2 x_{it2} + \ldots + \beta_k x_{itk} + a_i + u_{it}, \qquad t = 1,2,\ldots,T,$$

where

y_{it} is the value of dependent variable for *i*th entity in *t*th time period,
x_{itk} is the value of explanatory variable x_k for *i*th entity in *t*th time period,
β_k is the coefficient of explanatory variable x_k,
a_i is the time invariant characteristics for *i*th entity, and
u_{it} is the residual term.

The explanatory variables or regressors utilized in the analysis include the following: *IRR* represents gross irrigated area to gross area sown; *MANDI* represents the number of regulated markets (kishan mandis) per lakh hectares of net area sown; *COOP* represents the number of cooperative agricultural marketing centers per lakh of rural population; *BANK* represents the number of rural development and cooperative banks per lakh of rural population as a proxy for agricultural credit availability; *WATER* represents total remaining water to total underground water, as a proxy for groundwater exploitation; and *RAIN* represents annual rainfall (in millimeters). Data on the first five regressors are obtained from various issues of *District-Wise Development Indicators of UP* and *Statistical Diary of UP*. Data on *RAIN* comes from the *Rainfall Statistics of India* 2012 and 2013. Unfortunately, Amethi, Chitrakoot, Kasganj, Sambhal, and Shrawasti districts have been dropped from the second-stage analysis, as data on a couple of the regressors are missing. Thus, for the second-stage regression we use data for 68 districts leading to a balanced panel of 136 observations.

We have calculated TE scores for both CRS and VRS technologies to illustrate how the underlying technology assumption affects this score. The CRS assumption is imposed as the input–output quantities are district-level aggregates, and these aggregate input–output bundles are feasible only if the underlying technology is additive, that is, CRS (Ray and Ghose, 2014), whereas the VRS assumption has also its own merit since even if one could consider all the physical inputs, environmental factors, like rainfall, relative humidity, temperature, soil conditions, and so on are still usually not be included in the empirical model. In other words, CRS is the best if we can take into account all factors of production that we can think of. Since the VRS assumption is closer to the real-world situations, further analyses are based on VRS TE scores. The output-oriented TE scores for the models were calculated using DEAP Version 2.1.

Let us now look at the salient features of the results summarized in Table 3.1. It is easy to observe that the efficiency under VRS is higher than that under CRS, this is due to the restrictive nature of CRS assumption. We can see that, on average, districts are very inefficient though there is considerable improvement in efficiency from 2012–2013 to 2013–2014. If we consider the CRS assumption, then for 2012–2013, the mean efficiency is 0.586, which means that the DMUs are producing only 58.6% of the potential output, given the inputs. For 2013–2014, the mean TE has increased considerably to 0.837, but still there is much scope for improvement. The improvement in performance can also be inferred from the number of DMUs at the frontier, which shows a considerable increase over the study period. A cursory glance at the minimum scores, across all the measures, projects an encouraging trend. The performance of the worst performing DMUs is considerably better in 2013–2014 as compared to the previous year. Furthermore, the standard deviation measure shows that the dispersion of efficiency scores has decreased in the course of study years. A similar observation can be made for the VRSTE scores. Thus, there is still much

Table 3.1 Summary Statistics of TE Scores

	Mean	*Median*	*SD*	*Min*	*Max*	*DMUs on frontier*
2012–2013						
CRSTE	0.586	0.528	0.205	0.173	1	7
VRSTE	0.759	0.736	0.195	0.269	1	17
2013–2014						
CRSTE	0.835	0.839	0.133	0.482	1	13
VRSTE	0.874	0.888	0.129	0.511	1	26

Source: Authors' calculation

scope for improvement. The comparison of the mean and the median indicates that for 2012–2013 the relative concentration of TE scores is higher toward the frontier under both CRS and VRS assumptions. But as we progress to 2013–2014, the relative concentration increases towards the lower values. Table 3.2 lists districts with TE scores in specific ranges and displays the relative frequency of VRSTE scores for 2012–2013 and 2013–2014. The mass of the TE distribution is concentrated in the 0.7–1.0 range of the scores. The improvement in efficiency is readily evident from the table,

Table 3.2 DMUs in Different Ranges of VRSTE Scores, 2012–2013 and 2013–2014

	2012–2013	*2013–2014*
TE range	*Districts (frequency, percentage)*	*Districts (frequency, percentage)*
0–0.300	Unnao (1, 1.37%)	–
0.301–0.400	–	–
0.401–0.500	Sonbhadra, Balrampur, Mau, Moradabad, Mahoba, Faizabad, Amroha (7, 9.59%)	–
0.501–0.600	KanpurNagar, Basti, Sultanpur, Raebareli, Santkabir Nagar, Mirzapur, KushiNagar, Shrawasti, Pratapgarh (9, 12.32%)	Chitrakoot, Banda, Mahoba (3, 4.11%)
0.601–0.700	Varanasi, AmbedkarNagar, Kaushambi, Deoria, Gonda, Hamirpur, Etah, Maharajganj, Hardoi, Rampur, Kanpur Dehat, Chitrakoot (12, 16.44%)	Sultanpur, Sonbhadra, Hamirpur, Kaushambi (4, 5.48%)
0.701–0.800	Banda, Saharanpur, Pilibhit, Ballia, Auraiya, Ghazipur, Gorakhpur, Farrukhabad, Bareilly, Fatehpur, Bahraich (11, 15.07%)	Jhansi, Shrawasti, Lucknow, Faizabad, Balrampur, Ghazipur, Mirzapur, Pratapgarh, Ballia, Moradabad, Kanpur Nagar, Jalaun (12, 16.44%)
0.801–0.900	Mainpuri, SiddharthNagar, Chandauli, Jhansi, Barabanki, Etawah, Jaunpur, Allahabad, Jalaun, Bulandshahar (10, 13.70%)	Rampur, Jaunpur, Unnao, Gonda, Fatehpur, Varanasi, Mau, Saharanpur, Deoria, Bareilly, Bahraich, Sant Kabir Nagar, Allahabad, Kushi Nagar, Amroha, Basti, Maharajganj, Gorakhpur, Ambedkar Nagar (19, 26.03%)

(Continued)

Table 3.2 (Continued)

	2012–2013	2013–2014
TE range	*Districts (frequency, percentage)*	*Districts (frequency, percentage)*
0.901–1	Meerut, Azamgarh, Baghpat, Hathras, Bijnor, Lalitpur, Agra, Aligarh, Amethi, Badaun, Firozabad, G. B. Nagar, Ghaziabad, Kannauj, Kasganj, Kheri, Lucknow, Mathura, Muzaffarnagar, Sambhal, Sant Ravidas Nagar, Shahjahanpur, Sitapur (23, 31.51%)	Sitapur, Barabanki, Raebareli, Azamgarh, Etah, Farrukhabad, KanpurDehat, Badaun, Mainpuri, Hardoi, Pilibhit, Auraiya, SiddharthNagar, Chandauli, Etawah, Bulandshahar, Meerut, Baghpat, Hathras, Bijnor, Lalitpur, Agra, Aligarh, Amethi, Firozabad, G. B. Nagar, Ghaziabad, Kannauj, Kasganj, Kheri, Mathura, Muzaffarnagar, Sambhal, SantravidasNagar, Shahjahanpur (35, 47.95%)

Source: Authors' calculation

while in 2012–2013, there are 29 districts having VRSTE below 0.7, the number of such districts has come down to only 7 in the next year.

In DEA, the benchmark or the reference DMU for the given DMU is created as a combination of some other DMUs, which are called peers. Each peer is assigned a weight (λ) for the formation of the benchmark. The weight is the proportion of the peer DMU that is used for this purpose. Table 3.3 gives the peers and corresponding weights for the poorest three performers for the study years. For instance, if we consider the case of Unnao for 2012–2013, its peers are Aligarh, Lucknow, and Shahjahanpur, with associated weights being 0.313, 0.475, and 0.212, respectively. That means the benchmark DMU for Unnao is formed using 0.313 part of Aligarh, 0.475 portion of Lucknow, and 0.212 share of Shahjahanpur. To arrive at the efficiency measure of Unnao, its output is compared to the output of this benchmark.

The weights also help us find the output targets which the DMU should achieve to become fully efficient. For example, if we want to calculate the output targets for Unnao, then the required data is in Table 3.4. The table provides the weights and actual output values for the peers of Unnao. The output targets for Unnao are calculated thus:

- $\text{fgrain}^{\text{Target}} = (\lambda^{\text{Aligarh}} \times \text{fgrain}^{\text{Aligarh}}) + (\lambda^{\text{Lucknow}} \times \text{fgrain}^{\text{Lucknow}}) + (\lambda^{\text{Shahjahanpur}} \times \text{fgrain}^{\text{Shahjahanpur}})$
- $\text{scane}^{\text{Target}} = (\lambda^{\text{Aligarh}} \times \text{scane}^{\text{Aligarh}}) + (\lambda^{\text{Lucknow}} \times \text{scane}^{\text{Lucknow}}) + (\lambda^{\text{Shahjahanpur}} \times \text{scane}^{\text{Shahjahanpur}})$

Table 3.3 Peers and Corresponding Weights for the Poorest Three Performers in VRSTE Scores

Districts	*Peer 1*	*Peer 2*	*Peer 3*	*Peer 4*	*Peer 5*
2012–2013					
Unnao	Aligarh (0.313)	Lucknow (0.475)	Shahjahanpur (0.212)		
Sonbhadra	Lucknow (0.329)	Kasganj (0.283)	St. Ravidas Nagar (0.384)	Agra (0.005)	
Balrampur	Lucknow (0.704)	Aligarh (0.21)	Agra (0.044)	Shahjahanpur (0.042)	
2013–2014					
Chitrakoot	Kasganj (0.084)	Sambhal (0.164)	Chandauli (0.176)	Sant Ravidas Nagar (0.577)	
Banda	Sambhal (0.711)	Aligarh (0.289)			
Mahoba	Kasganj (0.114)	Sambhal (0.099)	Lalitpur (0.391)	Sant Ravidas Nagar (0.372)	Agra (0.024)

Note: Values in brackets are the decision-making unit weights (λ).

Table 3.4 Weights and Outputs of Peers for Unnao

	Weight (λ)	*fgrain*	*scane*	*Oseed*	*Pot*
Output Target	–	1,273,161	4,929,003	28,882	269,241
Aligarh	0.313	1,310,459	422,480	40,493	603,315
Lucknow	0.475	1,136,492	9,063,213	26,865	90,791
Shahjahanpur	0.212	1,524,529	2,307,228	16,283	176,607

Note: The outputs are in metric tons.

- $oseed^{Target} = (\lambda^{Aligarh} \times oseed^{Aligarh}) + (\lambda^{Lucknow} \times oseed^{Lucknow}) + (\lambda^{Shahjahanpur} \times oseed^{Shahjahanpur})$
- $pot^{Target} = (\lambda^{Aligarh} \times pot^{Aligarh}) + (\lambda^{Lucknow} \times pot^{Lucknow}) + (\lambda^{Shahjahanpur} \times pot^{Shahjahanpur})$

Substituting the values from table we get the following:

- $fgrain^{Target} = (0.313 \times 1{,}310{,}459) + (0.475 \times 1{,}136{,}492) + (0.212 \times 1{,}524{,}529) = 1{,}273{,}208$
- $scane^{Target} = (0.313 \times 422{,}480) + (0.475 \times 9{,}063{,}213) + (0.212 \times 2{,}307{,}228) = 4{,}926{,}395$

- $oseed^{Target}$ = (0.313 × 404,93) + (0.475 × 26,865) + (0.212 × 16,283) = 28,887
- pot^{Target} = (0.313 × 603,315) + (0.475 × 90,791) + (0.212 × 176,607) = 269,404

The target values so arrived at are slightly different from the DEAP output because of rounding off figures. The slacks give the difference between the maximum possible radial expansion output and the reference DMU output. In radial expansion, all the outputs are increased in the same proportion, but there is a limit to expansion—the production possibility frontier. So with the maximum possible radial expansion, the DMU may reach a point on the production possibility frontier where it is still possible to increase some of the outputs without an increase in inputs. For Unnao, as can be seen in Table 3.5, when the actual output is multiplied with radial expansion, it gives the radial expansion output. The output targets give us the maximum possible outputs for the production possibility frontier. No further radial expansion is possible, because if the actual outputs are increased in any larger proportion, then it will exceed the output targets for at least one output, which is not permissible. So this is the maximum radial expansion of outputs. If we compare such a radially expanded output and target output, there is still a possibility of increment in output for sugarcane and oilseed, which are the concerned output slacks. The presence of slacks signals a faulty allocation of inputs in the production process.

Now we shift our discussion to graph hyperbolic TE model. The results, summarized in Table 3.6, show that the trends are similar to what we observed in the case of CRSTE and VRSTE discussed previously. We can notice that the graph hyperbolic efficiency scores are higher than their orientation-specific counterparts. However, in comparing the two, one has to keep in mind that the CCR or BCC TE scores relate to either contraction of inputs or expansion of outputs, keeping the other unchanged, whereas the graph hyperbolic efficiency scores are in relation to simultaneous changes in both. If we consider the GCRS results in 2012–2013, then we observe that, on average, the firms have the possibility of increasing their output by 32.45% and at the same time reduce their inputs by 32.45% or more. For 2013–2014, the corresponding figure is about 10%. Thus, in case of graph

Table 3.5 Output Radial Expansion and Slacks for Unnao

	fgrain	*scane*	*Oseed*	*pot*
Actual Outputs	341,993	12,453	4,477	72,323
Radial Expansion	3.722769	3.722769	3.722769	3.722769
Radial Expansion Output	1,273,161	46,359	16,666	269,241
Output Targets	1,273,161	4,929,003	28,882	269,241
Slacks	0	4,882,644	12,216	0

Source: Authors' calculation

Table 3.6 Summary Statistics of Graph Hyperbolic TE Scores

	Mean	*Median*	*SD*	*Min*	*Max*	*DMUs on Frontier*
2012–2013						
GCRSTE	0.755	0.726	0.130	0.416	1	7
GVRSTE	0.862	0.852	0.108	0.606	1	17
2013–2014						
GCRSTE	0.910	0.916	0.075	0.694	1	13
GVRSTE	0.938	0.938	0.060	0.751	1	26

Source: Authors' calculation

hyperbolic measure also, we see considerable improvement, although there is still scope for further progress. The increase in the number of DMUs at the frontier also indicates enhanced performance. The results of GVRS measures are analogous to the GCRS results. Table 3.7 lists districts with

Table 3.7 DMUs in Different Ranges of GVRS TE Scores, 2012–2013 and 2013–2014

	2012–2013	*2013–2014*
TE Range	*Districts (frequency, percentage)*	*Districts (frequency, percentage)*
0–0.600	–	–
0.601–0.700	KushiNagar, Moradabad, Unnao (3, 4.11%)	–
0.701–0.800	Ambedkar Nagar, Amroha, Balrampur, Banda, Bareilly, Basti Deoria, Faizabad, Ghazipur, Gonda Gorakhpur, Hardoi, Kanpur Nagar Maharajganj, Mau, Mirzapur, Pilibhit Pratapgarh, Raebareli, Rampur, Saharanpur, Sonbhadra, Sultanpur (23, 31.51%)	Banda (1, 1.37%)
0.801–0.900	Auraiya, Bahraich, Ballia, Barabanki, Etah, Farrukhabad, Fatehpur, Hamirpur, Kanpur Dehat, Kaushambi Mahoba, Mainpuri, SantkabirNagar, Shrawasti, Siddharth Nagar, Varanasi (16, 21.91%)	Ballia, Balrampur, Chitrakoot, Faizabad, Ghazipur, Hamirpur, Jalaun, Jaunpur, Jhansi, Kanpur Nagar, Kaushambi, Lucknow, Mahoba, Mirzapur, Moradabad, Pratapgarh, Shrawasti, Sonbhadra, Sultanpur, Unnao (20, 27.40%)

(Continued)

Table 3.7 (Continued)

	2012–2013	*2013–2014*
TE Range	*Districts (frequency, percentage)*	*Districts (frequency, percentage)*
0.901–1	Agra, Aligarh, Allahabad, Amethi, Azamgarh, Badaun, Baghpat, Bijnor, Bulandshahar, Chandauli, Chitrakoot, Etawah, Firozabad, G. B. Nagar, Ghaziabad, Hathras, Jalaun, Jaunpur, Jhansi, Kannauj, Kasganj, Kheri, Lalitpur, Lucknow, Mathura, Meerut, Muzaffarnagar, Sambhal, SantravidasNagar, Shahjahanpur, Sitapur (31, 42.47%)	Agra, Aligarh, Allahabad, Ambedkar Nagar, Amethi, Amroha, Auraiya, Azamgarh, Badaun, Baghpat, Bahraich, Barabanki, Bareilly, Basti, Bijnor, Bulandshahar, Chandauli, Deoria, Etah, Etawah, Farrukhabad, Fatehpur, Firozabad, G. B. Nagar, Ghaziabad, Gonda, Gorakhpur, Hardoi, Hathras, Kannauj, Kanpur Dehat, Kasganj, Kheri, KushiNagar, Lalitpur, Maharajganj, Mainpuri, Mathura, Mau, Meerut, Muzaffarnagar, Pilibhit, Raebareli, Rampur, Saharanpur, Sambhal, SantkabirNagar, SantravidasNagar, Shahjahanpur, SiddharthNagar, Sitapur, Varanasi (52, 71.23%)

TE scores in specific ranges displays the relative frequency of GVRS TE scores for 2012–2013 and 2013–2014. The mass of the TE distribution is concentrated in the 0.9–1.0 class of the scores, which, from 31 DMUs in 2012–2013, increases to 52 in 2013–2014. Furthermore, the number of DMUs below the score of 0.8 has decreased from 27 to only 1 as we move from 2012–2013 to 2013–2014.

Benchmarking techniques are often used to rank the DMUs on the basis of performance. An interesting point to ponder upon is how the different efficiency measures affect the ranking of the firms under consideration.

To explore it further, we calculate Spearman's rank correlation coefficient (with ties) among the four efficiency measures. The results indicate that there may be a considerable difference in ranking if we shift from CRSTE to VRSTE ($\rho = 0.635$). The same is the case with the graph hyperbolic measures for CRS and VRS assumptions ($\rho = 0.732$). However, if we consider the CRSTE versus GCRSTE and VRSTE versus GVRSTE, then the concerned correlation structure is very high, with ρ values of 0.999 and 0.903, respectively. Thus, switching the scale assumption may have a considerably larger impact on the rankings than changing from traditional orientation-specific model to corresponding graph hyperbolic measure.

Table 3.8 presents the second stage regression results originating from VRSTE and GRSTE models. We observe that the set of variables that significantly influence performance remains the same. The result shows that the ratio of gross irrigated area to gross sown area (IRR) has an insignificant impact on TE. This is not on expected lines, as improvement in irrigation facilities should lead to an increase in agricultural productivity. The ratio of the total remaining water to the total underground water (WATER) has an inconclusive impact on efficiency, as the coefficient is statistically insignificant. This is against expectations, as earlier studies have argued, that excessive groundwater-based irrigation has led to a decline in agricultural productivity. The density of cooperative agricultural marketing centers (COOP) is seen as positive and a significant contributor to improvements in TE. This emphasizes the importance of marketing channels in improving agricultural performance. RAIN has a positive (although very small) and significant impact on TE, which supports the perception of dependency of agriculture on rain. The density of kisan mandis (MANDI) and rural development and cooperative banks (BANK) have no significant impact on TE.

Table 3.8 Second-Stage Regression Results

	VRSTE		*GVRSTE*	
	Coef.	*P-value*	*Coef.*	*P-value*
IRR	0.000	0.863	0.001	0.559
MANDI	0.035	0.280	0.016	0.388
WATER	0.002	0.290	0.001	0.559
COOP	0.027	0.063	0.017	0.058
BANK	0.514	0.346	0.245	0.500
RAIN	0.000	0.009	0.000	0.002
CONS	0.276	0.208	0.550	0.000
R^2 (within)	0.249		0.327	
F-value	5.740	0.000	6.82	0.000

Source: Authors' calculation

3.5 Concluding remarks

Agriculture is the mainstay of the state of UP. In this chapter, we analyze agricultural productive efficiency at the district level in the state of UP. As a fact finding exercise for the road map to agricultural development, this study provides the first estimates of technical efficiency using DEA methodology and data for 2012–2013 and 2013–2014. Performing below the state average of technical efficiency scores, more than half of the districts have a lot of catching up to do. Based on these scores, we identify the factors that explain the differences in agricultural productive efficiency at the district level. It can be observed that irrigation facilities, the presence of marketing places, have come out significant in all the models. This emphasizes the impact of infrastructural facilities on agricultural efficiency.

Of course, such two-staged procedure, involving estimation of DEA TE scores at the first stage while regressing such scores on a set of other explanatory variables at the second, has been criticized in the literature for two reasons. Firstly, so estimated TE scores are subject to upward bias since the estimated frontier (which is considered as the benchmark to have the TE score of a DMU) using the sample observations may not be the true one and there may be a significant gap between these two. Finally, such efficiency estimates are serially correlated also (via the common benchmark for all the DMUs, in general, and hence the fully efficient DMU[s], in particular). Unfortunately, however, it is very difficult to have an estimate of the nature of such serial correlation, since the true data-generating process is unknown for the sample data. Scholars argue that the presence of such unknown serial correlation makes the standard inferencing procedure (at the second stage) invalid. Simar and Wilson (2007) propose a simulation-based double-bootstrap procedure to have consistent second-stage inferences in this connection. However, we choose to follow the conventional procedure for at least two reasons: (a) As we have noted, this chapter is a humble effort for the beginners, an introduction of the methodology and a brief history of the concerned literature is our primary focus, keeping aside the recent developments and other advanced materials for the further interested readers, and (b) although our data have some sample characteristics, since we have to leave out few observations from our analyses for non-availability of information, we are dealing with almost the entire population. Nonetheless, our inferential findings may be tentative for the reason mentioned earlier.

Notes

1 For economic implications and interpretations of these axioms, see Ray (2004, p. 27).
2 Note that Y_i will be a scalar in case all firms produce a single and the same good.

3 Note that a feasible solution to the preceding problem is given by $\varphi = 1$ and $\lambda_s =$ a unit vector (the sth component being unity). Hence, the optimal value, φ_s^{M*}, will be greater than or equal to 1.
4 This is also known as Farrell efficiency.
5 Note that a feasible solution to the above problem is given by $\theta = 1$ and $\alpha_s =$ a unit vector (the sth component being unity). Hence, the optimal value, θ_s^{I*}, will be less than or equal to 1.
6 This is the CCR model.

References

Banker, R. D., A. Charnes, and W. W. Cooper. (1984). "Some Models for Estimating Technical and Scale Inefficiencies in Data Envelopment Analysis." *Management Science* 30, no. 9: 1078–1109.

Byerlee, D., De Janvry, A., Sadoulet, E., Townsend, R., & Klytchnikova, I. (2008). *World development report 2008: agriculture for development* (No. 41455, pp. 1–390). The World Bank.

Charnes, A., W. W. Cooper, and E. Rhodes. (1978). "Measuring the Efficiency of Decision Making Units." *European Journal of Operational Research* 2, no. 6: 429–444.

Charnes, A., W. W. Cooper, and E. Rhodes. (1981). "Evaluating Program and Managerial Efficiency: An Application of Data Envelopment Analysis to Program Follow Through." *Management Science* 27, no. 6: 668–697.

Coelli, T. J., D. S. P. Rao, C. J. O'Donnell, and G. E. Battese. (2005). *An Introduction to Efficiency and Productivity Analysis*. Springer Science & Business Media.

Debreu, G. "The Coefficient of Resource Utilization." *Econometrica* 19, no. 3 (1951): 273–292.

De Koeijer, T. J., G. A. A. Wossink, P. C. Struik, and J. A. Renkema. (2002). "Measuring Agricultural Sustainability in Terms of Efficiency: The Case of Dutch Sugar Beet Growers." *Journal of Environmental Management* 66, no. 1: 9–17.

Farrell, M. J. "The Measurement of Productive Efficiency." *Journal of the Royal Statistical Society: Series A, General* 120, no. 3 (1957): 253–281.

Førsund, F. R., and N. Sarafoglou. "On the Origins of Data Envelopment Analysis." *Journal of Productivity Analysis* 17, no. 1/2 (2002): 23–40.

Hoff, A. "Second Stage DEA: Comparison of Approaches for Modelling the DEA Score." *European Journal of Operational Research* 181, no. 1 (2007): 425–435.

Khan, W., and S. A. Ansari. "Does Agriculture Matter for Economic Growth of Uttar Pradesh (India)?" *Economy of Region/Ekonomika Regiona* 14, no. 3 (2018): 1029–1037.

Raman, R., and R. Kumari. "Regional Disparity in Agricultural Development: A District Level Analysis for Uttar Pradesh." *Journal of Regional Development and Planning* 1, no. 2 (2012): 71–90.

Ray, S. C. (2004). *Data Envelopment Analysis: Theory and Techniques for Economics and Operations Research*. New York, USA: Cambridge University Press.

Ray, S. C., and A. Ghose. "Production Efficiency in Indian Agriculture: An Assessment of the Post Green Revolution Years." *Omega* 44 (2014): 58–69.

Ray, S. C., S. C. Kumbhakar, and P. Dua (eds.). (2015). *Benchmarking for Performance Evaluation: A Production Frontier Approach*. Springer (2015).

Ray, S. C. (2019). "Data Envelopment Analysis with Alternative Returns to Scale." In T. ten Raa and W. H. Greene (eds.), *The Palgrave Handbook of Economic Performance Analysis*. Palgrave Macmillan, 145–188. https://media.economics.uconn.edu/working/2018-20.pdf

Simar, L., and P. W. Wilson. (2007). "Estimation and Inference in Two-Stage, Semi-Parametric Models of Production Processes." *Journal of Econometrics* 136, no. 1: 31–64.

4 Agricultural productivity in Bihar and its determinants

A district-level analysis

Amey Sapre, Mahendra Kumar Singh and Deep Mukherjee

4.1 Introduction

The concern about the sustainability of agriculture as an economic activity has several dimensions. For instance, in general, it is widely believed that agriculture productivity has shown improvements over the last decade. However, levels and the pace of such productivity gains have differed across regions and sectors of the economy. Similarly, increases in productivity have environmental implications where the challenge is to achieve increased productivity with lower environmental degradation. Alongside technological innovations, agricultural sector in developing and backward areas is replete with unsustainable and unviable economic practices (see OECD [2013] and FAO [2018] for a worldwide perspective on sustainable agriculture). These situations create a mix of conditions in which there are few success stories of increasing productivity gains and moving towards sustainability versus increasing disparities on account of several policy or coordination failures. Understanding productivity gains (or even disparity) in agriculture has a broader significance, especially in the context of developing economies or regions. Agriculture in such regions or economies is primarily a way of life in rural areas as it is deeply intertwined with traditional ways of cultivation and earning for livelihood. However, with the changing environment in business, technology, trade and commerce, new opportunities for innovation and modernization emerge. In the backdrop of several such traditional pull or push factors, questions on growth and sustainability of agriculture as an economic activity need to be answered.

In simple words, productivity growth can be understood as a result of the change in the productive efficiency with which farmers combine inputs to produce output and technological change. Total Factor Productivity (TFP) estimates provide a summary measure to objectively compare producers' performance across regions. In India and especially in major agricultural states like Bihar, the analysis of productivity gains in agriculture has important policy and practical implications. This chapter deals with the appraisal of TFP growth in the agricultural sector of Bihar and the identification of factors that explain variations in productivity. The premise of analysing productivity growth is based on the fact that agriculture and allied activities continues to be the backbone of the state economy and remains central to the state's economic progress and sustainability. To begin with, geographically, Bihar lies in the plains and basin of the river Ganges and is well endowed with several natural resources, particularly rich alluvial soil and water bodies. The state has three agro-climatic zones, namely, North-West, North-East and South, which makes it unique in the cultivation of a variety of crops. As of 2014, the agricultural sector contributed nearly 40% of the gross state domestic product (GSDP) and had nearly 75% of the state's population engaged in farming and allied activities (Government of Bihar 2015).

Despite agriculture being such a crucial sector, agricultural performance in the state has had a dismal record. Cultivation in the state continues to be affected by frequent droughts and floods which lead to extensive loss in agriculture produce and lower farm incomes. Coupled with policy and coordination failures, the under-performance of the sector pushes a large part of the rural population into socio-economic distress. Such situations of distress also lead to other problems such as unemployment, food insecurity, malnutrition and poverty. The evidence and persistence of these problems in Bihar have been well documented in various mediums over the years; see, for instance, Shah (2016), Sharma and Rodgers (2015), Ghosh (2015) and Mahajan (2015), among others. It is well established that the performance of the agricultural sector is directly associated with the availability of a conducive environment for cultivation, inputs, manpower and physical and financial capital. In recent years, the government of Bihar prepared a road map for accelerated development of the agricultural sector by implementing several programs and schemes. These included targeted schemes on rice cultivation, crop diversification, micro-irrigation and other enhanced facilities for soil and cultivation information. Correspondingly, in the last decade, a substantial increase in the annual average budget allocation was done from INR 200 crores from 2001 to 2005 to more than INR 2000 crores between 2007 and 2013 (Singh 2011). As a result, agricultural growth reached 3.7% between 2010 and 2014 from a negative growth between 2000 and 2005 (Government of Bihar 2015).

In recent years, the state government outlined some of the factors necessary for effective implementation of a road-map for future years, namely (1) to identify key determinants or the underlying factors that significantly contribute to agricultural growth and (2) assess district-level agricultural performance in order to identify progressive and laggard districts. The process will enable a shift policy focus towards correcting the factors that are unable to deliver in non-performing or economically lagging districts. Taking a cue from these factors, in this chapter, we conduct a systematic analysis of agriculture productivity for three major crops, viz. wheat, rice and maize, for all districts of Bihar over a period of 10 years (2004–2013). The purpose is to construct productivity indexes using crop-level data to identify high- and low-performing districts. The next step is to identify factors that contribute to such productivity growth at the district level and bridge the research gap by providing district-wise crop-level productivity indexes for a decade, which are presently unavailable. The arrangement is as follows: Section 4.2 covers some essential studies on the theory and findings of relevant empirical studies on agricultural productivity, Section 4.3 describes the theoretical model for computing the productivity index, Section 4.4 describes the data and sources, Section 4.5 outlines the empirical strategy and results and Section 4.6 concludes with a discussion on policy perspectives.

4.2 Literature on agricultural productivity in various parts of India

Agricultural TFP growth in Bihar and particularly at the district level has not been rigorously studied previously. We review a few studies that have addressed district level productivity for other states. For instance, Adams and Bumb (1979) in their cross-sectional district-wise analysis of Rajasthan argue that land productivity depends primarily on conventional inputs. Other factors such as cropping pattern, taken as the ratio of cash crop value to total crop value and cropping intensity taken as the ratio of multiple cropped areas to net area shown have a direct and positive relation to levels of productivity. They also find that the use of modern technologies, urban expansion and institutional set-up as facilitative or intermediary variables that could affect levels of productivity. Muragi (2001) computes productivity indexes for the state of Punjab for 1961 to 1994. The author studies district-level outputs of crops and fruits and take livestock into consideration for building productivity indexes. Twenty crops, fruits and livestock products were aggregated into a total output index using district-specific farm harvest prices for crops and state-level prices for livestock and fruits.

Kumar et al. (2004) calculate TFP growth estimates of the crop sector in Indo-Gangetic Plain for 94 districts between 1980–1981 and 1996–1997. They find an expenditure on research and development (R&D) and extension, literacy rate and levels of infrastructure (such as roads, electrification, educational institutions, health facilities and banking, among others) to be the most important sources contributing to TFP growth. The study finds that over-utilization of groundwater has led to the decline of TFP growth in agriculture. Bhushan (2005) uses Data Envelopment Analysis (DEA) to estimate productivity growth in wheat production in major wheat-producing states from 1982–1983 to 1999–2000. His results indicate that technological progress has contributed mainly to the TFP growth of wheat, but the trends are uneven among states. He argues that a rise or fall in TFP growth is mainly due to an increase or decrease in levels of technological change rather than changes in technical efficiency. The study claims that credit systems, irrigation and road networks are areas of policy priority in less developed areas. Kumar et al. (2008) study the performance of agricultural productivity growth in South Asian countries. For India, crop-specific TFP analysis is performed by using the micro farm–level data for all the major crops grown in states from 1970–1971 to 2000–2001. They argue that expenditure on R&D, literacy and market access have accounted for a significant part of the TFP growth.

In a similar study, Bhushan (2015) uses the DEA to estimate district-level agricultural productivity in Bihar. Using a Malmquist Index, the study decomposes TFP growth into efficiency and technological gains to identify the best performing districts in the state. The findings suggest that technical change contributed to productivity growth by 2.0%, while efficiency deterioration reduced productivity growth by 1.7% per year on average between 2000 and 2012. Recently, Bhushan (2019) linked biophysical, social, economic and health resources at the district level to re-estimate TFP growth rate for agricultural output. The author develops a Livelihood Resilience Index at the district level and explains its connection with agricultural TFP growth.

Empirically, the Malmquist Index is a popular choice for estimating TFP; however, theoretically, it has been shown that the index does not belong to the class of complete TFP indexes. For instance, O'Donnell (2008), Diewert and Fox (2005) and Grifell-Tatje and Lovell (1995) have shown that the Fisher and Tornquist Indexes can be derived from a basic aggregator input and output function and hence are part of a complete TFP index. The notion of completeness refers to the fact that the indexes can be completely decomposed into measures of technical and efficiency change. O'Donnell (2008) describes how TFP measures from the Malmquist Index can be unreliable due to its incompleteness. Part of the reason for its incompletes is because the index fails the factor reversal test and cannot be computed without the knowledge of the underlying production technology.

In terms of computation, the Malmquist Index can be obtained as a special case of the Tornquist Index under the following assumptions: (1) firms or production units are revenue maximizers (cost minimizers), (2) the index uses positive inputs and outputs and (3) the underlying functions are translog; then both Tornquist and Malmquist Indexes are similar. However, the Tornquist Index remains superlative as it requires less restrictive assumptions on production technology. Fare et al. (1992) and (1995) have shown that unless the underlying production technology exhibits constant returns to scale and the functions are inversely homothetic, the Malmquist Index results into biased estimates of changes in TFP. Thus, in an empirical set-up, it is more useful to proceed with less restrictive assumptions on functional forms, technology functions and given data constraints.

To build on the premise the analysis we outline a relatively simple setup to first obtain comparable TFP scores across districts in Bihar. Having obtained TFP scores, the second step would be to identify factors that can explain variation in productivity levels. Taken together, the two-step process would enable us to analyze agricultural performance in a much broader way and provide cues for policy intervention. To begin with, we first describe the theoretical framework by which TFP scores can be computed.

4.3 Total factor productivity: An index-based approach

The theoretical setup involves computing the TFP using data on inputs and outputs of crops. Theoretically, TFP is a variable that measures the total output not explained by the use of conventional inputs. In other words, this indicates a measure of the total technological change leading to an increase in output after accounting for all relevant inputs. Thus, to explain the growth of output, one has to decompose the growth into various sources. First, the output growth can be decomposed in two parts, namely, input growth (e.g. land, labor, fertilizer, seed, etc.) and, second, the residual or unexplained part, which is denoted as the TFP growth. Implicitly, the TFP growth is accounted by aspects such as R&D, innovation, improved techniques and better quality infrastructure, among others. These aspects are expected to indirectly contribute to a higher growth in output over and above the use of inputs. Analytically, the decomposition of TFP growth into its own sources also provides a strong policy implication to understand the hidden growth potential and areas of innovation (Bhushan 2005).

We use the Divisia–Tornqvist Index for TFP computation purpose. The Tornquist Index can be computed both as a price or quantity index and are a weighted geometric average of price relatives and are a discrete approximation of the continuous version of the Divisia Index. The index is built using two key components, namely, the Total Output Index (TOI) and the Total Input Index (TII). The expressions for computation are as follows (Kumar et al. 2004):

$$TOI = \frac{TOI_t}{TOI_{t-1}} = \pi_j \left(\frac{q_{jt}}{q_{jt-1}} \right)^{\sqrt{r_{jt}+r_{jt-1}}} = A_t \tag{4.1}$$

$$TII = \frac{TII_t}{TII_{t-1}} = \pi_j \left(\frac{x_{jt}}{x_{jt-1}} \right)^{\sqrt{s_{jt}+s_{jt-1}}} = B_t \tag{4.2}$$

where, r_{jt} is the *jth* output share in total revenue, q_{jt} is the *j*th crop output (quantity), s_{it} is the *i*th input share in total input cost, x_{it} is the *i*th input quantity and t denotes the time period. Using this, the TOI in period (t) is given by $(TOI)_t = A_1, A_2, A_3, \ldots, A_t$, where the A_is represent the output index values for each time period. Similarly, the TII in period (t) is given by $(TII)_t = B_1, B_2, B_3, \ldots, B_t$, where B_is correspondingly represent the input index values. Combining the two, the TFP Index for year (t) is given by

$$TFP_t = \left(\frac{TOI_t}{TII_t}\right) = \left(\frac{A_t}{B_t}\right) \forall t = 1, 2, \ldots, T. \tag{4.3}$$

Computationally, the index can also be represented as

$$\ln(TFPI_{st}) = \ln\left[\frac{(\text{Output Index})_{st}}{(\text{Input Index})_{st}}\right] \tag{4.4}$$

which can be simplified as $\ln(TFPI_{st})$ = ln (Output Index) – ln (Input Index). Expanding equation (4.4), we have

$$\frac{1}{2}\sum_{i=1}^{m}(r_{is} + r_{it})(\ln q_{it} - \ln q_{is}) - \frac{1}{2}\sum_{j=1}^{n}(s_{js} + s_{jt})(\ln x_{jt} - \ln x_{js}) \tag{4.5}$$

where (q) and (x) denote output and input quantity, respectively; (s) represents the share of input of the ith crop in the total input cost; and the other variables are as previously defined in equations (4.1) and (4.2). The growth rates of variables (prices or quantities) are expressed in the difference in their logarithms. The weight of each variable is taken as its share, i.e. the ratio of its value to the total for each period. The Tornqvist Index also corresponds to the Translog production (or cost) function and is considered a close approximation of the Fisher Index. The Tornqvist Index is a non-transitive index as it does not satisfy the circularity property. The difference between the transitive and non-transitive index is that the first type satisfies the circularity property. That is to say, for any three periods, say, s, t and r, the property requires that $P_{st} = P_{sr} \times P_{rt}$. This implies that a direct comparison using s and t yields the same index as an indirect comparison using r for the given periods. A further discussion on TFP and its measurement can be found in Coelli et al. (2005).

4.4 Empirical work

To empirically compute the TFP values, we use secondary data sources on various agricultural variables available from government bulletins, surveys, reports and handbooks of statistics. The data series are collected for the period 2004–2005 to 2013–2014 for most agricultural crops and primary inputs for various districts. The data on crop output were unavailable on a consistent basis for two districts, viz. Arwal and Sheohar. Thus, these districts are excluded from analysis. Data were collected on the following variables: net sown area; irrigated area by tube well; number of cultivators and agriculture labor; crop yields of rice, wheat and maize; number of electrified and total villages in a district; Kisan Credit Card (KCC) holders; literacy rate of the district; annual rainfall; and few other demographic variables. The list of variables and their description is presented in Table 4.1.

We compute TFP indexes for three major crops, namely wheat, rice and maize. Given data constraints, we choose the following variables for inputs and outputs for calculation. To begin with, consider the following expression:

$$\frac{Y_i}{L_i} = f\left(\frac{l_{1i}}{L_i}, \frac{l_{2i}}{L_i}, \frac{f_{1i}}{L_i}, \frac{f_{2i}}{L_i}\right) \tag{4.6}$$

where, for the ith district, Y_i is quantity of the jth crop in tones, that is, rice, wheat and maize; L_i is net sown area in the district in hectares; l_1 is the number of cultivators in the district; l_2 is the number of agricultural laborers in the district; f_1 is consumption of urea fertilizer in the district; and f_2 is consumption of Diammonium Phosphate (DAP) fertilizer in the district.

Equation (4.6) expresses output as a function of inputs per unit of land. This setup helps us express productivity per unit of land and take the relevant inputs in the same unit across districts. The basic inputs considered are number of cultivators and agricultural laborers and the consumption of types of fertilizers. Based on this property, we use the nontransitive version and compute the index for the three crops for each year for 36 districts. The indexes were computed using the TFPIP Ver. 1.0 software. Using the TFP values, we further develop a regression analysis to analyze the determinants of TFP at the district level. Let the regression equation be denoted by

$$TFP_{it} = \beta_0 + \sum_{k=1}^{4} \beta_k \ln R_{it} + \beta_5 X_{it} + v_{it} \tag{4.7}$$

where R_1 denotes the ratio of the total irrigated area upon total net sown area, R_2 as ratio of tube-well-irrigated area on total irrigated area, R_3 as ratio of number of villages electrified divided by the total number of villages in that district and R_4 as ratio of total area of wheat, rice and maize under high-yielding variety (HYV) seeds divided by total irrigated area. We also include other control variables such as dummy for droughts, floods and literacy rate at the district level. The list of variables used and their description are presented in Table 4.1.

The choice of ratios instead of level variables captures the heterogeneity and accounts for the size and geographical characteristics of across the districts. The inclusion of variables like the number of electrified villages serves

Table 4.1 Variables and description

Var. Name	*Definition*	*Unit*	*Source*
NTC	Non-transitive cumulative Tornqvist Index	–	Computed
NSA	Net sown area	Hectares	LUS, Bihar
Twirr	Net irrigated area by tube well	Hectares	LUS, Bihar
Irri	Total Net area irrigated	Hectares	LUS, Bihar
Rain	Rainfall	MM	ES, Bihar
Villages	Total no. of villages in a district	Number	Census 2001, 2011
Electrified	District wise number of electrified villages	Number	HB Stat, Bihar
KCC	Number of Kisan Credit Card holders	Number	ES, Bihar
Cultivators	Number of cultivators	Number	Census 2011
Drought	Drought year	Binary	NDRMA
Floods	Flood year	Binary	NDRMA
Ricehyv	Rice area under HYV seeds	Hectares	HB Stat, Bihar
Wheathyv	Wheat area under HYV seeds	Hectares	HB Stat, Bihar
Maizehyv	Maize area under HYV seeds	Hectares	HB Stat, Bihar
Ricearea	Area under rice cultivation	Hectares	HB Stat, Bihar
Wheatarea	Area under wheat cultivation	Hectares	HB Stat, Bihar
Maizearea	Area under wheat cultivation	Hectares	HB Stat, Bihar

Note: LUS = Land Use Statistics, Government of Bihar; Census = Census of India; ES = Economic Survey, Government of Bihar; HB Stat = Bihar Statistical Handbook, Government of Bihar; NDRM = National Disaster Relief Management Authority; HYV = high-yield variety.

a control for infrastructure facilities across the districts. The adoption and spread of HYV seeds in the state are captured through the area under its cultivation as a proportion to the irrigated area and total literacy rate at the district level serves as an indicator for awareness and information. In terms of data consistency, it may be noted that several of these variables are available with a considerable time lag. Thus, given the timelines of surveys in the state, some of these variables may be subject to revision as and when actual data become available. Before we analyze the results of the empirical model, some descriptive statistics of the variables of interest are necessary. Table 4.2 shows the averages of the basic variables for the period 2004–2005 to 2013–2014.

4.5 Results

First, we rank the districts as per their net sown area and TFP scores for the initial and last period to get a comparison of their position between the two periods. We can then identify the differences in the basic variables which would build a case for understanding the factors that may contribute to productivity growth. Using the two-period ranks, a Spearman's rank correlation (ρ) is computed to get a measure of association between the district's ranks on TFP and net sown area. One can also compute this statistic for the ranks of the same variable between two periods. The computed values are presented in Table 4.3. Comparing between relative positions of districts on TFP scores and net sown area, the rank correlation indicates a stronger association in case of net sown area. This suggests that ranks of districts on net sown area have more or less remained unchanged over the years. In the case of TFP ranks, positions of several districts have altered between the two time points. Taking both variables, the correlation between TFP and net sown area for both periods is negligible. The finding is consistent with the earlier correlation result on ranks as the relative position of districts on net sown area have not changed, while changes on TFP rank are considerable.

To build on this finding further, we select the top 10 districts based on net sown area in 2013–2014 to form a group, while the remaining 26 districts are assigned as the other group. A test of means (averages) is carried out for the two groups to analyze differences in fertilizer consumption (DAP), ratio of irrigated land to net sown land and few others. The result is presented in Table 4.4. The result shows a significant difference between the top 10 districts and the remaining in terms of consumption of fertilizer (DAP) and net irrigation area. The result suggests that the high or better performing districts have some qualitative difference in terms of these variables as compared to the others. These differences could be further explored and these variables could be taken as independent variables to explain the differences in levels of productivity.

Let us now focus on a comparative picture of change in district rankings in terms of average TFP in the first half of the period (2004–2008) and the second half of the period (2009–2013). Rohtas, Bhojpur, and Buxar are the top three performing districts in terms of productivity, and the difference in average TFP over two periods across these three districts is not much. Kishanganj, Jamui, and East Champaran are the lowest three performing districts in terms of productivity, and the difference in average TFP over two periods across these six districts is enormous. The first three districts are almost twice productive compared to the last three districts.

Using equation (4.7) and the variables defined earlier, we estimate a fixed effects (FE) and a population average model on a panel data of districts observed for 10 years. The FE model captures the unobserved time-invariant features at the district level, while the population average model allows for correlated variables across the panel is useful to analyze the differences at the average instead of subject- or entity-specific values. In particular, the model is preferable in cases of data clustering and with time-varying parameters, the population average model also provides a robust estimate of standard errors. In this case, we derive the

Table 4.2 Average values and ranks of selected variables for districts, 2004–2005 to 2013–2014

Districts	*TFP Index Score*	*NSA (Ht.)*	*TWI Area (Ht.)*	*Rainfall (mm)*	*Literacy rate (%)*	*TFP Rank 2004–2005*	*TFP Rank 2013–2014*	*NSA Rank 2004–2005*	*NSA Rank 2013–2014*
Araria	0.592	184,298.0	64,876.0	1,377.83	0.499	30	6	13	8
Aurangabad	0.739	199,737.6	49,578.4	803.56	0.668	9	7	9	6
Banka	0.716	135,737.6	43,055.2	1,039.76	0.541	26	14	20	25
Begusarai	0.574	116,984.1	77,775.0	858.49	0.597	17	25	27	26
Bhagalpur	0.638	136,013.3	46,540.4	940.07	0.595	14	27	22	23
Bhojpur	0.914	185,374.7	45,501.1	881.39	0.675	3	10	11	7
Buxar	0.877	137,923.9	43,387.7	777.09	0.666	4	8	23	21
Darbhanga	0.597	165,207.8	33,499.1	889.94	0.534	31	32	15	17
East Champaran	0.579	290,426.0	115,836.5	1,174.92	0.508	16	31	1	2
Gaya	0.694	162,489.2	101,847.4	741.96	0.602	8	13	8	13
Gopalganj	0.665	147,555.6	56,264.3	1,077.03	0.607	11	11	21	19
Jamui	0.552	55,739.0	13,173.7	777.39	0.552	32	36	31	34
Jehanabad	0.765	61,215.8	34,351.4	883.43	0.638	6	1	33	32
Kaimur	0.852	156,174.2	47,889.1	849.94	0.656	1	5	19	16
Katihar	0.644	170,654.6	82,731.3	1,182.54	0.476	27	3	16	9
Khagaria	0.627	87,979.8	69,873.3	914.48	0.535	28	26	30	30
Kishanganj	0.481	119,845.6	28,658.8	1,897.86	0.487	33	35	25	29
Lakhisarai	0.683	63,969.1	25,746.3	1,011.47	0.586	12	12	32	31
Madhepura	0.654	126,705.8	65,718.9	1,117.47	0.48	23	16	24	22
Madhubani	0.512	227,369.6	60,491.2	936.95	0.542	35	34	4	4
Munger	0.687	48,105.1	14,347.6	990.81	0.677	13	19	34	33

(Continued)

Table 4.2 (Continued)

Districts	*TFP Index Score*	*NSA (Ht.)*	*TWI Area (Ht.)*	*Rainfall (mm)*	*Literacy rate (%)*	*TFP Rank 2004–2005*	*TFP Rank 2013–2014*	*NSA Rank 2004–2005*	*NSA Rank 2013–2014*
Muzaffarpur	0.562	210,689.5	81,978.1	1,085.19	0.593	25	22	6	5
Nalanda	0.578	169,115.0	98,378.2	823.62	0.615	29	15	14	18
Nawada	0.675	103,037.1	60,053.3	844.41	0.564	10	21	28	28
Patna	0.743	183,397.7	69,549.5	888.88	0.687	5	4	7	15
Purnea	0.607	198,720.3	71,755.8	1,367.22	0.468	24	29	5	10
Rohtas	1	250,120.3	16,999	819.26	0.703	2	2	3	3
Saharsa	0.716	109,036.9	48,695.7	842.47	0.495	18	30	29	27
Samastipur	0.624	179,369.9	109,452.1	925.3	0.574	34	18	12	11
Saran	0.708	181,033.9	84,197	996.94	0.623	7	23	10	14
Sheikhpura	0.655	38,692.2	19,607	850.15	0.599	19	9	35	35
Sitamarhi	0.542	25,303.6	57,875.5	1,503.78	0.485	22	24	36	36
Siwan	0.643	167,340.1	88,826.2	916.18	0.648	20	20	17	12
Supaul	0.636	144,638.6	51,704.6	1017.3	0.522	21	33	18	20
Vaishali	0.649	125,794.3	63,000.6	969.51	0.624	36	17	26	24
West Champaran	0.722	276,767.3	43,457.2	1,233.66	0.512	15	28	2	1
Total	0.669	148,404.5	57,963.14	1,006.07	0.579				

Note: TFP = Total Factor Productivity; Ht. = area in hectares; NSA = net sown area; TWI = tube well irrigation.

Table 4.3 Rank Correlation of selected variables for 2005 and 2014

Rank Correlation (ρ)	*Value*
ρ (TFP, 2005 & 2014)	0.512
ρ (NSA, 2005 & 2014)	0.952
ρ (TFP & NSA 2005)	0.077
ρ (TFP & NSA 2014)	0.109

Note: TFP = Total Factor Productivity; NSA = net sown area.

Table 4.4 Test of Means of Basic Agricultural Variables for Top 10 Districts as Compared to the Rest, 2013–2014

Variable	*Top (10)*	*Bottom (26)*	*Pr (diff ≠ 0)*
Urea/Net Sown Area	300.43	359.50	0.05**
	(11.57)	(18.07)	
DAP/Net Sown Area	63.72	78.45	0.12
	(4.80)	(5.58)	
Net Irrigation/Net Sown Area	55.75	69.09	0.002*
	(1.97)	(2.55)	
No. of KCC/No. of Cultivators	20.98	18.58	0.24
	(1.72)	(1.10)	

Note: Standard error in parenthesis; * and ** indicate value significant at 5% and 1%, respectively. DAP = ; KCC = Kisan Credit Card.

average estimates for a 10-year period for each district. The estimated equation is as follows:

$$TFP_{it} = \beta_0 + \sum_{k=1}^{4} \beta_k R_{it} + \beta_5 \left(Drought\right)_{it} + \beta_6 \left(Flood\right)_{it} + \beta_7 \left(Size\right)_{it} + \beta_8 \left(LR\right)_{it} + v_{it} \tag{4.8}$$

where R_is are the ratios defined previously. In absence of any reliable data on floods and drought, we construct the two variables as follows. Based on the guidelines of the national disaster management, a drought year is declared when the annual rainfall is deficient by 20% of normal or more, and a severe drought year is declared when annual rainfall is deficient by 25% to 40% of normal or more (NDRM 2010). Using this criterion, we take deviations of actual annual rainfall from the 10-year average for each district. Using the deviation, we consider a negative deviation in excess of 20% and denote that as a drought situation. Thus, Drought is specified as a dummy variable, where it takes a value 1 for a negative deviation of

20% and above from the long period average. Similarly, we take a positive 20% deviation from the long period average and denote it as Flood with a dummy variable taking value 1 and 0 otherwise. In the equation (4.8), *LR* is the total literacy rate of the district in percentage, *Size* is net sown area per cultivator, *i* denotes districts, *t* denotes time in years, β_is represent parameters to be estimated and v is an error term (Table 4.5).

Our result of interest is Model 2 which is the population average model. The result shows that the ratio of irrigated area to net sown area R_1 has a positive and significant impact on productivity. This result is on expected lines as improvement in irrigation facilities is a major contributor to increasing agricultural productivity. We find the ratio of tube-well-irrigated area to total irrigated area R_2 to have a negative impact on productivity. The coefficient is close to being statistically significant, and this also supports earlier findings as studies in the past have argued that excessive tube-well irrigation has led to a decline in the quality of soil and fall in productivity. The electrification of villages, R_3, is seen as a positive and significant contributor to improvements in productivity. This effect indicates that improvements in infrastructure have had a considerable impact on improving the quality of agriculture. We find area under HYV seeds (R_4) to be a positive determinant for productivity growth. The coefficient, although close to being statistically significant, does indicate that use of HYV seeds has led to improvements in productivity for all major crops across the districts.

The overall literacy rate has shown a positive and significant impact on productivity levels. The result is indicative of the fact that higher literacy levels may translate into spreading awareness and information which eventually contribute to improvements in productivity. Droughts have an adverse impact on agricultural productivity through the loss of cultivable land and output. This effect is captured by the negative sign on the coefficient of the drought dummy. The coefficient is negative and significant, indicating the adverse impact on productivity. The result on Flood is similar, but the coefficient is statistically insignificant. We include size as the ratio of net sown area per person (or the land–man ratio) to control for the effect of per person cultivated land. We find this positively related to productivity level, suggesting that, on average, districts with higher land–man ratio have higher levels of productivity. The result on the effect of size could be explored more in detail as the changes in the land–man ratio can be decomposed into changes in the land brought under cultivation and the labor force engaged in agriculture. In evaluating the results, we have also attempted to re-estimate the model with a different set of control variables. In particular, we use the ratio of KCC holders to cultivators as an alternate variable to capture the effect of developmental progress across districts. In Model 3, we present the alternate result by including the ratio R_5 and omitting R_4. We find the effect of KCC holders as positive and significant while retaining all our previous findings. This effect lends us some credibility in our finding as we are able to retain our basic results with an alternate formulation.

Table 4.5 Regression Results

Dep. Var. ln(TFP)	*Model 1: FE*	*Model 2: PA*	*Model 3: PA*
ln R_1 (ratio of irrigated to NSA)	0.319***	0.178***	0.174***
	(0.095)	(0.053)	(0.051)
ln R_2 (ratio of tube-well-irrigated area to irrigated area)	-0.020	-0.042	-0.052
	(0.049)	(0.035)	(0.034)
ln R_3 (ratio of electrified villages to total villages))	0.079*	0.075**	
	(0.041)	(0.034)	
ln R_4 (ratio of HYV area to irrigated area)	0.061	0.074	0.065
	(0.092)	(0.054)	(0.051)
ln (literacy rate)	1.165***	0.873***	0.728***
	(0.198)	(0.143)	(0.153)
Drought	-0.050*	-0.048*	-0.058**
	(0.029)	(0.029)	(0.029)
Flood	0.002	-0.010	-0.002
	(0.032)	(0.031)	(0.032)
ln (Size) (NSA per cultivator)	0.139	0.109**	0.098**
	(0.123)	(0.051)	(0.047)
ln R_5 (Ratio of KCC holders to cultivators)			0.052***
			(0.019)
Constant	0.563***	0.284***	0.252***
District FE	Yes	No	No
N, Groups	360, 36	360, 36	360, 36
R^2	0.30	–	–

Note: Standard errors in parentheses; FE = fixed effects; PA = population average; NSA = net sown area; HYV = high-yielding variety seeds; KCC = Kisan Credit Card. ***$p < 0.01$, **$p < 0.05$, *$p < 0.1$.

In summary, the results indicate few important aspects for policy in leading to improvements in agricultural productivity. The first is the set of inputs that are used in estimating productivity levels at the crop level and, second, the factors that explain the differences in productivity across the district level. One of the findings suggests that, at the district level, the use of HYV seeds can lead to an increase in productivity for different crops. To see this phenomenon more clearly, we can compare the average area under cultivation under the three crops. Table 4.6 shows the district wise averages (for the period 2004–2013) of the area under HYV cultivation for districts ranked as per their TFP scores. For a comparison, we can group districts into three categories, viz. high (1–0.75), medium (0.74–0.6) and low (<0.6) TFP scores to get a sense of the variation across districts. First, the cultivation pattern clearly shows that most districts (24 out of 36) have a preference towards rice as, on average, the area under cultivation for rice is higher than other two crops. Second, for districts with higher TFP scores, the area under rice is much higher compared to other crops in other districts. In recent years, several policy initiatives have been made in the area of rice cultivation, and continuing similar policies in the future is more likely to lead to productivity gains. The result also highlights that alternate sources of irrigation (other than tube well) need to be explored as excessive tube-well irrigation can lead to a negative impact on agricultural productivity. Alternate sources such as canals, streams and rainwater harvesting, among others, may prove to be more beneficial in the long run in raising productivity levels of major crops.

Bihar experiences wide fluctuations in agricultural output due to natural disasters such as floods and droughts. Such disasters have a negative effect on production and push a large part of the population engaged in agriculture into economic distress. It is thus essential that more comprehensive flood and drought management programs are structured and implemented to minimize the impact of such calamities. Table 4.6 shows the number of flood instances across districts between 2004 and 2013. On average, the impact of floods can be seen on districts with lower productivity levels. These districts, despite having similar levels of area under cultivation, have been prone to floods with a higher frequency as compared to the rest. Another major determinant of agricultural productivity is the spread of awareness, information and capacity building of farmers to training and extension programs. Initiatives on R&D activities, crop information, soil testing, irrigation programs and other learning programs through agricultural universities and centers can aid in the long-term development of agriculture in the state. It is imperative that such activities are given policy focus and initiation.

4.6 Conclusion

In this chapter, we analyze agricultural productivity at the district level for the state of Bihar. Using the Divisia–Tornqvist Index for three major crops, we prepare TFP estimates for 36 districts for the period 2004–2005 to 2013–2014. Based on the TFP scores for each district, we identify some important factors that explain differences in TFP across districts. Our findings show that an increase in irrigation coverage has had a positive and significant impact on raising productivity. We find that the recent initiative of the electrification of villages has also contributed to raising productivity in agriculture. The results also suggest that use of HYVs for major crops has the potential to bring about productivity gains. The analysis also suggests that there are deeper linkages of agricultural productivity and the socio-economic environment in which agriculture is undertaken as an economic activity in the state. In this context, the disparity in levels and trajectory of productivity growth across districts can help in identifying factors that operate differently across the states.

The disparity in productivity gains across districts also has implications for state policy. Agriculture in the state is severely affected by frequent

Table 4.6 District-Wise Area under Cultivation Statistics and TFP Levels, 2004–2005 to 2013–2014

Districts	*Rice HYV*	*Wheat HVY*	*Maize HYV*	*Avg. TFP Score*	*N. Floods*	*TFP Status*
	Avg. Area (Lk. Ht)	*Avg. Area (Lk. Ht)*	*Avg. Area (Lk. Ht)*			
Rohtas	158.40	119.80	0.10	1.000	6	High
Bhojpur	91.80	54.90	2.30	0.914	0	High
Buxar	55.60	53.60	0.30	0.877	0	High
Kaimur	87.90	56.30	0.00	0.852	5	High
Jehanabad	28.20	19.20	0.30	0.765	7	High
Patna	46.50	38.20	2.20	0.743	3	Medium
Aurangabad	110.30	42.10	0.00	0.739	2	Medium
West Champaran	84.80	74.00	5.90	0.722	6	Medium
Saharsa	32.20	29.70	12.60	0.716	5	Medium
Banka	61.80	17.60	4.80	0.716	0	Medium
Saran	47.70	71.70	13.30	0.708	7	Medium
Gaya	50.70	32.50	1.90	0.694	0	Medium
Munger	9.60	7.50	1.70	0.687	0	Medium
Lakhisarai	11.30	16.20	0.70	0.683	0	Medium
Nawada	35.60	31.00	0.80	0.675	9	Medium
Gopalganj	54.30	79.20	6.40	0.665	9	Medium
Sheikhpura	9.10	9.10	0.10	0.655	6	Medium
Madhepura	39.20	30.10	19.50	0.654	2	Medium
Vaishali	27.90	31.20	15.60	0.649	5	Medium
Katihar	60.10	29.00	17.30	0.644	8	Medium
Siwan	63.80	76.40	8.50	0.643	1	Medium
Bhagalpur	19.20	25.10	15.30	0.638	10	Medium
Supaul	51.10	32.10	6.50	0.636	5	Medium
Khagaria	5.90	25.80	7.75	0.627	0	Medium
Samastipur	46.40	47.50	12.20	0.624	6	Medium
Purnea	66.10	37.50	22.50	0.607	8	Medium
Darbhanga	31.50	48.90	2.40	0.597	0	Low
Araria	63.90	42.30	8.20	0.592	10	Low
East Champaran	79.90	73.90	6.70	0.579	7	Low
Nalanda	71.90	59.40	3.20	0.578	3	Low
Begusarai	12.20	40.60	20.70	0.574	1	Low
Muzaffarpur	59.60	76.00	10.20	0.562	7	Low
Jamui	9.00	2.60	0.00	0.552	0	Low
Sitamarhi	47.90	45.10	1.90	0.542	10	Low
Madhubani	62.90	56.90	0.70	0.512	5	Low
Kishanganj	46.60	15.90	0.70	0.481	8	Low

Note: HVY = high-yielding variety; Lk. Ht. = lakh hectare; N. floods = number of floods between 2004 and 2013; TFP status: 1–0.75 high, 0.74–0.60 medium, and <0.60 low.

droughts and floods which leads to a significant loss of cultivable area and output. As a policy initiative, a comprehensive drought and flood management program can minimize the loss in agricultural output to a considerable extent. As the agricultural sector remains central for the progress of the state economy, higher productivity levels this sector can lead to a long-term sustainable and higher growth for the state. In particular, a well-directed policy on promoting HYV seeds, flood and drought management and the promotion of R&D activities through training and informative programs can bring about a significant change in agriculture. While this study is probably the first district-level analysis of Bihar's agricultural productivity performance, few caveats must also be mentioned. The results are highly contingent on the quality of data available. In our assessment, the data on basic agricultural and district-level variables suffer from several limitations, such as outliers, a lack of proper definition and time lag in compilation, among others.

Acknowledgment

This project was carried out under the aegis of the *Jigyasa* event under the "Badh Chala Bihar Campaign" and the results were presented in the concluding event of the campaign in August 2015. A team from IIT Kanpur, supervised by Dr. Deep Mukherjee, accomplished this project with following members: Amey Sapre, Mahendra Kumar Singh, Kritika Awasthi and Amit Kumar. In completing the project, we have received help and cooperation from several officials of the Bihar government. We thank Shri Narendra Sahani, Director, and Shri Shailesh Kumar, Deputy Director, of Dept. of Agriculture; and Krishi Bhawan, Patna, for their cooperation and for providing district-level data on agriculture on request. We also thank Shri Vishnu Dayal, Director, and Shri Nirbhay Kumar, Deputy Director, and all officers of the Directorate of Economics and Statistics, Patna, for providing data and publications required for this project.

References

Adams, J., and B. Bumb. "Determinants of Agricultural Productivity in Rajasthan, India: The Impact of Inputs, Technology, and Context on Land Productivity." *Economic Development and Cultural Change* 27, no. 4 (1979): 705–722.

Bhushan, S. "Sources of Productivity Growth and Livelihoods Resilience in Bihar in the Recent Decade: A District-Level Non-Parametric Analysis." In *Transition Strategies for Sustainable Community Systems*. Cham: Springer (2019), 231–248.

Bhushan, S. "Sources of Productivity Growth in Bihar in the Recent Decade a District Level Analysis." *The Indian Economic Journal* 63, no. 1 (2015): 41–51.

Bhushan, S. "Total Factor Productivity Growth of Wheat in India: A Malmquist Approach." *Indian Journal of Agricultural Economics* 60, no. 1 (2005): 32–48.

Coelli, T. J., D. S. P. Rao, C. J. O'Donnell, and G. E. Battese. *An Introduction to Efficiency and Productivity Analysis*. New York: Springer (2005).

Diewert, W. E., and K. J. Fox. "Malmquist and Tornqvist Productivity Indexes: Returns to Scale and Technical Progress with Imperfect Competition." Center for Applied Economics Research, Working Paper, 2005/03, University of New South Wales (2005).

FAO. *Transforming Food and Agriculture to Achieve the SDGs*. Rome: United Nations, Food and Agriculture Organization of the United Nations (2018).

Fare, R., S. Grosskopf, B. Lindgren, and P. Roos. "Productivity Changes in Swedish Pharmacies 1980–1989: A Non-Parametric Malmquist Approach." *Journal of Productivity Analysis no.* 3 (1992): 85–101.

Fare, R., S. Grosskopf, and C. A. K. Lovell. *The Measurement of Efficiency of Production*. Boston: Kluwer Academic Publishers (1995).

Government of Bihar. *Economic Survey, 2004–05*. Patna: Finance Department, Government of Bihar (2005).

Government of Bihar. *Economic Survey, 2014–15*. Patna: Finance Department, Government of Bihar (2015).

Government of India, *Economic Survey, 2014–15*. New Delhi: Department of Economic Affairs, Ministry of Finance, Government of India (2015).

Ghosh, S. "State of Bihar." *Economic and Political Weekly* 50, no. 11 (2015): 4.

Grifell-Tatje, E., and C. A. K. Lovell. "A Note on the Malmquist Productivity Index." *Economics Letters* 47, no. 2 (1995): 169–175.

Kumar P., A. Kumar, and S. Mittal. "Total Factor Productivity of Crop Sector in the Indo-Gangetic Plain of India: Sustainability Issues Revisited." *Indian Economic Review, New Series* 39, no. 1 (2004): 169–201.

Kumar, P., S. Mittal, and M. Hossain. "Agricultural Growth Accounting and Total Factor Productivity in South Asia: A Review and Policy Implications." *Agricultural Economics Research Review* 21, no. 2 (2008): 145–172.

Mahajan, S. "Decay and Darkness in Bihar." *Economic and Political Weekly* 50, no. 2 (2015): 78–79.

Muragi, R. "The Green Revolution and the Productivity Paradox: Evidence from the Indian Punjab." *Agricultural Economics* 25, no. 2–3 (2001): 199–209.

NDRM. *National Disaster Management Guidelines: Management of Drought*. New Delhi: National Disaster Management Authority, Government of India (2010).

O'Donnell, C. J. "An Aggregate Quantity-Price Framework for Measuring and Decomposing Productivity and Profitability Change." Centre for Efficiency and Productivity Analysis, Working Paper Series, No. WP07/2008, School of Economics, University of Queensland (2008).

OECD. *Compendium of Agri-Environmental Indicators*. Paris: OECD Publishing (2013).

Singh, R. K. P. "Changing Face of Agriculture in Bihar." *The Bihar Times*, November 8, 2011. Retrieved from http://www.bihartimes.in/articles/RKP/agriculture.html

Sharma, A. N., and G. Rodgers. "Structural Change in Bihar's Rural Economy." *Economic and Political Weekly* 50, no. 52 (2015): 45–53.

Shah, M. "Eliminating Poverty in Bihar," *Economic and Political Weekly* 50, no. 6 (2016): 56–65.

Part II

Energy and climate change

5 Indian youth's willingness to pay for climate change policies

A contingent valuation study

Debarun Sengupta, Abhishek Kumar and Sarani Saha

5.1 Introduction

Rapid industrialization and explosion in population have led to unjustifiable exploitation of natural resources by humans, which eventually created many problems. Climate change, which implies a statistically significant variation in the mean temperature of Earth, is one of the significant outcomes of such activities. Its impacts are observed through variations in temperature, shifts in rain patterns, and the rise of sea level (Rishi 2009). As a consequence of climate change, the number of natural disasters is continuously increasing (Banholzer et al. 2014). In 1979, the First World Climate Conference formally raised this issue among the scientific community, but the pace of climate change research and assessment increased with the establishment of Intergovernmental Panel on Climate Change (IPCC) in 1988. Later, the introduction of the United Nations Framework Convention on Climate Change (UNFCCC) at the Earth Summit (1992) drew the attention of the masses towards climate change issues. The objective of the Convention was to mitigate greenhouse gas concentrations in the atmosphere (United Nations 1992). The participating members of the UNFCCC met annually at the Conference of Parties (COP) to formulate and evaluate corresponding strategies. It announced the Kyoto Protocol (1997) which imposed legally binding emission reduction goals for its member countries (United Nations 1998). At COP21 (December 2015) of the UNFCCC, 195 nations signed the Paris Agreement to promote sustainable development and restrict the rise in global temperature under 2° Celsius of what it was in the pre-industrial era (United Nations 2015). Unlike the Kyoto Protocol, the Paris Agreement did not ask for any legally binding emissions reduction targets; instead, it allowed countries to set their voluntary commitments based on their national circumstances.

Being the fourth-largest CO_2 emitter in the world (Olivier et al. 2016), India has always been a critical player in the climate change negotiations. India ratified the Kyoto protocol in 2002 and launched different national and state-level programs to mitigate climate change which includes National Action Plan on Climate Change (NAPCC) and the Climate Change Action Programme (CCAP) under the 12th five-year plan. NAPCC covers eight different "National Missions" like the National Solar Mission, National Water Mission, and others. Recently, India signed the Paris Agreement and submitted its Intended Nationally Determined Contributions (INDCs), in which it aimed to reduce the emissions intensity of its gross domestic product by 33% to 35% from 2005 levels by 2030 (Ministry of Environment and Forests, Government of India 2015).

India's INDCs to Paris Agreement include ambitious targets and require a huge investment. According to the estimates of the government of India (GoI), it is expected to cost at least US$2.5 trillion at 2015 prices (Department of Economic Affairs, Economic Division 2016). These investments can be financed through national funds, like National Clean Energy Fund and National Adaptation Fund for Climate Change, or international funds like Adaptation Fund and Green Climate Fund. Still, the burden of investment is tremendously large, and India needs to diversify its sources of finance. Climate change–related research and development (R&D) is regarded as a public good, and hence, it could also be financed through other avenues such as surcharge in electricity bills, public donation to climate funds, increased taxes on fossil fuel energy sources (like petrol, coal, and others), and so on. However, these modes of financing depend on an individual's level of risk perception towards climate change. So, if an individual perceives that climate change has a negative impact, then he or she would be willing to contribute to these policies. Compared to the developed nations, people in developing countries like India have little awareness, which creates a barrier to the funding of these policies in developing countries. Thus, these nations need to first change the individual's motivation and awareness level, which affects their perception (Rishi 2009).

This study employed a single referendum format to evaluate an individual's willingness to pay (WTP) for a climate change–related R&D fund. This study is based on primary data. The focus group of the study consists of the final-year undergraduate students from a technical educational institution in Kanpur City, Uttar Pradesh. The respondents were asked questions regarding their awareness and knowledge on this issue, and a discrete choice referendum question was asked about their willingness to make a onetime contribution for the fund. This chapter used a logit model to estimate the parameters and found that gender, income group and the bid amount significantly influenced an individual's choice to make contributions. The rest of the chapter is organized as follows: Section 5.2 provides a review of corresponding literature for both developed and developing countries. Section 5.3 provides the methodology behind the referendum format, followed by the empirical analysis of the model in Section 5.4. Finally, Section 5.5 presents the conclusions from the whole study.

5.2 Related literature

In the last five decades, Contingent Valuation Method (CVM) has been developed as a method for elicitation for economic values of non-market commodities (Smith 2006). CVM was first used by Davis (1963) to value big game hunting in Maine, and then many other studies were conducted which employed this method. However, the critics of CVM argued that it gives a hypothetical result. This led to a controversy when it was used as a legal method to determine the non-use or passive value of goods and services. The controversy reached its high when CVM was used to determine the damage created by the *Exxon Valdez* oil spill in Alaska. In 1992, the National Oceanic and Atmospheric Administration (NOAA) set up a panel, known as Blue Ribbon Panel, to evaluate the CVM to estimate the non-use values. This panel provided a guideline which helps unify the CVM methodology for better estimation. After the NOAA panel report, the study related to the CVM grew rapidly (Carson and Hanemann 2005).

Wiser (2007) conducted a CVM study of 1,574 U.S. residents who pay their electric bills and an opinion survey of 202 U.S. residents to estimate their WTP for renewable energy and hence, environmental improvements. This paper focused on the differences in stated WTP of an individual under different payment schemes (i.e., collective or voluntary) and different provision arrangement (i.e., government or private) of the good. Under the collective payment method, all households and businesses will have a mandatory surcharge in electricity bills but voluntary method allowed individuals to choose whether to pay for renewable energy or not. The author employed a split-sample approach, whereby each respondent received a different set of CVM questions corresponding to payment approach and provision arrangement of the good (i.e. collective-government, collective-private, voluntary-government and voluntary-private) and used three different bid values, namely 50¢ per month, $3 per month and $8 per month, to be paid for a duration of three years. Each respondent received just one of these three bid levels. On comparing the empirical distributions of WTP responses, the author found that the collective payment method elicits a higher WTP than do voluntary ones and that private provision arrangements elicit a higher WTP than do governmental ones. It also noted that on an average, household's WTP for renewable energy is 25% (approx.) higher with collective payment method compared to voluntary ones. In all four cases based on the payment method and the provision arrangement of the good, higher household income and liberal views promoted higher WTP. The survey also indicated that females were more willing to pay for renewable energy and that respondents with children were less willing to pay.

Li et al. (2009) conducted a CVM study of U.S. households to estimate their annual WTP to support the creation of a National Energy Research and Development Fund (ERDF), for investments in R&D on energy sources not reliant on fossil fuels. The ERDF was proposed to be set up with the motive of reducing future dependence on foreign oil and emissions of carbon dioxide (CO_2). The authors used data from a set of national telephone (916 observations) and internet (1,417 observations) surveys. They used higher fees on the use of fossil energy (i.e., electricity and other products and services that rely on coal, oil, and gas) as the payment vehicle. The bid values used in the survey were chosen randomly from a set which varies from $6 (low value) to $2,400 (high value). Authors employed a split-sample treatment to determine the effect of a nuclear energy R&D option in the mix of energy sources, on the overall value of ERDF. Half of the respondents were asked about the significance of investments in new nuclear reactor R&D, while the other half were not asked about it. The referendum question was followed by a question to explore the certainty level of the respondent to the previous question. To obtain the maximum likelihood estimation (MLE) of WTP, this chapter used probit model, and it found that factors like perceptions about the importance of energy issues, the need to reduce dependence on foreign energy sources, and benefits of the development of crop-based fuels positively influence the respondent's support for the creation of a national ERDF. The study also concluded that if the respondent was female or having liberal views or belongs from a higher income group, then he or she contributed more.

Solomon and Johnson (2009) conducted a CVM study and a fair share[1] (FS) survey to investigate the public's valuation to climate protection, using their WTP for biomass or cellulosic ethanol. The survey was conducted among 745 households of Michigan, Minnesota, and Wisconsin to study the issue in the context of U.S. The authors also considered that increased demand for biofuels (made from corn and soybean) would eventually lead to increase in food prices, and hence, the survey recipients were asked about their FS charge for cellulosic ethanol in the case of a food emergency. They provided half of the respondents with an FS scenario and another half with basic CVM scenario and used a payment card approach to implement the WTP survey. This study used a single-bounded payment card approach with bid values ranging from $0.00 to $1.00 which represented higher prices per gallon for cellulosic ethanol. Authors found annual household income, political views, rural residence, and concern about climate change as the significant explanatory variables in basic CVM scenario and were positively related to an individual's WTP. Furthermore, from the FS study, annual household income, gender (being female), and concern about climate change were the significant variables which positively affected FS charge. It found that the WTP for FS bids were lower than the basic CVM scenario. Assuming all survey non-respondents had zero WTP, the authors found that the annual mean WTP was $252 per capita and the annual mean FS was $192 per capita.

Hong-Xia et al. (2014) conducted a CVM to estimate the WTP for supporting the emission reduction policy in China. The data for the study were collected from four different cities, namely Beijing, Fujian, Shandong, and Shanghai, from September 2009 to May 2010. Respondents were asked about their WTP to support the government actions toward climate change mitigation. The study used a payment card method to record the stated responses. The Cameron and Huppert (1989) WTP modeling approach was followed to estimate the coefficients using a maximum likelihood technique. The result indicated that the Chinese public was willing to pay CN¥

201.86 per year. The study also showed that among those four different cities Beijing had the highest WTP while Shanghai had the lowest WTP. The study concluded that factors like age, sex, education levels, political orientation, life satisfaction, local environmental condition, and knowledge of the environment have a positive impact on WTP.

Gupta (2016) used CVM to know the effectiveness of carbon tax in Indian road passenger transportation. The study was carried out in three different metropolitan cities in India, namely Delhi, Mumbai, and Bangalore. Only respondents who were 18 years old and who owned a vehicle were included in the study. Each respondent was asked about how much he or she would be willing to pay for a measure to mitigate CO_2 emission from road passenger transport. To avoid the negative values, WTP values were censored at zero. The probit and tobit models were used to analyze the data. The study concluded that factors like age, education, income, environmental awareness, and local environmental status have a positive impact on determining the WTP amount.

5.3 Theoretical framework

The aim of the study conducted in this chapter is to estimate an individual's WTP for climate change R&D funds. The R&D funding of government is similar to a public good. Essentially, here we estimate the WTP for provision of public good and also find the effect of covariates (basically individuals characteristic) on this WTP.

Let there be an individual, say individual i, whose utility depends on two factors:

1. Various marketable goods, he or she consumes (x)
2. The R&D investment on climate change by the government (q)

So, the utility function of individual i can be written as

$$U_i = U(x,q). \tag{5.1}$$

If an individual thinks that he or she will get benefits (loss) from this R&D investment, then U is an increasing (decreasing) function of q. The individual always tries to optimize (or maximize) his or her utility given his or her income (y) and the price of the marketable goods (p). Now, putting the optimal values of x and q in the utility function given in equation 5.1, we get the individual's indirect utility function which can be expressed as

$$V_i = V(p,q,y). \tag{5.2}$$

Now to know how an individual values this R&D investment, we have to make a contrast between two situations: a situation with the current provision of funds and a situation with a changed provision in funds (Carson and Hanemann 2005). Suppose that the fund provision increases from q_0 to q_1, and if the individual regards this change as good, then his or her utility will improve; that is,

$$V_i(p,q_1,y) > V_i(p,q_0,y). \tag{5.3}$$

Let C be WTP. To make a valuation of this increase in R&D funds, we try to estimate the maximum WTP amount which satisfies the following equation:

$$V_i(p,q_1,y-C) = V_i(p,q_0,y). \tag{5.4}$$

Essentially, the maximum WTP of an individual gives how much the individual is willing to pay for the increase in funds provision. Generally, the value of C varies among individuals. The value of C depends on three attributes (Carson and Hanemann 2005):

1. Initial investment amount (q_0)
2. Terminal investment amount (q_1)
3. The value of p and y at which this change occurred.

So, we can define the WTP for the ith individual as

$$WTP_i = C_i(q_0,q_1,p,y). \tag{5.5}$$

Equation 5.5 is known as the WTP function for the ith individual. A crucial point to note in this case is that the individual responses are not a direct measure of WTP, but these can be used to derive the WTP. The derivation of WTP involves a statistical analysis of the responses. Therefore, it is necessary for us to make our model stochastic. This stochastic model gives us the probability distribution of the responses. Generally, we add an error component (ε) in our utility model which makes our model stochastic. So the WTP function can be rewritten as

$$WTP_i = C_i(q_0, q_1, p, y; \varepsilon). \tag{5.6}$$

In the survey conducted in this chapter, final-year undergraduate students were asked for a onetime contribution towards the climate change–related R&D fund. It was stated in the survey that the contribution amount would be deducted as a percentage of their annual salary after a year they join the job.

There are various survey formats like open-ended survey, bidding game, payment card, referendum, or discrete choice method, among others, available for estimating the WTP for investment in R&D funds (Boyle 2003). Among these survey formats, the referendum or discrete choice method is considered as a presumptive method of elicitation for responses (Kolstad 2016), because the other methods suffer from an incentive compatibility problem; so, the respondent can influence the outcome by revealing a false WTP value (Haab and McConnell 2003). Hence, in our study, we choose the referendum method as elicitation format.

During the survey, we asked a similar question as mentioned above to each respondent, but the bid (WTP) amount varied among the respondents. The respondent says yes only if the amount is less than or equal to his or her WTP amount. Let D be the maximum amount of WTP at which the respondent says yes, so the probability of getting the response "yes" can be represented in terms of WTP as

$$\Pr(\text{response is YES}) = \text{pr}\left\{C_i\left(q_0, q_1, p, y; \varepsilon\right) \geq D\right\} = 1 - G(D), \quad (5.7)$$

where $G(D)$ represents the cumulative probability distribution (CDF) of a respondent's WTP. Let Y be the response of the individual. As the response is binary in nature, we get two discrete value of Y, one for yes and zero for no. This response does not depend only on the amount we ask in the survey but also on other individual attributes (like age, sex) and social attributes. So we can write

$$Y = \alpha + \beta X + \varepsilon, \quad (5.8)$$

where X is the matrix of independent variables and β is the corresponding coefficient matrix. The questionnaire used in the survey includes various questions regarding the individual's awareness, motivation, perceptions about climate change, and socio-demographic aspects. The WTP question in the survey reads as

> Suppose the government is considering to approach all new job joiners for one-time contribution towards an R&D fund for global warming related research. A percentage of your annual salary will be deducted after the first year in the job. The collected money will be spent on research for technology development to address global warming issues. Would you be willing to pay "S"% of your first year's salary?

where S is randomly assigned from the set of values $S \in \{0.5, 1, 2, 3, 4, 5, 7, 10\}$

The response to this question was recorded as either Yes (coded as 1) or No (coded as 0), and based on that, the model is estimated. A simple linear regression model corresponding to the above survey is

$$Y_i = \beta_0 + \beta_1 X_i, \quad (5.9)$$

where Y_i is a dummy variable which takes the value of 1 if the ith individual is willing to pay X percent of his or her annual salary at the completion of first year of his or her job. Here, X_i is the percentage of salary asked of ith individual. Here, the variable x_i can attain some distinct values only, based on the questionnaire. In Table 5.1, we show the frequency of individual response to the various bid amount asked in the questionnaire. The total number of responses is 732, out of which 533 said yes and 199 said no. In Figure 5.1, we plot the probability of yes response for the different bid amount. The locus of these points is downward-sloping, indicating that the probability of yes decreases as the bid amount increases. It also tells us that there is a sharp decrease in individual willingness after a significant increase in the bid amount.

However, the dependent variable in equation 5.9 is a dummy variable, so instead of using a dichotomous variable as the outcome variable, the study uses the expected value of outcome variable as the dependent variable. So, equation 5.9 can be written as

$$E(Y|X = x) = 1 \times X\,\Pr(Y = 1|X = x) + 0 \times X\,\Pr\left(Y = 0|X = x\right)$$
$$= \Pr(Y = 1|X = x) = \pi\left(x\right).$$

Table 5.1 Frequency of Responses to Various Bid Amounts

Responses	*Bid Amount (% of annual salary)*								*Total*
	0.5	*1*	*2*	*3*	*4*	*5*	*7*	*10*	*732*
[0] NO	11	10	12	16	14	24	54	58	199
[1] YES	86	85	81	78	78	70	31	24	533

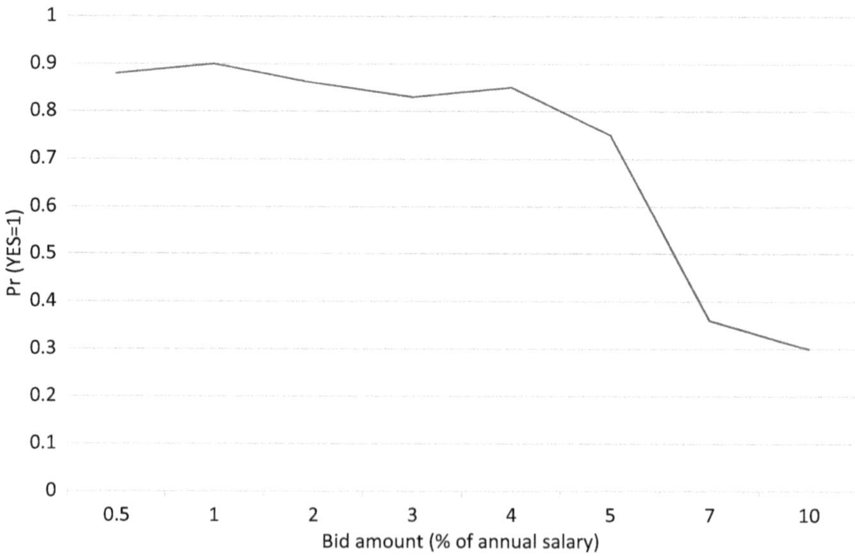

Figure 5.1 Linear Interpolation of Percentage of an Individual's Bid Amount for the Research and Development Fund

Therefore, we can write π (x) as follows:

$$\pi(x) = E(Y|X = x) = \beta_0 + \beta_1 x. \tag{5.10}$$

Now, the Left-Hand Side (LHS) can vary between 0 and 1, while the Right-Hand Side (RHS) can attain any value for different values of x. This mismatch in ranges of RHS and LHS can be corrected by transforming the LHS variable to use log odds ratio (OR) of $\pi(x)$,in place of $\pi(x)$ only. Here, $\pi(x)$ denotes the probability of a yes response for a given x. Log-OR is the log of OR, that is, the ratio of odds of the individual saying yes to the odds of individual saying no to the valuation question.

In equation 5.11, we use logistic transformation which makes the function linear in its parameter and also continuous:

$$g(x_i) = \ln\left[\frac{\pi(x_i)}{1-\pi(x_i)}\right] = \beta_0 + \beta_1 x_i. \tag{5.11}$$

Now, $\pi(x_i)$ can be solved using antilog of equation 5.11, and in general form, it can be written as in equation 5.12:

$$\pi(x_i) = \frac{e^{\beta_0 + \beta_1 x_i}}{1 + e^{\beta_0 + \beta_1 x_i}}. \tag{5.12}$$

The equation 5.12 gives us the probability that the *i*th individual would accept to pay the bid amount. Let there be k independent variables, where x_{ij} represent *j*th independent variable in ith observation. We can write $x_i' = (x_{i1}, x_{i2}, \ldots, x_{ik})$. Therefore, the model can be expressed as given in equation 5.13:

$$g(x_i) = \ln\left[\frac{\pi(x_i)}{1-\pi(x_i)}\right] = \beta_0 + \sum_{j=1}^{k} \beta_j x_{ij}, \tag{5.13}$$

where $\pi(\mathrm{x_i}) = \dfrac{e^{\beta_0 + \sum_{j=1}^{k} \beta_j x_{ij}}}{1 + e^{\beta_0 + \sum_{j=1}^{k} \beta_j x_{ij}}}.$

Since, the survey inherits close end format questions, the probability of a response being yes for the different values of WTP offered in the questionnaires given by equation 5.7:

$$\Pr(\text{Response is Yes}) = \Pr(C \geq B) = 1 - G(B) = \frac{1}{1 + e^{\beta_0 + \sum_{j=1}^{k} \beta_j x_{ij}}}. \tag{5.14}$$

Once this logit model is estimated, the parameter β_i can be used to evaluate the probability with which an individual is willing to pay for the fund.

5.4 Empirical framework

Let there be "n" independent observations for (x_i, y_i), that is, i = 1, 2, . . . , n, where y_i is the outcome variable and $x_i$is the independent variable. Here, the method of MLE will be used to obtain the unknown parameters for the logistic regression model. In order to do so, we first construct the joint probability function of obtaining the observed set of data, assuming the parameters are given, then maximize it to obtain the estimates of the parameters. This joint probability function is called the likelihood function. From equation 5.10 we can write

$$\pi(x_i) = Pr(y_i = 1 | X = x_i) \text{ for all } i;$$
$$\text{Or, } 1 - \pi(x_i) = Pr(y_i = 0 | x_i) \text{ for all } i. \tag{5.15}$$

Using these probabilities, the likelihood of a particular observation (x_i,y_i) can be written as

$$\pi(x_i)^{y_i}\left(1 - \pi(x_i)\right)^{(1-y_i)}. \tag{5.16}$$

Note that for the univariate model, $x_i = (x_1)$ while, for a multivariate model, $x_i' = (x_{i1}, x_{i2}, \ldots, x_{ik})$. The observations are independent, so the likelihood function for the given data can be expressed as

$$l(\beta) = \prod_{i=1}^{n} \pi(x_i)^{y_i} (1 - \pi(x_i))^{(1-y_i)}. \tag{5.17}$$

Equation 5.17 can be simplified by using log and is known as a log-likelihood function:

$$L(\beta) = \log(l(\beta)) = \sum_{i=1}^{n} \{y_i \log[\pi(x_i)] + (1 - y_i)\log[1 - \pi(x_i)]\}. \tag{5.18}$$

This log-likelihood equation is maximized to estimate the parameters. It is differentiated with respect to the parameters, that is, β_i, and first-order conditions can be solved using iterative techniques in various statistical software.

5.4.1 Results

Once the parameters of the model are estimated, the next step is to interpret the coefficients and its significance in the model (Table 5.2). Unlike linear regression model, where the coefficients represent the slope, the coefficient's interpretation is not that simple in a logit model. This section analyzes and

Table 5.2 Estimation Results for Univariate Model

Response	*Coef.*	*Std. Err.*	*Odds Ratio*	*Std. Err.*	$\frac{dy}{dx}$	*Std. Err.*
WTP	−0.371***	0.035	0.690***	0.024	−0.057**	0.003
Constant	2.645***	0.194	14.087***	2.735	–	–

Note: N = 732.*** $p < 0.01$, ** $p < 0.05$, * $p < 0.1$.

explains the estimation results of the survey. First, a simple model is estimated to study the impact of WTP on the individual's response:

$$g(x) = Y = \beta_0 + \beta_1 \text{ WTP}, \tag{5.19}$$

where $Y = 1$, *if* individual's response is YES.
0, Otherwise.

The results show that WTP amount significantly influences the individual's decision to contribution or not. The median WTP amount is defined as the bid amount (i.e., the percentage of annual salary) which is associated with a 50% probability of yes (Hanmman 1984). Using equation 5.14 and putting the value of the coefficients $\Pr(\textit{Response is} \text{ “Yes”}) = \frac{1}{1+e^{2.645-0.371\,WTP}} = 0.5$, we obtain that the median WTP amount is 7.13%; that is, for the bid amount, 7.13%, the probability that response is yes equals 0.5. The plot for fitted values of probability of yes responses with bid amount is given in Figure 5.2.

However, this probability of a yes response by an individual also depends on factors other than the bid amount (WTP). Controlling for other factors (like the income class of the individual, the individual's awareness of climate change issues, and sociodemographic aspects of the individual), the multivariate model can be developed as

$$\begin{aligned} g(x) = Y = \beta_0 + \beta_1\text{WTP} + \beta_2\text{Aware} + \beta_3\text{WTP}\times\text{Aware} \\ + \beta_4\text{Male} + \beta_5\text{WTP}\times\text{Male} + \beta_6\text{Rich} + \beta_7\text{WTP}\times\text{Rich}. \end{aligned} \tag{5.20}$$

where WTP indicates the bid amount and Aware is a dummy variable, which takes a value of 1 if the respondent has prior knowledge about climate change issue and 0 otherwise. Similarly, Male is an indicator variable to indicate whether the respondent is male or not. Rich is a categorical variable that indicates whether the respondent belongs to a higher income group.

In our study, we used two alternative models to know the impact of the rich variable. In Model 1, the variable Rich1 indicates those respondents who belong to the first upper quartile of the income group. Similarly, in Model 2, we use Rich2 which specifies whether the individual belongs to

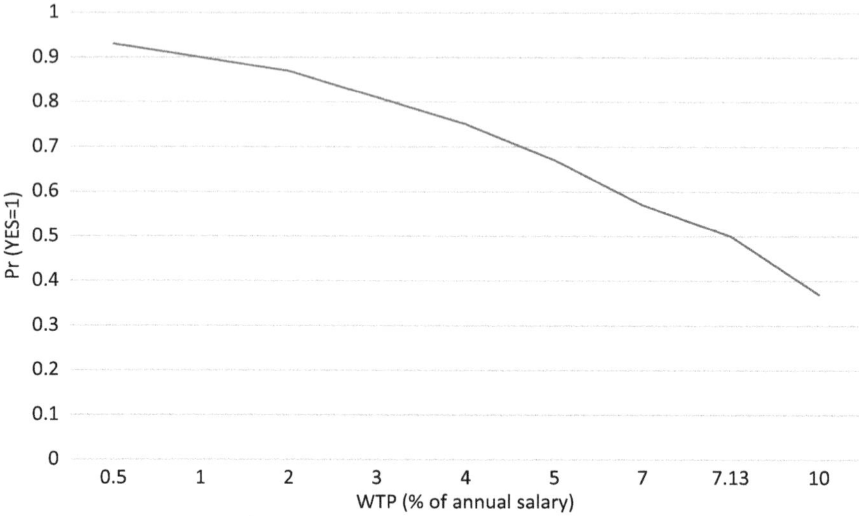

Figure 5.2 Probability of Response Being Yes with Different Bid Amounts (predicted model)

the second or higher quartile of the income group. However, the methodology followed to estimate this model will remain the same as was explained earlier. Here, the model is estimated using statistical software (Stata). The estimation results are given in Table 5.3.

In the first model specification (i.e., Model 1), the estimation result shows that the coefficients of WTP, Rich1, WTP*Aware, and WTP*Rich1 are statistically significant. In the logistic regression setup, the coefficients of the regressors are the log of OR in favor of payment. The negative coefficient for WTP supports the theoretical outcome that increasing the bid value decreases the corresponding probability to make the payment. From Table 5.3, it can be seen that when all the variables, except WTP, have zero values, then the log OR corresponding to an increase of one unit in bid amount (WTP) is –0.531. In the logistic regression model, OR is a crucial measure and has a much clearer interpretation, so in terms of OR, we can say that the individual's probability of making payment reduces to 0.588 times on an increase of 1 unit in WTP.

Most of the literature suggests that rich people are more likely to make the payment compared to people who are not rich. However, the negative coefficient of Rich1 in Model 1 (also Rich2 in Model 2) suggests otherwise. This is a surprising outcome that the rich people are less likely to make payment compared to the non-rich. The reason for this anomaly is that the coefficient of rich denotes the log OR for rich versus non-rich in favor of making the payment when the bid amount equals zero. At low-bid amounts, the probability of making payment does not get influenced much by the individual's income status, but as the WTP amount rises, the rich

Table 5.3 Estimation Results for Multivariate Model

Response	*Model 1*		*Model 2*	
	Coeff	*Odds Ratio*	*Coeff*	*Odds Ratio*
WTP	−0.531***	0.588***	−0.884***	0.413***
	(0.128)	(0.075)	(0.147)	(0.061)
Aware	−0.529	0.589	−0.505	0.603
(Yes = 1)	(0.415)	(0.245)	(0.435)	(0.263)
WTP*Aware	0.127*	1.135*	0.111	1.117
	(0.074)	(0.084)	(0.076)	(0.085)
Male	−0.296	0.743	−0.406	0.666
(Yes = 1)	(0.664)	(0.493)	(0.674)	(0.449)
WTP*Male	0.007	1.007	0.063	1.065
	(0.124)	(0.125)	(0.119)	(0127)
Rich1	−1.067***	0.344***	-	-
	(0.409)	(0.141)		
WTP*Rich1	0.271***	1.311***	-	-
	(0.073)	(0.096)		
Rich2	–	–	−1.919***	0.147***
			(0.578)	(0.085)
WTP*Rich2	–	–	0.549***	1.732***
			(0.103)	(0.180)
_cons	3.557***	35.049***	4.74***	114.445***
	(0.712)	(24.967)	(0.814)	(93.179)
Pseudo R^2	20.45%		25.25%	
Log pseudo-likelihood	-340.71		-320.15	

Note: Rich1 divides the population in 1st quartile and Rich2 divides it in 2nd quartile; Number of Observations = 732.***$p < 0.01$, **$p < 0.05$, *$p < 0.1$.

people are more likely to make payments, as evident from the margin effects table (Table 5.4). The margin effects will be discussed in details later in the chapter. We have also obtained significant coefficients for the interaction terms between WTP and indicator variables aware and Rich1. The coefficient of interaction variable between WTP and awareness variable indicates that if the WTP value increases by one unit, then the probability of making a payment by an aware individual and an individual who is not aware differs by 0.127 unit. Similarly, the coefficient of interaction variable between WTP and indicator variable Rich1 specifies that due to one unit increase in WTP value the probability of making payment by a rich and a non-rich person differs by 0.271 unit.

In Model 2, WTP, Rich2, and interaction between WTP and Rich2 are statistically significant and their direction is similar to coefficients obtained in Model 1. From Table 5.3, it can be seen that when all the variables, except WTP, have zero values, then the log OR corresponding to an increase of one unit in bid amount (WTP) is −0.884. In terms of OR, we can say that the individual's probability of making payment reduces to 0.413 times

Table 5.4 Margin Effects

WTP	*Model 1*		*Model 2*	
	Rich1	*Aware*	*Rich2*	*Aware*
1	-0.075	-0.033	-0.089	-0.029
2	-0.063	-0.029	-0.077	-0.028
3	-0.037	-0.021	-0.036	-0.023
4	0.003	-0.004	0.050	-0.011
5	0.059	0.023	0.183	0.010
6	0.129	0.055	0.328	0.035
7	0.200	0.087	0.437	0.055
8	0.264	0.111	0.488	0.068
9	0.311	0.123	0.492	0.080
10	0.339	0.125	0.468	0.092

due to an increase of 1 unit in WTP. In this model also, the negative coefficient of Rich2 can be explained using the margin effects of Rich2 given in Table 5.4. The coefficient of WTP*Rich2 (0.549) indicates the difference in the log-OR corresponding to a change in the WTP by one unit between rich people and the non-rich people.

Another way to explain the dependence of outcome variable on various explanatory variables is to utilize the average marginal effects of the regressors. It provides the change in outcome variable corresponding to a unit change in a particular regressor, keeping other explanatory variables fixed. It can be obtained by taking derivatives of $\pi(x_i)$ with respect to a particular explanatory variable, say, x_{ij}.

$$\frac{d\pi(x_i)}{dx_{ij}} = \beta_j \pi(x_i)\big(1-\pi(x_i)\big). \tag{5.21}$$

Table 5.4 shows the marginal effects of Rich1 and Aware (Model 1), and Rich2 and Aware in Model 2 for values of WTP ranging from 1 to 10. At each WTP value, the margins indicate the difference between the likelihood of an individual to make the payment at the given WTP amount for different levels of the categorical variable. So, here the margins of Rich1 (or Rich2) indicates the difference between the likelihood of a rich and a non-rich individual to make payment at the given WTP amount. Similarly, the margins of Aware indicate the difference between the likelihood of an aware and an unaware individual to make payment at the given WTP amount.

The margins in Table 5.4 indicate an interesting result about the effect of the individual's income status and the awareness level on his or her likelihood to make payment changes the direction after a certain level of bid amount.

We can see that for the lower bid amounts the likelihood to make payment by the individual is negatively related with his or her income status (similarly for awareness level), but after the bid amount crosses 4% of the individual's annual salary, the relation changes direction and becomes positive. This is in accordance with literature and our expectations that the rich people and more aware people are more willing to make payment for the fund.

5.5 Conclusion

This chapter employed a referendum format to estimate the WTP of the young generation for a climate change–related R&D fund and studied various attributes that affect this WTP. It found that the sign on the variables is in accordance with literature, that is, being rich and aware, increases the individual's WTP while the gender was not statistically significant. It also showed that the median WTP for the fund is about 7.13% of the individual's annual income. However, it is to be noted that this is a sample-specific outcome. This study was conducted with students from an educational institution in Kanpur (Uttar Pradesh) only, and hence, we cannot generalize the study for the pan-India sample. However, this study indicates that the Indian youth perceives the global warming issue as a threat and is motivated to participate in climate change mitigation policies and donate for the R&D fund for this purpose.

Notes

1 The Fair Share (FS) scenario asks questions of the respondents considering the various outcomes of the policy. For example, here, the use of biofuels could lead to increase in food prices and food shortages. So the respondents were asked about their FS in the case of a food emergency.

References

Banholzer, Sandra, James Kossin, and Simon Donner. 2014. "The Impact of Climate Change on Natural Disasters." In A. Singh and Z. Zoomers (eds.), *Reducing Disaster: Early Warning Systemys For Climate Change*. Springer (2014), 21–49.

Boyle, Kevin J. "Contingent Valuation in Practice." In Patricia A. Champ, Kevin J. Boyle, and Thomas C. Brown (eds.), *A Primer on Nonmarket Valuation*. New York: Springer (2003), 111–169.

Cameron, Trudy Ann, and Daniel D. Huppert. "OLS versus ML estimation of non-market resource values with payment card interval data." *Journal of Environmental Economics and Management* 17, no. 3 (1989): 230–246.

Carson, Richard T., and W. Michael Hanemann. 2005. "Contingent Valuation." In K. G. Miller and J. R. Vincent (eds.), *Handbook of Environmental Economics, Volume 2*. Elsevier, 821–936 .

Department of Economic Affairs, Economic Division. *Economic Survey 2015–16*. New Delhi: Ministry of Finance, Government of India (2016).

Davis, R. "The Value of Outdoor Recreation: An Economic Study of the Marine Woods", *Doctoral Dissertation in Economics*, Harvard University. (1963).

Gupta, Monika "Willingness to Pay for Carbon Tax: A Study of Indian Road Passenger Transport." *Transport Policy* (2016): 46–54.

Haab, Timothy C., and Kenneth E. McConnell. 2003. "Parametric Models for Contingent Valuation." In Timothy C. Haab and Kenneth E. McConnell (eds.), *Valuing Environmental and Natural Resources: The Econometrics of Non-Market Valuation*. Edward Elgar Publishing Limited (2003), 16–58.

Hanmman, W. Michael. " Welfare Evaluations in Contingent Valuation Experiments with Discrete Responses." *American Journal of Agricultural Economics* (1984): 332–341.

Hong-Xia, Duan, Lu Yan-Li, and LI Yan. "Chinese Public's Willingness to Pay for CO_2 Emissions Reductions: A Case Study from Four Provinces/Cities." *Advances in Climate Change Research* (2014): 100–110.

Hosmer, David W., and Stanley Lemeshow. *Applied Logistic Regression. A Wiley-Interscience Publication* (2000).

Jaccard, James, Choi K. Wan, and Robert Turrisi. "The Detection and Interpretation of Interaction Effects between Continuous Variables in Multiple Regression." *Multivariate Behavioral Research* 25, no.4(1990): 467–478.

Kolstad, D. Charles. "Stated Preferences, Experiments and Referenda." In D. Charles Kolstad (ed.), *Intermediate Environmental Economics*. New Delhi: Oxford University Press (2016), 194–211.

Li, Hui, Hank C. Jenkins-Smith, Carol L. Silva, Robert P. Berrens, and Kerry G. Herron. "Public Support for Reducing US Reliance on Fossil Fuels: Investigating Household Willingness-to-Pay for Energy Research and Development." *Ecological Economics* (2009): 731–742.

Ministry of Environment and Forests, Government of India. "India's Intended Nationally Determined Contribution is Balanced and Comprehensive: Environment Minister." Press Information Bureau. October 02, 2015. Retrieved from http://pib.nic.in/newsite/PrintRelease.aspx?relid=128403

Olivier, J. G. J., G. Janssens-Maenhout, M. Muntean, and J. A. H. W. Peters. *Trends in Global CO_2 Emissions: 2016 Report*. PBL Netherlands Environmental Assessment Agency European Commission, Joint Research Centre (2016).

Rishi, Parul. 2009. "Climate Change and Human Psyche A Relook at Choice." In Parul Rishi (ed.), *Environmental Issues Behavioural Insights*. New Delhi: Rawat Publications (2009), 40–60.

Smith, V. Kerry. "Fifty Years of Contingent Valuation." In Anna Alberini and James R. Kahn (eds.), *Handbook on Contingent Valuation*. Boston: Edward Elgar Pablishing Limited (2006), 7–65.

Solomon, Barry D., and Nicholas H. Johnson. "Valuing Climate Protection through Willingness to Pay for Biomass Ethanol." *Ecological Economics* (2009): 2137–2144.

United Nations. *United Nations Framework Convention on Climate Change* (1992).

United Nations. *Kyoto Protocol to the United Nations Framework Convention on Climate Change* (1998).

United Nations. *Paris Agreement* (2015).

Wiser, Ryan H. "Using Contingent Valuation to Explore Willingness to Pay for Renewable Energy: A Comparison of Collective and Voluntary Payment Vehicles." *Ecological Economics* (2007): 419–432.

Yip, Paul S. L., and Eric W. K. Tsang. "Interpreting Dummy Variables and Their Interaction Effects in Strategy Research." *Strategic Organization* 5, no. 1 (2007): 13–30.

6 An advance methodology for estimating the elasticities and rebound effect

Debalina Chakravarty and Priyanka Chatterjee

6.1 Introduction

Improvement in energy efficiency (EE) can deliver a range of benefits, not only to individual consumers but also to achieve sustainability goals at the national and international levels. Countries, particularly those with relatively low EE, can consider this as an attractive option because up to a certain extent the economy does not need to choose between averting climate change and promoting growth and development if it considers EE measures. Technically, it is expected that 1% increase in EE reduces energy demand by 1%. However, this is usually not the case because technological improvements evoke behavioral responses (Khazzoom, 1980). Often, an increase in efficiency by 1% will cause a reduction in resource use that is far below 1% or, sometimes, it can even cause an increase in resource use. Actually, gains in the efficiency of energy service consumption will result in an effective reduction in the per unit price of energy services. As a result, the consumption of energy services may increase, partially offsetting the impact of the efficiency gain in energy input use. For instance, a study by Roy (2000) states that the difference between technical and economic potential is 50% for lighting services in rural households in India. Literature argues that, this is due to the presence of the "rebound effect" (RE; Khazzoom, 1987). In the EE–RE debate, the whole argument is that it is uncertain whether EE improvement will reduce aggregate energy consumption in the consumers' context. The answer to this problem is not straightforward. Therefore, it is important to know the magnitude of RE to estimate the final consumption level. If the development strategy through EE improvement takes place in the next 10–20 years to achieve sustainable development, public and private investment should be directed at effective and credible policies. Therefore, if the policy makers can capture RE thoroughly, only then can they get the actual effectiveness of the energy efficiency policy. Since RE is widely estimated by using price elasticities (PE), there is a need to determine the appropriate methodology for estimating correct elasticities. In the literature, several econometric models have been used to estimate RE based on the estimation of PE parameters, for example, linear/log-linear regressions, Almost Ideal Demand System (AIDS), household production functions, and others (Davis, 2008; Guertin et al., 2003, Dubin et al., 1986). The AIDS model by Deaton and Muellbauer (1980) was applied by Brannlund et al. (2007), Wang et al. (2012) and Lin and Liu (2015) for determining the effect on consumption choice of Swedish households and Chinese households, respectively, after an increase in EE and for the estimation of RE. For the properties and wide applicability of the AIDS model in empirical research on household consumption behavior, this study discusses, explores and analyzes its framework as an advanced methodology for estimating the elasticities and RE with a case study from rural West Bengal, India.

This chapter has been organized into four sections: Section 6.2 briefly discusses theoretical discourses about estimating RE and price elasticities based on past literature. Section 6.3 describes the household behavior and the AIDS methodology. A case study for estimating RE and price elasticities using AIDS framework has been discussed in Section 6.4. Section 6.5 presents a summary of the study and concluding remarks.

6.2 Theoretical discourse on price elasticities and rebound effect

The gains in the efficiency of energy service consumption will result in an effective reduction in the per unit price of energy services. As a result, consumption of energy services may increase or partially offset the impact of the efficiency gain in energy service consumption (Roy, 2000). RE in energy is nothing but the increase in energy service consumption demand (corresponding to energy input demand) due to an increase in EE (Binswanger, 2001; Greene, 1992; Brookes, 1990a, 1990b; Khazzoom, 1980). Khazzoom first triggered the RE debate based on the theoretical logic supplemented by empirical evidence in early 1980s. Khazzoom, argued that improvement in the efficiency of one energy service influences not only the demand for own end use but also the demand for others (Khazzoom, 1980).

Khazzoom (1987) considers that the long-run service demand depends on the energy service price which is a ratio of energy price (P) to appliance efficiency (ε) (Khazzoom, 1987). Thus, energy service demand S= $S\left(\frac{P}{\varepsilon}\right)$, and therefore, S′= $\frac{\partial S}{\partial P}$ = $s'/\varepsilon<0$. Energy demand (E) is a ratio of energy service demand and EE.

Therefore, $E = \frac{S}{\varepsilon} = \frac{S\left(\frac{P}{\varepsilon}\right)}{\varepsilon}$

$$\frac{\partial E}{\partial \varepsilon} = \frac{\varepsilon \frac{\partial S}{\partial \varepsilon} - S}{\varepsilon^2} = \frac{-\varepsilon\, sP \frac{1'}{\varepsilon^2} - S}{\varepsilon^2} = -P\frac{s'}{\varepsilon^3} - \frac{S}{\varepsilon^2} < 0$$

Therefore,

$$\left|\eta_\varepsilon(E)\right| = \left|\frac{\partial E}{\partial \varepsilon}\frac{\varepsilon}{E}\right| = \left|\left[-P\frac{s'}{\varepsilon^3} - \frac{S}{\varepsilon^2}\right]\frac{\varepsilon}{E}\right| = \left|-\frac{S}{E\varepsilon} - \frac{Ps'}{E\varepsilon^2}\right| = \left|-\frac{S}{S} - \frac{Ps'}{S\varepsilon}\right|$$

$$= \left|-1 - \frac{Ps'}{S\varepsilon}\right| = \left|-1 + \left[-\frac{P}{S}s'\right]\right| = \left|-1 - \eta_p(S)\right| = 1 + \eta_p(S) \qquad (6.1)$$

Here, $\eta_\varepsilon(E)$ is the change in energy demand due to a change in EE (Khazzoom, 1987). $\eta_p(S)$ is the absolute value of the elasticity of energy service demand with respect to energy price, and this is the measure of RE suggested by Khazzoom (1987). In practice, demand elasticity with respect to efficiency is difficult to estimate since efficiency is an unobservable variable. According to the definition of EE ($\varepsilon = \frac{P_E}{P_s}$), the price (cost) of the energy service (P_s) is proportional to the energy price (P_E). If ε is considered constant, the RE can also be defined as equal to the PE of energy consumption, that is $\eta_p(E)$ (Sorrell, 2007; Saunders, 2005; Sorrell and Dimitropoulos, 2008; Frondel et al., 2007; Binswanger, 2001). If and only if EE(ε) is constant, the PE of energy service is equal to PE of energy consumption (Frondel et al., 2007). RE, therefore, can be estimated through the expression given next under the assumption that the end users respond in the same way to decrease in the energy prices as to increase in EE (Sorrell and Dimitropoulos, 2008). Therefore,

$$\text{RE} = \eta_\varepsilon^S = \left|\eta_{p_s}(S)\right|, \qquad (6.2)$$

where $\eta_{ps}(S)$ is own PE of energy services, provided the sign of the elasticity is negative. When the PE is positive or perverse,

$$\text{RE} = \eta_\varepsilon^S = -\eta_{p_s}(S) \qquad (6.3)$$

This estimation procedure of RE is an academically sound and well-accepted method (Bentzen, 2004). Several econometric models are available to estimate RE through price elasticities that are explored and summarized in Chakravarty et al. (2013). In the theoretical and empirical literature, although the basic mechanism of RE is widely accepted, its magnitude is highly debated (Greening et al., 2000). While some of the studies found minor consequences of RE (Schipper and Grubb, 2000; Laitner, 2000; Howarth, 1997; Henly et al., 1988a, 1988b; Lovins, 1988), others found that the RE, especially economy-wide, could be sufficiently large to completely offset the energy savings arising from improvement in energy efficiency (Khazzoom, 1980, 1987; Brookes, 1990a; Saunders, 1992; Herring, 1999; Sanne, 2000; Birol and Keppler, 2000; Roy, 2000). Those who have found large RE tend to believe that energy efficiency measures have limited importance in meeting international obligations to reduce fuel consumption for environmental or climate reasons (Brookes, 2004). RE is found to differ across sectors and across countries, even from one household to another depending on the price responsiveness and energy intensity (Allan et al., 2007; Turner, 2009). Most of the empirical studies have explicitly neglected the cross-substitution possibilities between energy service and non-energy service consumption (Bernt and Wood, 1975). The absence of such studies on substitution possibilities becomes particularly apparent if one is interested in deriving the appropriate magnitude of RE. Therefore, it is important to discuss the mechanism of RE and its corresponding methodologies to estimate the magnitude.

6.3 Household behavior and the AIDS framework

Estimation of demand functions consistent with economic theory has been a highly published area (Barnett and Seck, 2008) in the last 60 years. The majority of those papers have considered the flexible functional forms, which rely heavily on duality theory. The Generalized Leontief (Diewert, 1971), the Translog (Christensen et al., 1975), the Rotterdam (Theil, 1965, 1975; Barten, 1964, 1968, 1977), and the AIDS (Deaton and Muellbauer, 1980) are examples of popular demand system models (Barnett and Seck, 2008).

Any behavioral change due to efficiency improvement might lead to an alteration of the household consumption pattern for all goods and services. To represent such intra-consumption category trade-offs between end uses, expenditure minimizing modelling approach is adopted, in which household (i) attempts to allocate its available money Ei to various end uses (j) to maximize the total utility Ui. It should be noted that most of the existing estimates of RE ignore the possibility of change in demand for other goods and services due to EE improvement (Girod, 2009).

In this study, AIDS, the linear version of a demand system which was proposed by Deaton and Muellbauer (1980) is discussed. The AIDS is accepted as an advanced methodology for the following advantages. The model provides the first-order approximation for the demand system. It provides definite estimations of axioms of choice. It aggregates consumers perfectly. It has a functional form that is consistent with household budget data. It is easier to make estimations in the form of a linear approach. It can be easily used to test homogeneity and symmetry constraints (Blanciforti and Green, 1983). In this framework, the basic assumption is that the demand functions for the goods within the subgroups have the same functional form as the demand functions for the main group (Brännlund et al., 2007).

Deaton and Muellbauer start with a Price Independent Generalised Log-Linear (PIGLOG) indirect cost function:

$$\log c(u,p) = (1-u)\log\left[a(p)\right] + u\log\left[b(p)\right], \tag{6.4}$$

where *a(p)* interpreted as the cost of subsistence and *b(p)* the cost of bliss. Deaton and Muellbauer (1980) consider *a(p)* is of translog form as

$$\log\left[a(p)\right] = \alpha_0 + \sum_k \alpha_k \log p_k + \frac{1}{2}\sum_k \sum_j \gamma_{kj}^{*} \log p_k \log p_j \tag{6.5}$$

$$\text{and } \log\left[b(p)\right] = \log a(p) + \beta_0 \prod p_k^{\beta_k} \tag{6.6}$$

Therefore,

$$\log c(u,p) = \alpha_0 + \sum_k \alpha_k \log p_k + \frac{1}{2}\sum_k \sum_j \gamma_{kj}^{*} \log p_k \log p_j + u\beta_0 \prod p_k^{\beta_k s} \tag{6.7}$$

Now,

$$\frac{\delta \log c(u,p)}{\delta \log p_i} = \frac{p_i q_i}{c(u,p)} = w_i \tag{6.8}$$

Therefore,

$$w_i = \alpha_i + \Sigma \gamma_{ij} \ln p_j + \beta_i u \, \beta_0 \prod p_k^{\beta_k} \tag{6.9}$$

Now, by inversion of the indirect utility function, if $x = c(u,p)$ then, $u = f(p,x)$.
Substituting the value of u in the expenditure share equation, it will give the general non-linear form of AIDS is as follows:

$$w_i = \alpha_i + \Sigma \gamma_{ij} \ln p_j + \beta_i \ln(Y / P) \tag{6.10}$$

where w_i represents the budget share of *i*th good, p_j is price of the *j*th good, *Y* is the total expenditure on all goods, and *P* is Stone's price index (Deaton and Muellbauer, 1980) given by

$$\ln P = \varphi \Sigma w_i \ln p_i \tag{6.11}$$

The general form of P, suggested by Deaton and Muellbauer (1980), is

$$\ln P = \alpha_0 + \sum_k \alpha_k \log p_k + \frac{1}{2}\sum_k \sum_j \log p_k \log p_j \tag{6.12}$$

In this equation, the aggregate price index is a weighted sum of the price of individual goods and services (p_i) at time point *t*, where weights are the expenditure share of goods and services (w_i). φ is the proportion parameter. The model that uses Stone's index is called the "Linear Approximate AIDS" (LA-AIDS), following Blanciforti and Green (1983). If prices are highly collinear, *P* may be approximately proportional to *p*, that is, $P \approx p$. In the extreme case, when *P* is exactly (linearly) proportional to *p*, the LA-AIDS model can be used to estimate the parameters of the AIDS model (Green and Alston, 1990). Here, the shares (w_i) must sum to 1, as $\Sigma w_i = 1$.

Typically, the parameters are constrained such that, the following conditions are satisfied:

Adding up condition, $\sum_i \alpha_i = 1, \sum_i \gamma_{ij} = 0, \sum_i \beta_i = 0$

Homogeneity condition, $\sum_j \gamma_{ij} = 0$

Symmetry condition, $\gamma_{ij} = \gamma_{ji}$ for all i, j

With the estimated parameters, one can calculate elasticities by the following expressions (Edgerton et al., 2012),

$$\text{Expenditure elasticity, } \eta_{iy} = \left(\beta_i / w_i\right) + 1 \tag{6.13}$$

$$\text{Compensated PE, } \eta_{ij} = \frac{\gamma_{ij} + \beta_i \beta_j \ln\left(Y / P\right) - \beta_i w_j}{w_i} - \delta_{ij}, \tag{6.14}$$

where $\delta_{ij} = 1$ when, $i = j$ and $\delta_{ij} = 0$ when $i \neq j$;

$$\text{Uncompensated PE, } \eta_{ij}^U = \eta_{ij} - \eta_{iy}.w_j. \tag{6.15}$$

So the uncompensated PE is the compensated PE minus the expenditure share multiplied by expenditure elasticity. In equation 6.14, i = j implies own price elasticities and $i \neq j$ implies elasticity across prices. The elasticity of demand implies the responsiveness of the quantity demanded to factors influencing it, such as own price, price of other related goods and service and income. These elasticities can be derived either by using the demand function or the Engel Function.[2] It is convenient to use these elasticities in demand analysis because they are independent of the units of measurement of the good, prices, income and so on, and hence, comparisons of the same can be made across goods as well as a cross section.

If the basic model (equation 6.10) is written for each commodity group, for example, food articles (1), fuel and lighting (2) and other miscellaneous goods and services (3), then the following system of demand equations for *i*th households might be obtained:

$$w_{1i} = \alpha_1 + \gamma_{11} \ln P_{1i} + \gamma_{12} \ln P_{2i} + \gamma_{13} \ln P_{3i} + \gamma_{14} \ln P_{4i} + \beta_1 \ln\left(\frac{Y}{P}\right)_i + \varepsilon_{1i} \tag{6.16}$$

$$w_{2i} = \alpha_2 + \gamma_{21} \ln P_{1i} + \gamma_{22} \ln P_{2i} + \gamma_{23} \ln P_{3i} + \gamma_{24} \ln P_{4i} + \beta_2 \ln\left(\frac{Y}{P}\right)_i + \varepsilon_{2i} \tag{6.17}$$

$$w_{3i} = \alpha_3 + \gamma_{31} \ln P_{1i} + \gamma_{32} \ln P_{2i} + \gamma_{33} \ln P_{3i} + \gamma_{34} \ln P_{4i} + \beta_3 \ln\left(\frac{Y}{P}\right)_i + \varepsilon_{3i}, \tag{6.18}$$

where, i = 1, 2, . . . , n and j = 1, 2, 3.

The slope coefficient shows expenditure and price effects on the expenditure share of the consumption category. Since additivity and negativity constraints of demand system are automatically met by the model AIDS, no test is required for these constraints under this framework.

If the correlations between the error terms in the individual equation had been zero, then the seemingly unrelated regression on the system of equations would have been equivalent to running separate ordinary least squares (OLS) regressions on each equation (Brookes, 2004). This would be the case if all of the values of the explanatory variables were the same in all equations (Brookes, 2004; Cameron and Trivedi, 2009), as in the equations 6.16 through 6.18. Deaton and Muellbauer (1980) also suggested a simple equation to equation regression method to estimate the system of demand equations for consumption categories (equation 6.10) under basic AIDS framework with the Stone price index (Majumder, 1986). The problem of heteroscedasticity and multicollinearity is also checked through the Breusch-Pagan test and Variance Inflationary Factor (VIF), respectively, for cross-section data. Regression analysis has been carried to discern causal relation and relative strength of determining variables to help in deriving policy implications.

6.4 The case study: Consumption expenditure of households with rural non-farm employment in West Bengal, India

To understand the applicability of the AIDS framework for estimating the price elasticities and RE, a survey has been conducted on the consumption expenditure of households earning from rural non-farm employment in West Bengal, India.

Historically, agricultural activity from rural India contributes a substantial part of the country's gross domestic product. However, over time, the employment growth rate of the agriculture sector has gradually decreased (Ranjan, 2006), pushing more employment towards alternative employment opportunities in rural sectors leading to the expansion of the rural non-agricultural sector or the rural non-farm sector (RNFS; Kashyap and Mehta, 2007). The RNFS sector consists of mining and quarrying, household and non-household manufacturing, processing, repair, construction, trade and commerce, transport and other services in villages and rural towns that means that it includes all the sectors, other than agriculture and fisheries. The expansion of this sector is expected to increase rural employment, improve income distribution, alleviate poverty and ensure well-being of the households. However, with increase in household income and well-being, demand for energy services in rural areas will also increase at a much faster pace because of very high unmet demand (Roy, 2000). As per National Sample Survey Organisation (2012) data, 40% of monthly total consumption expenditure of an average rural household is on primary energy sources like firewood and chips, among others. Census statistics from the Government of India (2011) shows that in the rural area electricity is the second-largest (22%) energy input to provide energy services preceded by firewood and chips (47%). Moreover, demand for energy services in rural households is rising over the years (approximately 9% per annum from 2012–2013 to 2013–2014 estimated by MoSPI, 2014).

6.4.1 Data

To analyze the objective of the study, a primary survey has been carried out for getting consumption expenditure of households in two districts of West Bengal, which has dominance of rural non-farm employment—Bankura and Malda. Two blocks, Bishnupur from Bankura and Kaliachak II from Malda block, which again have high rural non-farm employment, were selected (Census, Government of India, 2011). After consulting with the respective block headquarters or BDO (Block Development Office), villages where more people work in the non-farm sector than in agriculture were selected. Thus, four villages, two from each district—(1) Satangapara from Mothabari Panchayat, (2) Masumchak from Gangaprasad Panchayat of Malda district, (3) Kakila from Radhanagar Panchayat and (4) Paikpara Dharapat from Ajodhya Panchayat of Bankura district—were chosen. A total of 369 households were surveyed by a systematic random sampling method. In this method, every fifth house of the village was selected for the survey. If household members of the fifth household were not available, the next house was surveyed, and accordingly, the next fifth house was selected.

After the selection of households, the survey was carried out with a designed and detailed questionnaire. Four separate sections of the questionnaire dealt with the identification of the sample household, their socioeconomic characteristics, a profile of each household member and the last section gives a detail description of the consumption expenditure done by the households in the last 30 days. The consumption expenditure of the household includes various food and food products (cereals, pulses, milk, fish, meat, vegetables and other food products), beverages, tobacco products, fuel and gas, electricity charges, medical expenses, educational expenses and other miscellaneous expenses. The questionnaire also includes a clothing, bedding and footwear category, but during the survey, it was found that items under this category were not generally purchased by the villagers. It's a tradition or custom in the sample households that generally, every year they get clothing from relatives during various functions and bedding from the block offices and various non-governmental organizations. Thus, while collecting data, we found that the villagers could not specify the expenditure done under this category. Hence, this expenditure is not considered for further analysis.

To understand the basic structure of the sample households, socio-religious structure, age-wise distribution, household types depending on the principal activity of the households, consumption expenditure and so on of the households in both the districts and villages are obtained. The average household size, that is, the number of family members in a household, in both the districts is five. The sample surveyed consists of both Hindu

(49.71%) and Muslim (50.29%) families, with Malda being the Muslim-dominated area and Bankura being a Hindu-dominated area. Out of the total Hindus, the majority were Scheduled Castes followed by General and Other Backward Caste (OBC). The principal activity of the maximum proportion of households were casual non-farm labor followed by self-employment in rural non-farm activities. In Bankura, 69.05% males and 59.71% females are literate whereas in Malda, 82.46% of the males and 81.07% females are literate.

6.4.2 Estimation of elasticities and RE

In this section, the expenditure function of the consumption categories such as food articles, fuel and lighting and other miscellaneous goods and services have been estimated (equations 6.16 to 6.18) for rural households that are in non-farm employment. Estimated parameters for the expenditure share equations are shown in Table 6.3 along with the standard errors and *t*-values. In this study, the OLS method is used for each equation (equations 6.16, 6.17 and 6.18) instead of estimating the system of equations as suggested by Deaton and Muellbauer (1980, Majumder, 1986).

In this cross section analysis, we had initially considered a total of seven variables: Expenditure Share of Food & Food Articles, Expenditure Share of Fuel & Lighting, Expenditure Share of Other Miscellaneous Goods & Services, Price of Food & Food Articles, Price of Fuel & Lighting, Price of Other Miscellaneous Goods & Services and Total Expenditure in expenditure functions for Food & Food Articles, Fuel & Lighting and Others Miscellaneous Goods & Services. The summary statistics of these seven variables are shown in Table 6.1. Since there is no variability in Price of Food & Food Articles, Price of Fuel & Lighting and Price of Other Miscellaneous Goods & Services over the cross-sectional data, we have dropped the variables. Therefore, the expenditure share function specified in equations 6.16, 6.17 and 6.18 reduced to the expenditure share function of households, which depends only on the real expenditure of the households.

The pairwise correlations between useful variables are shown in Table 6.2 where all the correlation coefficients are statistically significant at 99% confidence level. The sign of the correlation coefficient shows that there is an inverse relation between total expenditure and expenditure share of food and food articles as well as with expenditure share of fuel and lighting. However, the relation between total expenditure and expenditure share of Other Miscellaneous Goods & Services is positive. This observation is quite consistent with Engel's Law which states that the proportion of income consumers spend on necessary goods & services such as food decreases as their income increases (Engel, 1857). Subsequently, Wright in 1875 reconsidered the Engels' law as the higher the income, the lower will be the relative expenditure for subsistence and as income increases, the percentage of expenditure on various items increases (Wright, 1889). Here, like the Food

Table 6.1 Summary Statistics

Variables	*Mean*	*Standard Deviation*	*Minimum*	*Maximum*
Expenditure Share of Food & Food Articles	0.64	0.08	0.44	0.85
Expenditure Share of Fuel & Lighting	0.14	0.04	0.03	0.32
Expenditure Share of Others Miscellaneous Goods & Services	0.22	0.09	0.03	0.47
Price of Food & Food Articles	125.00	0.00	125.00	125.00
Price of Fuel & Lighting	127.40	0.00	127.40	127.40
Price of Others Miscellaneous Goods & Services	120.70	0.00	120.70	120.70
Total Expenditure	5,970.85	2,049.53	2,061.00	19,027.00
No. of Observation	369			

Source: Estimated from primary data.

Table 6.2 Pairwise Correlation between Variables

	Expenditure Share of Food & Food Articles	*Expenditure Share of Fuel & Lighting*	*Expenditure Share of Others Miscellaneous Goods & Services*	*Total Expenditure*
Expenditure Share of Food & Food Articles	1			
Expenditure Share of Fuel & Lighting	0.0326 (0.5320)	1		
Expenditure Share of Others Miscellaneous Goods & Services	−0.8877* (0.000)	−0.4891* (0.000)	1	
Total Expenditure	−0.2778* (0.000)	−0.3292* (0.000)	0.394* (0.000)	1

Source: Estimated from primary data.
Note: p-value in parenthesis; * indicates significant at 1%.

& Food Articles, Fuel & Lighting is also a necessary consumption category in rural households with non-farm employment.

6.4.3 *Regression analysis*

Before running the regression,[3] the Breusch–Pagan (B-P) test is conducted to test for heteroscedasticity in a linear regression model (Breusch and Pagan, 1979). It tests whether the estimated variance of the residuals from a regression are dependent on the values of the independent variables. If there is no heteroscedasticity, then ordinary least square should be the most appropriate tool for estimation. In this model, on the basis of Breusch–Pagan test for all expenditure share stochastic equations, no heteroscadasticity was found, thus indicating that the OLS method is most appropriate for these data sets. The R^2 and adjusted R^2 values which are obtained from each of the estimated OLS models, are in the range of 0.10 to 0.19, implying relatively low goodness of fit of the model specification. The mean VIF of each homoscedastic regression implies that there is no multicollinearity between the dependent variables. The estimates of Table 6.3 show a systematic pattern, broadly consistent with the basic theory of consumption. All estimated coefficients are statistically significant at 1% confidence level. Here, the slope coefficient shows expenditure effects on the expenditure share of the consumption category. Even in the regression analysis, the expenditure share is negatively associated with real expenditure for food articles (slope coefficient of equation 6.16) and fuel and lighting (slope coefficient of equation 6.17).

6.4.4 *Elasticity parameter estimates*

Table 6.4 provides the estimated uncompensated and compensated own-price elasticities (equation 6.14 and 6.15) and expenditure elasticities (equation 6.13). The elasticities are non-linear functions of the estimated

Table 6.3 Parameter Estimates of Demand System for Households

	Constant	*Real expenditure*	*R-squared*	*Adj R-squared*
Food & Food Articles	0.93* (0.04)	-0.09* (0.01)	0.10	0.10
Fuel & Lighting	0.30* (0.02)	-.042 * (0.00)	0.11	0.11
Other miscellaneous goods & services	-0.24* (0.04)	0.12* (0.01)	0.19	0.19

Source: Estimated from primary data.
Note: Standard errors in parentheses.
* Significant at the 1% level.

Table 6.4 Estimated PE for Fuel and Lighting in Residential Sector of India

	Expenditure Elasticity	*Own PE*	
		Uncompensated	*Compensated*
Food & Food Articles	0.88 (0.02)	−1.44 (0.12)	−0.88
Fuel & Lighting	0.71 (0.04)	−1.01 (0.30)	−0.91
Other Miscellaneous Goods & Services	1.56 (0.00)	−0.86 (0.33)	−0.86

Source: Authors' estimation from Table 6.3.
Note: Standard errors are in parentheses.

parameters so that standard errors for their estimates cannot be calculated exactly. The approximated estimates of the standard errors of elasticities are estimated under the assumption that the expenditure shares are constant and equal to the means (over the estimation time bounds) of the estimated values. It has also assumed that the estimated coefficients are independent of each other.

Expenditure remains the most important variable in explaining consumption demand, so expenditure elasticity is estimated by using the equation 6.13. The estimate suggests that expenditure elasticities are positive and relatively high, which implies the consumption of these goods and services is normal for rural households in India with non-farm employment. This indicates that if the real expenditure or income increases, households will not decrease their consumption and vice versa (Lewbel, 2006). The expenditure elasticities of first two consumption categories are less than one which is consistent with the Engels' (1857) Law However, although Engel identified food articles as the only necessary consumption category in a household, in this analysis, it is indeed interesting to note that fuel and lighting also emerge as a necessary consumption category for rural households of West Bengal. In the case of other miscellaneous goods and services, the expenditure elasticity is greater than 1, implying the category is a luxury consumption category as the increase in expenditure for this consumption is proportionally more than the increase in total expenditure.

The uncompensated price elasticities provide the "gross" effects that include both the substitution effect and the income effect, where the compensated price elasticities represent the pure price effects (i.e., only the substitution effect) or the "net" effects of price change on demand. The uncompensated and compensated price elasticities are estimated from equation 6.14 and 6.15 which are shown in the Table 6.4. All price elasticities are of negative sign implying that the consumption category is normal consumption of goods and services for the average household in the sample survey. This is also consistent with the "Law of Demand", whereby demand falls with the increase in price and vice versa. The uncompensated or Marshallian own PE for fuel and lighting service consumption demand is 1.01 whereas the Hicksian compensated PE is 0.91. The estimated own-price elasticities indicate if, for example, the price falls by 100%, then demand for fuel and lighting would increase by 101%. Of this total increase in consumption, 91% is purely due to price effect (i.e., the substitution effect) as the compensated elasticity suggests. The income effect of the price fall accounts for the remaining 10% (i.e., 101%–91%) increase in fuel and lighting consumption demand due to the increase in real income, although the absolute amount of money income remains unchanged. The income effect is relatively negligible in this case because it has a small share in a household budget; hence, their price changes have only minimal effects on real income. If per capita income also increases by 10% accompanied by a 10% fall in fuel and lighting price, then the consumption for fuel and lighting service would increase by 10.1 %. However, the increase in per capita income represents a shift in the fuel and lighting service consumption demand curve that also leads to an increase in the price level.

The cross-price elasticities can also be estimated if one could get reliable data on unit price of each consumption category for individual households by using equation 6.14. For the current study, we have failed to estimate that due to inadequate variability in data as mentioned in the previous sections.

6.4.5 Rebound effect

The estimated RE magnitudes from PE parameters of fuel and lighting are from 91% to 101%. When PE is negative or less than zero, RE is equal to PE, and when PE is perverse or greater than zero, the RE is negative compared to PE, as described in the literature (equation 6.2 and 6.3). The estimated RE is high, partial to low backfire as the magnitudes may varies between 91% to 101%. That implies that if EE of fuel and lighting increases by 100%, then it could further increase the demand for fuel and lighting by 101%. In other words, if EE policy can reduce 100 units of energy inputs given the energy service constant, and then it will actually increase by 101 units. An additional one unit of energy input consumption will rebound through end users' behavioral response.

The observed rebound may be explained by the presence of unmet demand in fuel and lighting of residential end users. In the presence of unmet demand, EE improvement in energy services is expected to generate a high rebound (even it could generate backfire) in energy service consumption (Roy, 2000, Schipper & Grubb 2000, Sorrell et al., 2009) in developing countries. In this context, it should be noted that the per capita energy consumption in India is 684 kWh, which is much less than the per capita energy consumption of developed countries. For example, the per capita energy consumption in the US is 20 times higher than the figure for India (World Bank Database, 2014).

6.5 Summary and conclusion

To achieve maximum EE gains, it is absolutely imperative to consider what proportion of potential energy savings could be achieved if one "net out RE" from EE gain. There has been a wide variety of literature attempting to measure RE. The AIDS model is one widely accepted advanced methodology for determining the effect on households' consumption choice after increases in EE and for estimating RE. This model is elaborated in this chapter with a case study on rural households, primarily engaged in non-farm employment in West Bengal, India, for determining PE for all goods and services. The results show that price elasticities are with the usual negative sign, food and food articles and fuel and lighting are the necessary consumption categories for these households and the other miscellaneous goods and services are the luxury consumption. Estimated RE is high-partial to low backfire (91%–101%). It should be noted that this partial rebound is high, but there is still a potential to save energy through EE measures. In this study, the LA-AIDS methodology is explained, particularly, in the context of consumption expenditure of all consumption categories, with a special focus on RE estimation; however, it could be used in explaining any kind of parametric or non-parametric household behavioral study through household expenditure data.

Notes

1 We would like to thank Professor Joyashree Roy, Department of Economics, Jadavpur University, Kolkata for her unceasing support, valuable comments and suggestions on this study.

2 An Engel curve is the function describing how a consumer's expenditure on some goods or services relates to the consumer's total resources, holding prices fixed (Lewbel, 2006).

3 We have used the statistical software STATA.

References

Allan, Grant et al. "The Impact of Increased Efficiency in the Industrial Use of Energy: A Computable General Equilibrium Analysis for the United Kingdom." *Energy Economics* 29, no. 4 (2007): 779–798.

Barnett, William A., and Ousmane Seck. "Rotterdam Model Versus Almost Ideal Demand System: Will the Best Specification Please Stand Up?" *Journal of Applied Econometrics* 23, no. 6 (2008): 795–824.

Barten, A.P. "Estimating demand equations." *Econometrica: Journal of the Econometric Society* (1968): 213–251.

Barten, A.P. "Maximum likelihood estimation of a complete system of demand equations." *European Economic Review* 1, no. 1 (1969): 7–73.

Barten, A.P. "The systems of consumer demand functions approach: a review." *Econometrica: Journal of the Econometric Society* (1977): 23–51.

Bentzen, Jan. "Estimating the Rebound Effect in US Manufacturing Energy Consumption." *Energy Economics* 26, no. 1 (2004): 123–134.

Berndt, E.R. and Wood, D.O. "Technology, prices, and the derived demand for energy." *The Review of Economics and Statistics* (1975): 259–268.

Binswanger, Mathias. "Technological Progress and Sustainable Development: What about the Rebound Effect?" *Ecological Economics* 36, no. 1 (2001): 119–132.

Birol, Fatih, and Jan Horst Keppler. "Prices, technology development and the rebound effect." *Energy Policy* 28, no. 6–7 (2000): 457–469.

Blanciforti, Laura, and Richard Green. "An Almost Ideal Demand System Incorporating Habits: An Analysis of Expenditures on Food and Aggregate Commodity Groups." *The Review of Economics and Statistics* (1983): 511–515.

Brännlund, Runar, Tarek Ghalwash, and Jonas Nordström. "Increased Energy Efficiency and the Rebound Effect: Effects on Consumption and Emissions." *Energy Economics* 29, no. 1 (2007): 1–17.

Breusch, Trevor S., and Adrian R. Pagan. "A simple test for heteroscedasticity and random coefficient variation." *Econometrica: Journal of the Econometric Society* (1979): 1287–1294.

Brookes, Len. "The Fifth Fuel and the Greenhouse Effect." *Energy World* 18, no. 2 (1990a): 199–201.

Brookes, Len. "The Greenhouse Effect: The Fallacies in the Energy Efficiency Solution." *Energy Policy* 18, no. 2 (1990b): 199–201.

Brookes, Leonard. "Energy Efficiency Fallacies—A Postscript." *Energy Policy* 32, no. 8 (2004): 945–947.

Brooks, Chris. *Introductory Econometrics for Finance*. Cambridge University Press, 2019. Cambridge, UK. 2008. ISBN-13 978-0-511-39848-3

Cameron, Adrian Colin, and Pravin K. Trivedi. *Microeconometrics using Stata*. Vol. 5. College Station, TX: Stata Press (2009).

Census, Government of India. (2011). Retrieved 15 March 2015 from http://censusindia.gov.in/. Weblink: https://www.censusindia.gov.in/2011census/hlo/Data_sheet/India/Source_Lighting.pdf

Chakravarty, Debalina, Shyamasree Dasgupta, and Joyashree Roy. "Rebound Effect: How Much to Worry?" *Current Opinion in Environmental Sustainability* 5, no. 2 (2013): 216–228.

Christensen, Laurits R., Dale W. Jorgenson, and Lawrence J. Lau. "Transcendental logarithmic utility functions." *The American Economic Review* 65, no. 3 (1975): 367–383.

Davis, Lucas W. "Durable Goods and Residential Demand for Energy and Water: Evidence from a Field Trial." *The RAND Journal of Economics* 39, no. 2 (2008): 530–546.

Deaton, Angus, and John Muellbauer. "An Almost Ideal Demand System." *The American Economic Review* 70, no. 3 (1980): 312–326.

Diewert, W. Erwin. "An application of the Shephard duality theorem: A generalized Leontief production function *Journal of Political Economy* 79, no. 3 (1971): 481–507.

Dubin, Jeffrey A., Allen K. Miedema, and Ram V. Chandran. "Price Effects of Energy-Efficient Technologies: A Study of Residential Demand for Heating and Cooling." *The Rand Journal of Economics* (1986): 310–325.

Edgerton, David L., et al. *The Econometrics of Demand Systems: With Applications to Food Demand in the Nordic Countries*. Vol. 34. Springer Science & Business Media (2012) Germany.

Engel, Ernst. "The production and consumption conditions of the Kingdom of Saxony." *J. Stat. Bur. R. Sax. Minist. Inter* (1857): 8–9.

Frondel, Manuel, Jorg Peters, and Colin Vance. "Identifying the rebound: evidence from a German household panel." *The Energy Journal* 29, no. 4 (2008).

Girod, Bastien. *Integration of Rebound Effects into LCA*. Doctoral Dissertation, EidgenössischeTechnische Hochschule ETH Zürich, Nr. 18741 (2009).

Green R. and J. M. Alston. "Elasticities in AIDS Models." *American Journal of Agricultural Economics* 72, no. 2 (1990): 442–445.

Greene, David L. "Vehicle Use and Fuel Economy: How Big Is the 'Rebound' Effect?" *The Energy Journal* 13 (1992): 117–143.

Greening, Lorna A., David L. Greene, and Carmen Difiglio. "Energy Efficiency and Consumption—the Rebound Effect—a Survey." *Energy Policy* 28, no. 6 (2000): 389–401.

Guertin, Chantal, Subal C. Kumbhakar, and Anantha Kumar Duraiappah. "Determining Demand for Energy Services: Investigating Income-Driven Behaviours." *International Institute for Sustainable Development* (2003).

Henly, John, Henry Ruderman, and Mark D. Levine. "Energy Saving Resulting from the Adoption of More Efficient Appliances: A Follow-Up." *The Energy Journal* 9, no. 2 (1988a): 163–170.

Henly, John, Henry Ruderman, and Mark D. Levine. "Energy Saving Resulting from the Adoption of More Efficient Appliances: A Follow-Up." *The Energy Journal* 9, no. 2 (1988b): 163–170.

Herring, Horace. "Does Energy Efficiency Save Energy? The Debate and its Consequences." *Applied Energy* 63, no. 3 (1999): 209–226.

Hicks, J. *Value and Capital*. 2nd ed. Oxford: Oxford University Press (1946).

Howarth, Richard B. "Energy Efficiency and Economic Growth." *Contemporary Economic Policy* 15, no. 4 (1997): 1–9.

Kashyap, S. P., and Niti Mehta. "Non-Farm Sector in India: Temporal and Spatial Aspects." *The Indian Journal of Labour Economics* 50, no. 4 (2007): 611–632.

Khazzoom, J. Daniel. "Economic Implications of Mandated Efficiency in Standards for Household Appliances." *The Energy Journal* 1, no. 4 (1980): 21–40.

Khazzoom, J. Daniel. "Energy Saving Resulting from the Adoption of More Efficient Appliances." *The Energy Journal* 8, no. 4 (1987): 85–89.

Laitner, John. "Structural Change and Economic Growth." *The Review of Economic Studies* 67, no. 3 (2000): 545–561.

Lewbel A.. "Engel Curves." In the *New Palgrave Dictionary of Economics* (2006). Retrieved from https://bing.bc.edu/arthur-lewbel/palengel.pdf.

Lin, Boqiang, and Hongxun Liu. "A Study on the Energy Rebound Effect of China's Residential Building Energy Efficiency." *Energy and Buildings* 86 (2015): 608–618.

Lovins, Amory B. "Energy Saving Resulting from the Adoption of More Efficient Appliances: Another View." *The Energy Journal* 9, no. 2 (1988): 155–162.

Majumder, Amita. "Consumer Expenditure Pattern in India: A Comparison of the Almost Ideal Demand System and the Linear Expenditure System." *Sankhyā: The Indian Journal of Statistics, Series B* (1986): 115–143.

Ministry of Statistic and Program Implementation, Government of India. Energy Statistics, 2014. http://mospi.nic.in/mospi_new/upload/Energy_stats_2014.pdf.

National Sample Survey Organisation. "*Energy Sources of Indian Households for Cooking and Lighting, 2009–10*." Report by Ministry of Statistics and Program Implementation, Government of India (2012).

Ranjan, Sharad. "Occupational Diversification and Access to Rural Employment: Revisiting the Non Farm Employment Debate." MPRA Paper No. 7870 (2006): 102–131.

Roy, Joyashree. "The Rebound Effect: Some Empirical Evidence from India." *Energy Policy* 28, no. 6 (2000): 433–438.

Sanne, Christer. "Dealing with Environmental Savings in a Dynamical Economy—How to Stop Chasing Your Tail in the Pursuit of Sustainability." *Energy Policy* 28, no. 6 (2000): 487–495.

Saunders, Harry D. "A Calculator for Energy Consumption Changes Arising from New Technologies." *Topics in Economic Analysis and Policy* 5, no. 1 (2005): 1–31.

Saunders, Harry D. "The Khazzoom-Brookes postulate and neoclassical growth." *The Energy Journal* 13, no. 4 (1992).

Schipper, Lee, and Michael Grubb. "On the Rebound? Feedback between Energy Intensities and Energy Uses in IEA Countries." *Energy Policy* 28, no. 6 (2000): 367–388.

Sorrell, Steve, and John Dimitropoulos. "The Rebound Effect: Microeconomic Definitions, Limitations and Extensions." *Ecological Economics* 65, no. 3 (2008): 636–649.

Sorrell, Steve, John Dimitropoulos, and Matt Sommerville. "Empirical Estimates of the Direct Rebound Effect: A Review." *Energy Policy* 37, no. 4 (2009): 1356–1371.

Sorrell, Steve. "*The Rebound Effect: An Assessment of the Evidence for Economy-Wide Energy Savings from Improved Energy Efficiency*." (2007). https://www.semanticscholar.org/paper/The-rebound-effect%3A-an-assessment-of-the-evidence-Sorrell/8e5260a35163402b6ada126baddc023966252618?p2df

Theil, Henri. "The information approach to demand analysis." *Econometrica* 33 (1965): 67–87.

Theil, Henri. "Theory and Measurement of Consumer Demand", volume I, North-Holland (1975).

Turner, Karen. "Negative Rebound and Disinvestment Effects in Response to an Improvement In Energy Efficiency in the UK Economy." *Energy Economics* 31, no. 5 (2009): 648–666.

Wang, H., P. Zhou, and D. Q. Zhou. "An empirical study of direct rebound effect for passenger transport in urban China." *Energy Economics* 34, no. 2 (2012): 452–460.

Wright, C. D. "Comparative Wages, Prices and Cost of Living." Reprint of 16th Annual Report of the Massachusetts Bureau of Statistics and Labor for 1885. Wright and Potter Printing Co. (1889). Boston

World Bank. "*World Bank Database*" (2014). Retrieved 24 May 2015 from http://data.worldbank.org/data-catalog/world-development-indicators.

7 Index decomposition analysis of energy use in India

Shyamasree Dasgupta, Nandini Das and Joyashree Roy

7.1 Introduction

Enhanced energy efficiency has a multidimensional influence on the overall sustainable development: it influences a nation's energy demand and energy security, supports domestic industry, plays a crucial role in enhancing the energy access and helps in poverty reduction (Roy et al., 2013; GEA 83). Climate goals are often aimed to be achieved through the implementation of policies in which energy efficiency plays a pivotal role. Recent trends of growth in the global economy and energy-related emissions suggest that there is a sign of decoupling between activity growth and energy use. In the World Energy Outlook Special Report on Energy and Climate Change published in 2015, the International Energy Agency mentioned that the decline in energy intensity by 2.3% in 2014 played a significant role in this decoupling (2015, pp. 3, 22). Against this backdrop, analysis of energy intensity trends, its role in decoupling of energy use and emission from growth in output/income has emerged as one of the major areas of research. The challenge is twofold—developing an appropriate methodological framework capable of dealing with both simple and complex issues of energy intensity and its role in decoupling; and applications of these methods for empirical analysis. Index Decomposition Analysis (IDA) is one of the most widely applied frameworks to identify the factors that influence the change in energy use and therefore to understand the decoupling phenomenon. The beauty of the method lies in its flexibility. While the basic model of IDA can be applied when only aggregate-level data are available, one can move on towards more sophisticated decomposition techniques when data are available with greater disaggregation. The objective of this chapter is to introduce the readers to the framework of a particular type of IDA based on Log Mean Divisia Index (LMDI) and its applications. Since the construction of variables and identification of data sets are crucial components of empirical research, we also touch on these two issues in relevant sections.

7.2 Measuring energy intensity

The first challenge that an empirical researcher faces is to identify and measure variables under consideration. Declining energy intensity (the amount of energy used per unit of activity/output produced) is a simple and effective way of measuring enhanced energy efficiency. Sector-specific energy intensities are also considered to be good indicators of the vintage of plants and equipment in the industries (IAEA, 2005, p. 18). United Nations Expert Group on Indicators of Sustainable Development, published in 2001, identified energy intensity as one of the three[1] energy-use-related indicators for sustainable development in the context of consumption and production (UNDESA, 2001, p. 25).

Energy intensity can be calculated in multiple ways depending on how energy use and activity level are measured. Empirical research in the area of energy economics has a tradition of using both energy inputs and economic activities measured in monetary terms so that energy intensity becomes a ratio of energy costs and product sales or production value (both measured in constant prices). We denote this indicator as EI_1 (refer to Equation 7.1).

$$\begin{aligned} EI_1 &= \frac{Fuel\ expenditure\ at\ constant\ price}{Value\ of\ output\ at\ constant\ price} \\ &= \frac{Fuel\ used}{Output\ produced} \times \frac{Constant\ price\ of\ fuel}{Constant\ price\ of\ output} \end{aligned} \tag{7.1}$$

This indicator has been extensively used in the studies on Indian industries and is useful for industrial performance analysis (Roy et al., 1999, p. 42; Dasgupta et al., 2011, p. 6; Sahu and Narayanan, 2011, p. 19; Dasgupta and Roy, 2017, p. 6). Another advantage of defining energy intensity in this manner is as follows. It can be used as an aggregate measure for a group of industries producing a range of outputs measured in different units. Since both the numerator and the denominator are measured in monetary units, this indicator is free of units. But the decline of EI_1 can be attributed to two sources—a decline in the proportion of fuel use to output produced and/or decline in the price of fuel relative to the price of output (in constant prices; refer to Equation 7.1). Any downward trend in EI_1 will overestimate the gain in energy efficiency when $\frac{Constant\ price\ of\ fuel}{Constant\ price\ of\ output}$ is also declining. It, therefore, gives a relatively less accurate account of the change in energy technology. Second, this indicator is not comparable with global technology benchmarks as all the variables are measured in monetary units.

The limitations of EI_1 lead us to the second indicator—EI_2, where activity/ production as well as energy use are measured in physical units (Bhattacharya and Cropper, 2010, p. 3):

$$EI_2 = \frac{Energy\ consumed}{Physical\ quantities\ of\ output\ produced} \tag{7.2}$$

EI_2 is expressed as, for example, kilo Calorie (kCal) of thermal power per ton of clinker in the cement industry or giga Joule (GJ) of energy per ton of crude steel and so on. It has its basis in production engineering and is referred to as the specific energy consumption (SEC). But EI_2, cannot be constructed for an aggregate set of industries producing different types of final products, although it represents the state of the technology in a more appropriate manner and can be compared with the global benchmarks. In the literature of industrial management, the term "benchmark" is interpreted mainly in two ways: "internal benchmark" and "external/competitive benchmark" (Raa, 2009, p. 7). While "internal benchmark" can be derived based on the energy-saving potential of a particular production unit, "external benchmark" is competitive and can be conceptualized in terms of the best-practice energy-use technology at a global level. Potential energy efficiency that could be achieved by setting an "external/competitive benchmark", representing the existing technological limit in the form of best available technology (BAT), is always larger as compared to the internal one. The use of EI_2 is recommended in this context. A review by Freeman et al. (1997), however, shows that while there are limitations associated with all the measures, all of them are in use in literature (p. 703). The choice in favor of a particular approach in an empirical study should be made based on the objective of the study and the availability of data.

7.3 IDA: Understanding the drivers of change in energy use

Analyzing energy intensity trends is the first step to understand the status of energy efficiency. There are, however, empirical evidence showing that in spite of a substantial gain in energy efficiency, the absolute level of energy use and related carbon dioxide (CO_2) emissions continued to rise in the past (Allcott and Greenstone, 2012, p. 5; Roy et al., 2013, p. 287; Vogit et al., 2014, p. 47; Dasgupta and Roy, 2017, p. 2). Given this, it becomes important to understand the scale of energy efficiency gain vis-à-vis other drivers of energy use and their interaction to explain the relationship between activity growth and energy use. As mentioned, there are several methodological frameworks to empirically find out the answer to this. In this chapter, we discuss IDA, which has been developed as one of the most important tools to address this issue.

The basic framework of IDA is as follows:

Let us assume an aggregate indicator and a set of k factors/drivers contributing to its change over time. The purpose of IDA is to distribute the change in this aggregate indicator in a consistent manner to a set of predefined factors/drivers. This way, the IDA becomes an effective framework to systematically assess the immediate sources of change in energy use in a sector or in an aggregate economy. For example, in the context of industries, the aggregate indicator can be considered as total energy use by the industries; for the whole economy, it can be the total energy consumption by the country, among others.

7.3.1 LMDI to understand activity, structure and energy intensity-based drivers of change in energy use

To elaborate on the concepts, let us begin with an example of the manufacturing industries in India. Under a simple formulation, change in energy use in aggregate manufacturing can be attributed to three sources: activity, structure and energy intensity (Ang and Xu, 2013). Therefore, energy use at period t can be expressed as

$$E_t = \sum_i E_{it} = \sum_i Y_t \frac{Y_{it}}{Y_t} \frac{E_{it}}{Y_{it}} = \sum_i Y_t S_{it} I_{it} \tag{7.3}$$

where $E_t = \sum_i E_{it}$ = energy use of aggregate manufacturing, where E_{it} = energy use in sector I; $Y_t = \sum_i Y_{it}$ = production of aggregate manufacturing, Y_{it} = production of sector i; $S_{it} = Y_{it}/Y_t$= production share of sector i in aggregate manufacturing; $I_{it} = E_{it}/Y_{it}$= energy intensity of sector i; and period is denoted by the subscript t.

The change in energy use over a period ($E_T - E_0$) can be decomposed, based on additive or multiplicative formulations. In additive form, change in energy use is expressed as

$$E_T - E_0 = \Delta E_{TOT} = \Delta E_{AE} + \Delta E_{SE} + \Delta E_{IE} \tag{7.4a}$$

In multiplicative form, the change in energy use is expressed as

$$\frac{E_T}{E_0} = D_{TOT} = D_{AE}.D_{SE}.D_{IE} \tag{7.4b}$$

where ΔE_{TOT} and D_{TOT} = Total effect = Change in energy use; ΔE_{AE} and D_{AE} = Activity Effect, that is, change in energy use due to the change in activity; ΔE_{SE} and D_{SE} = Structural Effect, that is, change in energy use due to change in relative production share of energy-intensive sectors in aggregate manufacturing; and ΔE_{IE} and D_{IE} = Intensity Effect, that is, change in energy use due to the change in energy intensity.

The index decomposition methodology has been evolved over the past decades to determine the expressions for different components of change in energy use (ΔE_{AE}, ΔE_{SE}, ΔE_{IE} *or* D_{AE}, D_{SE}, D_{IE}). The method used in late 1970s and early 1980s was mainly based on Laspeyres Index[2] to estimate these components. The Divisia Index was first introduced as an alternative to Laspeyres Index by Boyd et al. (1987) to decompose the changing composition of US manufacturing production and separately identify the effects of energy efficiency improvements (pp. 77–96). It was then followed by Liu et al. (1992, pp. 161–177) and Ang and Liu (2001, pp. 537–548). In a continuous-time framework, Divisia Index was created by solving the integral path problems of changes in variables by Liu et al. (1992, pp. 161–177). This integral path problem could be transformed into a parametric estimation problem to derive the discrete time equivalent of Divisia Index. Such transformations are important as data are available at discrete time points (e.g., yearly or monthly), leading to undefined integral paths. For making such transformations, specific assumptions are made on the integral paths that lead to different parametric values to be applied in the general parametric Divisia method. Based on this, Ang and Lee (1994) showed that although there are an infinite number of ways to make assumptions leading to an infinite number of decomposition methods and results, a large number of methodologies applied to decompose the industrial energy demand are special cases of two general parametric methods based on Divisia Index: Method 1 (Equation 7.5) and Method 2 (Equation 7.6) as mentioned in the following (pp. 83–92):

Method 1:

$$\Delta E_{AE} = \left[E_0 + \alpha\left(E_T - E_0\right)\right]\ln\frac{Y_T}{Y_0} \tag{7.5}$$

$$\Delta E_{SE} = \sum\nolimits_i \left[E_{i0} + \beta_i\left(E_{iT} - E_{i0}\right)\right]\ln\frac{S_{iT}}{S_{i0}}$$

$$\Delta E_{IE} = \sum\nolimits_i \left[E_{i0} + \gamma_i\left(E_{iT} - E_{i0}\right)\right]\ln\frac{I_{iT}}{I_{i0}}$$

Method 2:

$$\Delta E_{AE} = \left[I_0 + \alpha\left(I_T - I_0\right)\right]\left(Y_T - Y_0\right) \tag{7.6}$$

$$\Delta E_{SE} = \sum\nolimits_i \left[I_{i0}Y_0 + \beta_i\left(I_{iT}Y_{iT} - I_{i0}Y_{i0}\right)\right]\left(S_{iT} - S_{i0}\right)$$

$$\Delta E_{IE} = \sum\nolimits_i \left[Y_{i0} + \gamma_i\left(Y_{iT} - Y_{i0}\right)\right]\left(I_{iT} - I_{i0}\right)$$

where $0 \le \alpha, \beta_i, \gamma_i \le 1$(in both method 1 and method 2)

Liu et al. (1992) showed that specific values of the parameters α, β_i, γ_i would imply assigning different weights to base (t = 0) and terminal (t = T) year, which is, in effect, equivalent to defining specific integral paths in the Divisia Index (pp. 161–177). However, given this kind of decomposition, ΔE_{AE} + ΔE_{SE} + ΔE_{IE} is theoretically equivalent to neither ΔE_{TOT} nor D_{AE}. D_{SE}. D_{IE} is to D_{TOT}. In the relevant literature, this difference, in both cases, is identified as the residual term *d*. Since different values of α, β_i, γ_i lead to different sets of decomposition results, there arises an issue of choice. Ang and Lee (1994) suggested certain criteria to choose the appropriate methodology in a context-specific manner. Theoretical foundation, ease of use, adaptability, result interpretation and most importantly lower value of the residual term are crucial for such criteria (Ang and Lee, 1994, pp. 83–92). Methods that pass the factor reversal test will not leave any residual term and hence are preferred.[3] Log-Mean Divisia Index[4] (LMDI) (Ang and Choi, 1997, pp. 59–73), which gives perfect decomposition (results do not contain any residual term),[5] satisfies the factor reversal test and is consistent in aggregation, is thus one of the most widely accepted measures of index decomposition.

In LMDI, the drivers of the change in energy use are formulated as follows:

In the additive form

$$\Delta E_{AE} = \sum_i w_i \ln\left(\frac{Y_T}{Y_0}\right)$$
$$\Delta E_{SE} = \sum_i w_i \ln\left(\frac{S_{iT}}{S_{i0}}\right)$$
$$\Delta E_{IE} = \sum_i w_i \ln\left(\frac{I_{iT}}{I_{i0}}\right), \tag{7.7a}$$

where $w_i = \dfrac{E_{iT} - E_{i0}}{\ln E_{iT} - \ln E_{i0}}$

In multiplicative form:

$$D_{AE} = \exp\left[\sum_i \widetilde{w_i} \ln\left(\frac{Y_T}{Y_0}\right)\right]$$
$$D_{SE} = \exp\left[\sum_i \widetilde{w_i} \ln\left(\frac{S_{iT}}{Si_0}\right)\right]$$
$$D_{IE} = \exp\left[\sum_i \widetilde{w_i} \ln\left(\frac{I_{iT}}{I_{i0}}\right)\right] \tag{7.7b}$$

where $\widetilde{w_i} = \dfrac{(E_{iT} - E_{i0})/(\ln E_{iT} - \ln E_{i0})}{(E_T - E_0)/(\ln E_T - \ln E_0)}$

Given this formulation, to decompose the change in energy use in the aggregate manufacturing industries in India, the first task is to define the "structure" of the aggregate manufacturing sector in terms of energy-intensive and less energy-intensive industries. It is evident that other things remaining constant, a shift in the production/activity in favor of energy-intensive industries vis-à-vis less energy-intensive industries will increase the total energy use of the aggregate manufacturing sector. In this example, we have identified the structure of the aggregate manufacturing industries in the following manner. The industrial energy efficiency policy in India, Perform Achieve and Trade (PAT), during its first phase of implementation

(2012–2015) had identified seven manufacturing industries to be the most energy-intensive ones. These industries are aluminum, cement, iron and steel, chlor alkali, fertilizer, pulp and paper and textile. In our analysis, we take a group of seven industries almost similar to the coverage under PAT and identify this group as the energy-intensive manufacturing industries (Group 1). The rest of the manufacturing industries are identified as the less energy-intensive manufacturing industries (Group 2). Once the structure is identified in this manner, the next task is to identify data on the energy use and output of these broad groups over a period of time (or at the least for two relevant time points—initial and terminal). In this chapter, we have considered 1990–1991 as the initial year and 2010–2011 as the terminal year. Our objective is to identify the role of the change in activity (i.e., production), structure and energy intensity of the manufacturing industries to the change in aggregate energy demand of the sector.

The Annual Survey of Industries (ASI), published by Central Statistical Organization (CSO), Ministry of Statistics and Programme Implementation (MoSPI), Government of India, documents yearly data on energy use and production levels of various organized manufacturing units in the country. For the decomposition analysis, we required data on the energy use and output produced by energy-intensive (Group 1) and less energy-intensive (Group 2) manufacturing industries in India. We have first identified the industry codes of relevant industries under Group 1 and Group 2 for 1990–1991 and 2010–2011 from the National Industrial Classification (NIC). As the industry classifications have changed over the study period, the matching of codes assigned by CSO to various industries was based on the Table of Concordance published along with the NIC.[6] Based on ASI data, "Value of Output" is taken as the measure of activity/output for each of the manufacturing industries. This is inclusive of the ex-factory value of products and by-products manufactured during the accounting year as well as the receipt for industrial and non-industrial services rendered to others, the value of semi-finished goods of the previous year sold in the current year, the value of electricity sold and the sale value of goods sold in the same conditions as purchased. To eliminate the inflationary trend that is, the effect of the change in the price level, "Value of Output" is deflated by the Wholesale Price Index (WPI) of the respective industrial products. Then, these values are aggregated for energy-intensive and less energy-intensive industry groups. The item "Fuels Consumed" in ASI reports represents energy use. Fuel Consumed represents the total purchase value of all items of fuels, lubricants, electricity, water and so on consumed by the factory during the accounting year, including gasoline and other fuels for vehicles except those that directly enter into products as materials consumed. The monetary value of Fuel Consumed for each industry is deflated by the weighted average of the WPIs of coal, mineral oil and electricity, where the weights are proportional to the contribution of each fuel in that particular industry. The contributions are determined based on the input–output matrix

(2007–2008) for India. WPI indices are obtained from "Volumes of Index Number of Wholesale Prices in India: A Time Series Presentation, Office of Economic Advisor, Ministry of Finance, Government of India" for various years. Based on these data we construct Table 7.1.

Based on data in Table 7.1, the next step is to construct Table 7.2, which contains information on energy intensity and industry group-wise production share.

Based on Equations 7.7a and 7.7b, this leads us to the following numbers under additive and multiplicative framework:

7.3.1.1 *Additive framework*

$$w_1 = \frac{E_{1T} - E_{10}}{\ln E_{1T} - \ln E_{10}} = 7{,}013.5$$

$$w_2 = \frac{E_{2T} - E_{20}}{\ln E_{2T} - \ln E_{20}} = 5{,}527.6$$

$$Activity\ Effect = \Delta E_{AE} = \sum_i w_i \ln\left(\frac{Y_T}{Y_0}\right) = 22{,}904$$

$$Structural\ Effect = \Delta E_{SE} = \sum_i w_i \ln\left(\frac{S_{iT}}{S_{i0}}\right) - 1{,}976$$

$$Energy\ Intensity\ Effect = \Delta E_{IE} = \sum_i w_i \ln\left(\frac{I_{iT}}{I_{i0}}\right) - 12{,}547$$

Table 7.1 Industry Group-Wise Production and Energy Use (both in constant prices) in Aggregate Manufacturing Sector in India in 1990–1991 and 2010–2011(values are in Rupees Lakhs)

Industry groups	*1990–1991 (t = 0)*		*2010–2011 (t = T)*	
	Energy Use (E_{i0})	*Output (Y_{i0})*	*Energy Use (E_{iT})*	*Output (Y_{iT})*
Energy-intensive manufacturing industries (Group 1)	5,105	47,563	9,347	194,572
Less energy-intensive manufacturing industries (Group 2)	3,714	86,342	7,853	637,131
Total	8,819	133,905	17,200	831,703

Table 7.2 Computation for LMDI

Industry groups	*1990–1991 (t = 0)*				*2010–2011 (t = T)*			
	E_{i0}	Y_{i0}	$S_{i0} = Y_{i0}/Y_0$	$I_{i0} = E_{i0}/Y_{i0}$	E_{iT}	Y_{iT}	$S_{iT} = Y_{iT}/Y_T$	$I_{iT} = E_{iT}/Y_{iT}$
Group 1	$E_{10} = 5,105$	$Y_{10} = 47,563$	$S_{10} = 0.36$	$I_{10} = 0.11$	$E_{1T} = 9,347$	$Y_{1T} = 194,572$	$S_{1T} = 0.23$	$I_{1T} = 0.05$
Group 2	$E_{20} = 3,714$	$E_{20} = 86,342$	$S_{20} = 0.64$	$I_{20} = 0.04$	$E_{2T} = 7,853$	$Y_{2T} = 637,131$	$S_{2T} = 0.77$	$I_{2T} = 0.01$
Total	$E_0 = 8,819$	$Y_0 = 133,905$	$S_0 = 1$	$I_0 = 0.07$	$E_T = 17,200$	$Y_T = 831,703$	$S_T = 1$	$I_T = 0.02$

7.3.1.2 Multiplicative framework:

$$\widetilde{w_1} = \frac{(E_{1T} - E_{10}) / (\ln E_{1T} - \ln E_{10})}{(E_T - E_0) / (\ln E_T - \ln E_0)} = 0.559$$

$$\widetilde{w_2} = \frac{(E_{2T} - E_{20}) / (\ln E_{2T} - \ln E_{20})}{(E_T - E_0) / (\ln E_T - \ln E_0)} = 0.441$$

$$Activity\ Effect = D_{AE} = \exp\left[\sum_i \widetilde{w_i} \ln\left(\frac{Y_T}{Y_0}\right)\right] = 6.20$$

$$Structural\ Effect = D_{SE} = \exp\left[\sum_i \widetilde{w_i} \ln\left(\frac{S_{iT}}{Si_0}\right)\right] = 0.85$$

$$Energy\ Intensity\ Effect = D_{IE} = \exp\left[\sum_i \widetilde{w_i} \ln\left(\frac{I_{iT}}{I_{i0}}\right)\right] = 0.37$$

Results indicate that between 1990–1991 and 2010–2011 (Figure 7.1), increased activity has been driving the increase in energy use in Indian manufacturing industries. But both structural and energy-intensity effects have helped in neutralizing a part of this rising energy use. These results have interesting messages to deliver—both the structure of manufacturing sector in India and energy-using technologies are becoming less energy-using. Let us first consider the results under additive framework. The actual increase in the value of energy use in Indian manufacturing sector in 2010–2011 as compared to 1990–1991 is Rs. 8,381 lakhs (in constant prices). The decomposition analysis tells us that had there been no change in the structure of the manufacturing sector and energy intensity of the sector, with the increase in activity/output (AE) alone, the value of energy use would have been Rs. 22,904 lakhs, which is 2.5 times higher than the actual increase. Due to the changing structure (SE), with increase in the production share of less energy-intensive manufacturing industries, there has been a reduction in energy use worth Rs. 1,976 lakhs. A more significant contribution, however, has emerged from energy-intensity effect (IE). There has been a decline in energy intensity in the overall manufacturing sector from 0.066 to 0.021[7] (Table 7.2). It implies that on an average, the use of energy input per unit of manufacturing output in India has declined from 0.066 (1990–1991) to 0.021 (2010–2011). This improvement in energy efficiency has helped bringing down the value of energy use by a sum of Rs. 12,547 lakhs (at constant price). This implies that actual energy use has increased at a much slower rate in the manufacturing sector than the rate of growth of output. One may observe that the figure 8381matches the figure AE (22,904) + SE (−1,976) +IE (−12,547). This demonstrates how LMDI-based IDA can be used to analyze decomposition using two data points. However, in the presence of time-series data, a chained decomposition can be carried out for

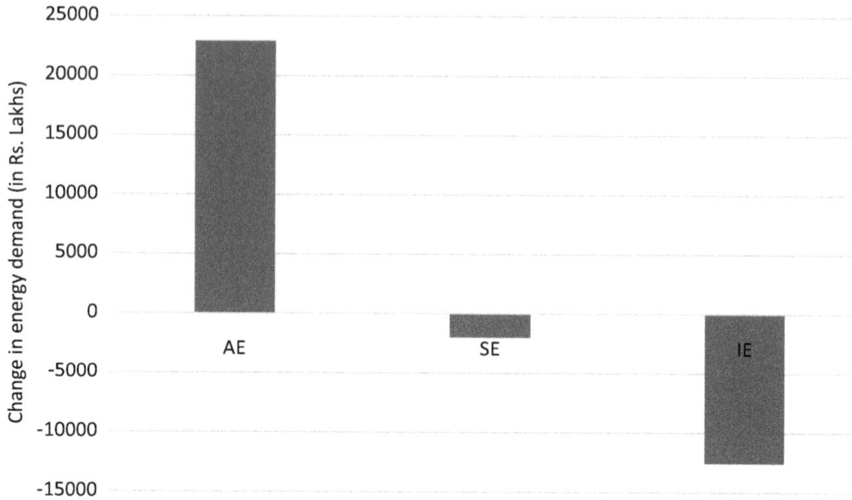

Figure 7.1 Representation of activity effect (AE), structural effect (SE) and energy intensity effect (IE) under additive (left) and multiplicative (right) LMDI

each year to show the trend of annual changes in energy use, AE, SE, and IE (Dasgupta and Roy, 2017, pp. 1–19).

The multiplicative framework is another way of representation. Using this method, our data show that the energy demand of the manufacturing sector has increased by a factor of 1.95 from 1990–1991 to 2010–2011. Had there been no change in the structure and energy intensity, this increase, induced by AE would have been by a factor of 6.2. It has been scaled down by structural effect and energy-intensity effect. The choice between additive and multiplicative form is inconsequential since multiplicative LMDI possesses the additive property in log form, and hence, one can be easily converted to the additive form.

7.3.2 LMDI to understand the relative strength of the drivers

While decomposition of total energy use is the first step in understanding the contribution of drivers of energy use, there is a whole range of literature that shows that during the past decades, the trend of decoupling has become much evident, with energy intensity playing a significant role. Given this, from the policy perspective, the next important step is to identify the drivers of change in "energy intensity". Based on the same example, one can observe that overall energy intensity of the manufacturing sector (E_t/Y_t) can be attributed to two drivers—shift in production toward less energy-intensive industries, that is, towards Group 2, and decline in energy-intensity of Group 1 and/or Group 2 (E_{it}/Y_{it}). LMDI can be used to identify the role of these two drivers in the following manner. Since it has already been argued that the choice between the additive and the multiplicative framework is rather inconsequential, here we discuss only the additive method.

Let, I_t denote the energy intensity of the aggregate manufacturing sector in India:

$$I_t = \frac{\mathrm{E_t}}{\mathrm{Y_t}} = \sum_i \frac{\mathrm{E_{it}}}{\mathrm{Y_t}} = \sum_i \frac{Y_{it}}{Y_t}\frac{E_{it}}{Y_{it}} = \sum_i S_{it} I_{it}, \tag{7.8}$$

In additive form, change in energy intensity is expressed as follows:

$$I_T - I_0 = \Delta I_{TOT} = \Delta \mathrm{I}_{SE} + \Delta \mathrm{I}_{IE}, \tag{7.9}$$

where $\Delta \mathrm{I}_{SE}$ is the change in energy intensity due to SE, and $\Delta \mathrm{I}_{IE}$ is the change in energy intensity due to sectoral energy-intensity effect.

Using LMDI,

$$\Delta \mathrm{I}_{SE} = \sum_i \widehat{w}_i \ln\left(\frac{S_{iT}}{S_{i0}}\right) \tag{7.10}$$

$$\Delta \mathrm{I}_{IE} = \sum_i \widehat{w}_i \ln\left(\frac{I_{iT}}{I_{i0}}\right),$$

where $\widehat{w}_i = \dfrac{\left(\dfrac{\mathrm{E_{iT}}}{\mathrm{Y_T}} - \dfrac{\mathrm{E_{i0}}}{\mathrm{Y_0}}\right)}{\ln\dfrac{\mathrm{E_{iT}}}{\mathrm{Y_T}} - \ln\dfrac{\mathrm{E_{i0}}}{\mathrm{Y_0}}}.$

Based on data in Table 7.2, $\widehat{w}_1 = 0.022$, $\widehat{w}_2 = 0.017$ and

$$\Delta \mathrm{I}_{SE} = \sum_i \widehat{w}_i \ln\left(\frac{S_{iT}}{S_{i0}}\right) = -0.0063$$

$$\Delta \mathrm{I}_{IE} = \sum_i \widehat{w}_i \ln\left(\frac{I_{iT}}{I_{i0}}\right) = -0.0389.$$

This shows, for the aggregate manufacturing industries in India, the decline in energy intensity is mainly because of the declining sectoral energy intensities. The magnitude of the change in energy intensity during the study period is 0.045, out of which only 0.0063, that is, 14% is due to the shift in production in favor of less energy-intensive sectors. The remaining 86% is due to decline in sectoral energy intensity that can be regarded as "pure" energy-intensity effect.

7.3.3 LMDI with more drivers and more sectors

One of the most important characteristics of LMDI is that it is easy to extend the formulation for any level of aggregation of sectors and drivers. In this section, we demonstrate the use of LMDI to decompose energy demand of the aggregate Indian economy with four drivers: activity, structure, energy intensity and fuel mix. We have taken the national gross domestic product (GDP) as an indicator of the macroeconomic activities and have divided the economy into four sectors: agriculture (Sector 1), industry (Sector 2), power generation (Sector 3) and services (Sector 4). Power generation is treated as a separate sector as the most significant amount of the total greenhouse gas emissions comes from the power sector (Raghuvanshi et al., 2006, p. 428; Mittal et al., 2012, p. 2). Since power generation is included as a separate sector in the analysis, we have considered only the use of primary energy in all the other sectors to avoid double counting. However, instead of using "energy" as an aggregate concept, in this example, we consider fuel-wise energy use, that is, the use of coal, oil and natural gas, for each of the sectors between 1990–1991 and 2012–2013. Data on GDP (in billion rupees) is collected from the *Handbook of Statistics on Indian Economy 2014–15*, published by Reserve Bank of India (at constant prices of 2004–2005). The energy use data has been obtained from the various issues of the *Statistical Yearbook of India* published by the MoSPI. For the energy data, final use data for coal, oil and natural gas for all four sectors are quoted in million tonnes oil equivalent (Mtoe). One may notice, given the change in data source, the way of defining energy intensity will also change. While in the case of the previous example, both energy use and output produced were measured in monetary terms, leading to the use of E_1 as the relevant indicator of energy intensity, in this example, we use E_2 (see Section 7.2).

So, here in this example, we have four sectors in the Indian economy where three types of fuels are being used as sources of energy. Since we have fuel-wise energy use data for each sector, we can identify four drivers of energy demand at any period t in the following manner:

$$E_t = \sum_{i,j} E_{ijt} = \sum_{i,j} Y_t \frac{Y_{it}}{Y_t} \frac{E_{it}}{Y_{it}} \frac{E_{ijt}}{E_{it}} \tag{7.11}$$

where E_{ijt} = Use of jth fuel in ith sector, E_{it} = Total energy use in ith sector, Y_{it}=GDP of ith sector, Y_t= GDP of the whole economy, $\frac{E_{ijt}}{E_{it}} = F_{ijt}$ = Proportion of jth fuel used in ith sector, $\frac{E_{it}}{Y_{it}} = I_i^t$ = Energy intensity of the ith sector, and $\frac{Y_{it}}{Y_t} = S_{it}$ = Share of ith sector in total output of the economy.

Based on additive formulation of LMDI, we have

$$\begin{aligned}\text{Change in energy use} = E_T - E_0 = \Delta E_{TOT} &= \Delta E_{OE} + \Delta E_{SE} \\ &+ \Delta E_{IE} + \Delta E_{FE}\end{aligned} \tag{7.12}$$

where $\Delta E_{OE} = \sum_{i,j} w_{ij}\text{In}\left(\frac{Y_T}{Y_0}\right)$ = Activity Effect, that is, change in energy use due to activity growth; $\Delta E_{SE} = \sum_{i,j} w_{ij}\text{In}\left(\frac{S_{iT}}{S_{i0}}\right)$ = Structural Effect, that is, changes in energy use due to change in structure of the economy; ΔE_{IE} = $\sum_{i,j} w_{ij}\text{In}\left(\frac{I_{iT}}{I_{i0}}\right)$ Energy Intensity Effect, that is, changes in energy use due to change in energy intensity; and $\Delta E_{FE} = \sum_{i,j} w_{ij}\ln\left(\frac{F_{ijT}}{F_{ij0}}\right)$ = Change in energy use due to change in fuel mix,where, $w_{ij} = \frac{E_{ijT} - E_{ij0}}{\ln E_{ijT} - \ln E_{ij0}}$.

Based on data for the base and terminal year, we construct Tables 7.3 and 7.4.

There is a methodological issue that needs to be addressed here. In Table 7.4, the use of coal in both agriculture and service and natural gas in the service sector are zero. It leads to a unique problem in the calculation of weights as it involves taking natural logs of these values. But dropping coal in the aggregate calculation cannot be done as two other sectors are major coal users. To overcome this problem, a simple practical way for computational ease without violating the nature of the relation is to replace zero with some very small positive fractions so that the natural logs are no longer undefined (Ang, 2012, p. 37).

This leads us to the following numbers under additive framework:

$w_{11} = 0$	$w_{21} = 75.22$	$w_{31} = 162.52$	$w_{41} = 0$
$w_{12} = 0.62$	$w_{22} = 6.01$	$w_{32} = 1.99$	$w_{42} = 40.28$
$w_{13} = 0.11$	$w_{23} = 0.88$	$w_{33} = 6.57$	$w_{43} = 0$

Total Change in Energy Demand= $E_T - E_0 = \Delta E$= 471.29 – 178.67 = 292.62
Activity Effect = ΔE_{AE} = +420.72
Structural Effect = ΔE_{SE} = −11.54
Energy Intensity Effect = ΔE_{IE} = −117.61
Fuel Mix Effect = ΔE_{FE} = + 1.05

Similar to the previous example from manufacturing industries, including all economic sectors between 1990–1991 and 2012–2013 (Figure 7.2), activity growth was the main driver of the increase in energy use in India. However, both the structural and the energy intensity effects have contributed in neutralizing a part of this increase. The total change in energy use, which is 292.62 Mtoe during the study period, can be decomposed into AE (+420.72), SE (−11.54), IE (−117.61) and FE (+1.05). This implies that had there been no change in the structure of the economy, IE and FE during these two decades, then the amount of energy use would have increased by

Table 7.3 Sectoral GDP (in billion rupees) and Energy Use Data (in Mtoe) for Indian Economy in 1990–1991 and 2012–2013

		1990–1991 (t = 0)		*2012–2013 (t = T)*	
Sectors		*Energy Use* (E_{ij0})	*Output* (Y_{i0})	*Energy Use* (E_{ijT})	*Output* (Y_{iT})
	Coal (E_{11})	0		0	
	Oil (E_{12})	0.54		0.70	
Agriculture (Sector 1)	Natural gas (E_{13})	0.08	3,361.76	0.16	6,494.24
	Total (E_1)	0.62		0.86	
	Coal (E_{21})	66.77		84.38	
	Oil (E_{22})	7.82		4.50	
Industry (Sector 2)	Natural gas (E_{23})	0.74	2,271.26	1.03	10,751.26
	Total (E_2)	75.33		89.93	
	Coal (E_{31})	76.19		297.67	
	Oil (E_{32})	2.75		1.38	
Power Generation (Sector 3)	Natural gas (E_{33})	3.27	269.71	11.57	1,029.23
	Total (E_3)	82.21		310.62	
	Coal (E_{41})	0		0	
	Oil (E_{42})	20.51		69.90	
Services (Sector 4)	Natural gas (E_{43})	0	6,686.49	0	36,424.75
	Total (E_4)	20.51		69.90	
Total		178.67	13,089.72	471.29	54,699.48

Table 7.4 Computation for Decomposition (four sectors and four drivers)

Sectors	*1990–1991 (t = 0)*					*2012–2013 (t = T)*				
	$I_{i0} = E_{i0}/Y_{i0}$	$S_{i0} = Y_{i0}/Y_0$	$F_{i10} = E_{i10}/E_{i0}$	$F_{i20} = E_{i20}/E_{20}$	$F_{i30} = E_{i30}/E_{30}$	$I_{iT} = E_{iT}/Y_{iT}$	$S_{iT} = Y_{iT}/Y_T$	$F_{i1T} = E_{i1T}/E_{iT}$	$F_{i2T} = E_{i2T}/E_{2T}$	$F_{i3T} = E_{i3T}/E_{3T}$
Sector 1	0.00018	0.257	0	0.87	0.13	0.00013	0.119	0	0.81	0.19
Sector 2	0.02717	0.212	0.87	0.10	0.01	0.00836	0.197	0.94	0.05	0.01
Sector 3	0.30480	0.021	0.93	0.033	0.04	0.30179	0.019	0.96	0.004	0.04
Sector 4	0.00307	0.511	0	1	0	0.00192	0.666	0	1	0

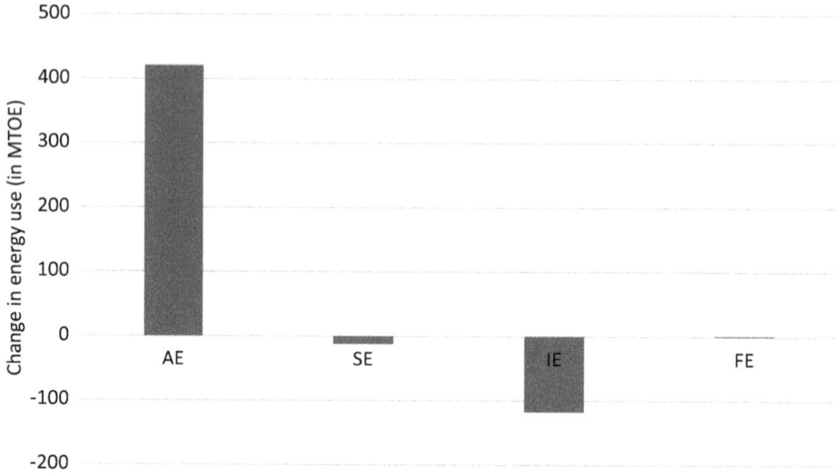

Figure 7.2 Representation of Activity Effect (AE), Structural Effect (SE), Energy Intensity Effect (IE) and Fuel Mix Effect (FE) under additive LMDI.

420.72 Mtoe, which is much higher than the actual increase. Due to the change in structure of the Indian economy (SE), with the increasing share of the less energy-intensive services sector and the decreasing share of the high energy-intensive industrial sector, there has been a decrease in energy use by 11.54 Mtoe. The impact of the change in FE, despite having a positive impact on energy, is very small in magnitude. Due to the change in FE, energy use has been increased by 1.05 Mtoe. It is noteworthy that IE, that is, the use of energy in Mtoe per billion rupee value of output in all four sectors—agriculture, industry, power generation and services—has declined in the terminal year compared to initial year (Table 7.4). Here also, one can easily derive that total change in energy use (292.62 Mtoe) matches the figure AE (420.72 Mtoe) + SE (–11.54 Mtoe) +IE (–117.61 Mtoe) + FE (1.05 Mtoe). Therefore, one can suggest that the decoupling of GDP growth and energy use growth in the Indian economy over 1990–1991 and 2012–2013 is attributable primarily to falling energy intensity due to technological improvements in various economic activity sectors (IE) and the changing composition of economic sectors in total GDP (SE).

7.4 Concluding remarks

This chapter demonstrates the use of LMDI-based IDA framework, its analytical strength and how it can be applied in empirical research. However, readers need to be aware of other methods applied in analysis of drivers of energy use. The structural decomposition analysis (SDA) is also a popular method. But IDA and SDA are applied in the context of two different types of data sets. While IDA is used in case of sectoral or macro data over time compiled from various sources, SDA uses input–output data, which is generally not available for consecutive years. While both the methods have a strong theoretical foundation, IDA has advantage over SDA in empirical research as the data requirement for the former is less. However, when data are available, SDA is capable of distinguishing between indirect and direct effects in a more detailed manner (Hoekstra and Bergh, 2003, pp. 39–64) as both forward and backward linkages are considered. Also, one must be careful in interpreting the effects of technological advancement under these two different frameworks. In the context of energy use, while the intensity effect in IDA represents the input/output ratio of any economic activity and time-series data can capture changing technology over time, SDA represents the technological effect of changes in the intermediate input structure. However, recent methodological developments in IDA have encouraged more complex economic modeling to understand the drivers in deeper detail. One such area of advancement is the use of multidimensional and multilevel index decomposition (Ang and Wang, 2015, pp. 67–76). A comparison of these two methods with a more detailed analysis can be a part of future research.

Notes

1 Other two being annual energy consumption per capita and share of consumption of renewable energy resources.
2 See references mentioned in the article published in *Energy Policy* in 2004: "Decomposition Analysis for Policymaking in Energy: Which Is the Preferred Method?" by Beng Wah Ang (p. 1133).
3 Conventional Laspeyres Index–based methods leave large residuals as they do not qualify factor reversal text.
4 The log mean of two numbers a and b is defined as $L(a, b) = (a - b)/(ln\ a - ln\ b)$
5 For proof of perfect decomposition see "A Simple Guide to LMDI Decomposition Analysis", published by Beng Wah Ang in 2012 (pp. 15–16).
6 For this information, you can follow these links: http://www.mospi.gov.in/sites/default/files/main_menu/national_industrial_classification/nic_2008_17apr09.pdf and http://www.mospi.gov.in/sites/default/files/main_menu/national_industrial_classification/nic_2004_concor_tab2.pdf.
7 This measure is unit-free since both the numerator and denominators are measured in monetary units.

References

Allcott, Hunt, and Michael Greenstone. "Is There an Energy Efficiency Gap?" *Journal of Economic Perspectives* 26, no. 1 (2012): 3–28.

Ang, Beng Wah. "Decomposition Analysis for Policymaking in Energy: Which Is the Preferred Method?" *Energy Policy* 32, no. 9 (2004): 1131–1139.

Ang, Beng Wah. "A Simple Guide to LMDI Decomposition Analysis." *Department of Industrial and Systems Engineering*, National University of Singapore, Singapore (2012).

Ang, Beng Wah, and S. Y. Lee. "Decomposition of Industrial Energy Consumption: Some Methodological and Application Issues." *Energy Economics* 16, no. 2 (1994): 83–92.

Ang, Beng Wah, and Feng Ling Liu. "A New Energy Decomposition Method: Perfect in Decomposition and Consistent in Aggregation." *Energy* 26, no. 6 (2001): 537–548.

Ang, Beng Wah, and H. Wang. "Index Decomposition Analysis with Multidimensional and Multilevel Energy Data." *Energy Economics* 51 (2015): 67–76.

Ang, Beng Wah, and X. Y. Xu. "Tracking industrial energy efficiency trends using index decomposition analysis." *Energy Economics* 40 (2013): 1014–1021.

Bhattacharya, S., and M. L. Cropper. "Options for Energy Efficiency in India and Barriers to Their Adoption: A Scoping Study." Discussion Paper, Resources for Future. RFF DP 10-20 Washington DC. 2010a. Retrieved from http://www.rff.org/files/sharepoint/WorkImages/Download/RFF-DP-10-20.pdf

Bhattacharya, Soma, and Maureen Cropper. "Options for Energy Efficiency in India and Barriers to their Adoption: A Scoping Study" (2010b). *Available at SSRN* 1590510.

Boyd, Gale, John F. McDonald, Marc Ross, and Donald A. Hanson. "Separating the Changing Composition of US Manufacturing Production from Energy Efficiency Improvements: A Divisia Index Approach." *The Energy Journal* 8, no. 2 (1987): 77–96.

Dasgupta, Shyamasree, and Joyashree Roy. "Analysing Energy Intensity Trends and Decoupling of Growth from Energy Use in Indian Manufacturing Industries during 1973–1974 to 2011–2012." *Energy Efficiency* 10, no. 4 (2017): 925–943.

Dasgupta, S., Roy, J., Bera, A., Sharma, A., and Pandey, P. "Growth Accounting for Six Energy Intensive Industries in India." *Journal of Industrial Statistics* 1, no. 1 (2011): 1–15.

Freeman, Scott L., Mark J. Niefer, and Joseph M. Roop. "Measuring Industrial Energy Intensity: Practical Issues and Problems." *Energy Policy* 25, no. 7–9 (1997): 703–714.

Hoekstra, R. and Bergh, J. "Comparing structural decomposition analysis and index." *Energy Economics*, 25, no. 1 (2003): 39–66.

IAEA. *Energy Indicators for Sustainable Development*. Vienna: International Atomic Energy Agency (2005).

IEA. "Energy and Climate Change, World Energy Outlook Special Report" (2015). Retrieved from https://www.iea.org/publications/freepublications/publication/WEO2015SpecialReportonEnergyandClimateChange.pdf

Liu, X. Q., B. W. Ang, and H. L. Ong. "The Application of the Divisia Index to the Decomposition of Changes in Industrial Energy Consumption." *The Energy Journal* 13, no. 4 (1992): 161–177.

Mittal, Moti L., Chhemendra Sharma, and Richa Singh. "Estimates of Emissions from Coal Fired Thermal Power Plants in India." *International Emission Inventory Conference* (2012): 13–16.

Raa, Thijs ten. *The Economics of Benchmarking: Measuring Performance for Competitive Advantage/T. ten Raa*. New York: Palgrave Macmillan (2009).

Raghuvanshi, Shiv Pratap, Avinash Chandra, and Ashok Kumar Raghav. "Carbon Dioxide Emissions from Coal Based Power Generation in India." *Energy Conversion and Management* 47, no. 4 (2006): 427–441.

Roy, Joyashree, Shyamasree Dasgupta, and Debalina Chakravarty. "Energy Efficiency: Technology, Behavior, and Development." In Edited by Andreas Goldthau *The Handbook of Global Energy Policy* (2013), Willey Blackwell. 282–302.

Roy, Joyashree, Jayant Sathaye, Alan Sanstad, Puran Mongia, and Katja Schumacher. "Productivity Trends in India's Energy Intensive Industries." *The Energy Journal* 20, no. 3 (1999): 33–61.

Sahu, Santosh, and Krishnan Narayanan. "Determinants of Energy Intensity in Indian Manufacturing Industries: A Firm Level Analysis." *Eurasian Journal of Business and Economics* 4, no. 8 (2011):13–30.

UNDESA. *Indicators of Sustainable Development: Guidelines and Methodologies, 2nd edition.* United Nations Department of Economic and Social Affairs, United Nations (2001).

Voigt, Sebastian, Enrica De Cian, Michael Schymura, and Elena Verdolini. "Energy Intensity Developments in 40 Major Economies: Structural Change or Technology Improvement?" *Energy Economics* 41 (2014): 47–62.

8 A framework for renewable energy policy modeling

A multistate model for India

Anoop Singh and M. Sandeep Chowdary

8.1 Introduction

Concerns for energy security and the environment have led to an emphasis on the development of renewable energy sources (RESs). Apart from environmental concerns, increasing import of fossil fuel—oil, coal and Liquified Natural Gas (LNG)—is posing a greater challenge to meet the long-term growing electricity demand in India. To address the existing energy shortages and to meet the growing energy needs to support economic development, RESs are expected to play an important role in future. While the development of RESs is gaining attention in the Indian policy making, the adoption of various policy measures has not shown encouraging results as a large number of states failed to fulfill renewable purchase obligation (RPO) set by the respective state electricity regulatory commissions (SERCs). High costs[1] and variability due to dependence on natural resources make it an unattractive option for capacity addition unless supported by appropriate policies and regulatory environment.

The renewable energy (RE) generation is often encouraged with policies like subsidies, fiscal incentives and RPO with feed-in tariffs (FiTs). The impact of these policies on the deployment of RESs for electricity generation is governed by the resource adequacy, access to technology, grid connectivity, cost and taxation structure, as well as overall investment climate in the country. Inadequate attention to the design of such policies may either lead to suboptimal policies leading to the failure of such policies or giving away too much advantage to investors for long time. Due to inadequate care in design of policy and regulatory framework, obligated entities in most of the state fail to meet their RPO. Apart from faulty policy design, these policies disregard economic efficiency. There is little incentive to develop cost-effective technology to develop the most cost-effective RES first.

The choice amongst the available policy options depends on the resource endowments and technological constraints as well as the cost of deployment of these resources. A modeling framework can be used to simulate the outcome with alternate policy scenarios in different market environments. This should help guide policy and regulatory initiatives in the country.

In a multistate nation like India, the State Electricity Regulatory Commissions (SERCs) set the RPO and the preferential FiT to be paid for the RE generated based on the technology used. Renewable Energy Certificates (RECs) have been introduced as an alternate revenue stream to support investment in RE and support compliance towards renewable purchase obligation. Singh (2009a) proposed the implementation of a market for RECs in India to achieve economic efficiency. Nishio and Asano (2006), and Mozumdar and Marathe (2004) discuss the economic benefits of RECs. The relative importance and efficacy of alternate policy measures such as FiTs and RECs have not been duly addressed in the literature in an applied framework. Singh (2010) identified and discussed some of the practical implementation aspects while designing regulations for implementing the REC framework in India. This chapter develops a modeling framework for evaluating such alternate policy measures under different market environments. We evaluate the economic efficiency in terms of cost of compliance and welfare impacts due to the adoption of alternate policies in state-specific and national market contexts.

8.2 Policies for renewable energy sources

Launched in 2009, Jawaharlal Nehru National Solar Mission, one of the eight missions identified under the National Action Plan for Climate Change (NAPCC) has initially set a target of 20,000 MW of solar energy–based power generation capacity (including 2,000 MW off-grid projects) by 2022. Cost reductions of solar photovoltaic technology led the government, in 2015, to set a more aggressive target of 100 GW of solar power generation capacity, including 60 GW of large and medium-scale grid-connected solar power projects and 40 GW of decentralized projects in the form of solar rooftops.

Support policies for RESs can be broadly classified into those supporting investment in plants harnessing RESs or electricity generation from such plants (Mendonça, 2007). The support schemes for RESs include the following:[2]

1. RPO
2. FiTs
3. Fiscal/financial incentives including subsidies
4. Demonstration projects and information dissemination
5. Market-based instruments like RECs

RPO mandates the obligated entities including the distribution utilities to buy a certain proportion of from RESs. Section 86 (e) of the Electricity Act 2003 (Government of India, 2003) mandates a certain proportion of electricity consumed in the area of the distribution licensee to be procured from the RESs. The SERCs in India have set out RPO for the obligated entities. The act also provides for SERCs to set preferential FiTs for electricity purchases by obligated entities from RESs. RE-based decentralized distributed generation projects have been are funded by the Ministry of New of Renewable Energy.

The cost and price barriers that are faced by RE generation can be addressed by setting a FiT which establishes a favorable price for the RE power generation compared to the conventional energy technologies. The capital subsidies are generally used during the initial stages to reduce capital cost of such projects, especially for technology demonstration projects. This, however, leads to excess capital costs and provides no incentive for cost reduction. Tax incentive policies provide accelerated depreciation benefits for investment in RESs, certain exemptions from customs duty and sales tax and income tax holidays for projects investment in the development of RESs. RE investments currently qualify for 40% accelerated depreciation in the first year of operation, down from 80% earlier. It also qualified for income tax exemption for 10 years, which was withdrawn from April 2017. Further, in lieu of 100% custom and excise duty exemption available to all components used for solar projects, post GST regime, these were bracketed into lower tax bracket. In 2009, the government had introduced a generation-based incentive (GBI), for wind generators[3] not availing accelerated depreciation, at the rate of Rs. 0.50 per unit of electricity fed into the grid, subject to a total limit of Rs. 10 million per MW. The GBI scheme was extended for the 12th five-year plan (i.e. from 1 April 2012 to 31 March 2017) for a minimum period of 4 years and a maximum period of 10 years. Furthermore, it was subject to an annual limit of Rs. 2.5 million for the first four years. As on 31 March 2020, total RE capacity in India reached 87 GW, including 34.6 GW solar and 37.7 GW wind (CEA (2020)). RE capacity share now accounts for over 23.5% of total installed capacity of 370 GW. A significant decline in solar energy cost, driven down further by competitive bidding, has brought solar energy into the focus and is soon expected to drive past wind power in terms of installed capacity.

The traditional policy choices often disregard economic efficiency. There is little incentive to develop the cost-effective resources as higher FiTs are offered to compensate for higher costs and low capacity utilization factor (CUF) in unfavorable conditions in some states.[4] Market-based instruments like RECs (Figure 8.1) are often proposed as alternatives that provide a promise of a competitive market outcome to reduce the cost of RE procurement by the obligated entities.

8.2.1 Market for renewable energy certificates

RECs are green attributes of electricity generated from RESs and which can be traded separately from the electricity (Figure 8.1). Singh (2009) aproposed the implementation of a market for RECs to achieve economic efficiency for RE promotion in India. The implementation issues, as well as institutional framework for India and suggestions for developing an efficient REC market, have been discussed in detail in Singh (2010). In 2010, the Central Electricity Regulatory Commission (CERC) issued regulations for introducing a national-level REC mechanism in India CERC (2010a). Under these regulations, one REC is issued for 1 MWh electrical energy fed into the grid. The RE generator may sell electricity to the distribution company at an average power purchase cost (APPC), and the associated RECs on power exchanges (PXs; Figure 8.1). RECs can be purchased and credited by a distribution company or other obligated entities towards meeting their RPO. Two kinds of RECs are defined separately for solar and non-solar RESs.

8.3 Methodology

This chapter sets up a static modeling framework to characterize the regulatory framework for RESs across states under different market environment for policy analysis. In the current version of the model, we consider four states alluding to Andhra Pradesh, Karnataka, Maharashtra and Tamil Nadu. The models define the supply and demand functions for REs/REC and the equilibrium condition. The model minimizes the overall cost of compliance as an objective function. We consider two types of RESs: solar and non-solar. Five different models are set up to evaluate policy options under an alternate market environment. In three such models, the obligated entities are to meet their RPO from the purchase of electricity generated from RESs. Two of these models allow for the trade of such electricity across the states. Two separate models allow trade in RECs towards RPO.

The models are set up using the General Algebraic Modelling System (GAMS). The RPO characterizes vertical demand function for the two types of RESs. Separate logistic supply-cost curves are estimated for solar and non-solar sources for each of the four states. This is further discussed in the following sub-section.

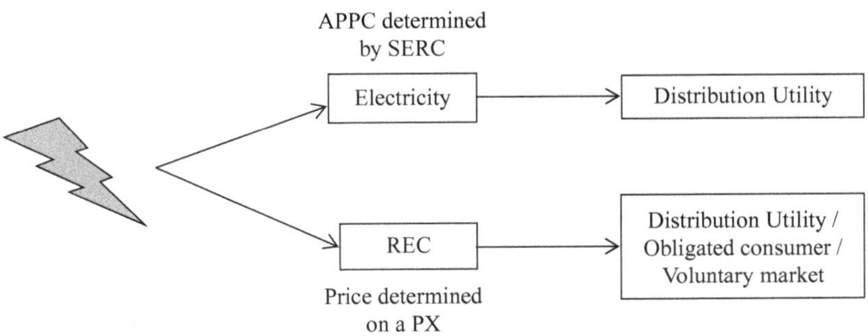

Figure 8.1 Concept of Renewable Energy Certificate

8.3.1 *Logistic supply function for renewable energy/renewable energy certificates*

We presume that investor's response to develop RESs depends on the return perceived by them over the life of a project. We use the estimated levelized tariff for solar and non-solar RESs for each state to derive the respective supply functions. Since the level of FiT declared by the respective SERCs change across time, investors make an assessment of the investment based on the expected levelized tariff. The long-term behavior of the supply function is assumed to be logistic; that is, the relationship between capacity added and the expected levelized tariff is assumed to be 'S'-shaped or logistic curve (Figure 8.2). This assumption follows the work by Olsina et al. (2006). The plausibility of this assumption is also confirmed by the results presented by Morthorst (1999).

The logistic curve for the supply function is specified as[5]

$$q = \frac{a}{\left(1 + b \times \exp\left(-p \times c\right)\right)}, \tag{8.1}$$

where a, b and c are the parameters to be estimated; p is the expected levelized tariff; and q is electricity generated from RESs.

The parameter a is the total potential for electricity generation from RESs for a technology in a state. The potential capacity is expressed in MW. Under a given capacity utilization factor (CUF), the RE electricity supply (MU) from this capacity can be calculated.

For non-solar technology, the maximum potential was assumed to be that for wind energy. For solar technology, the maximum potential is calculated from the average annual radiation received by the four states assuming that

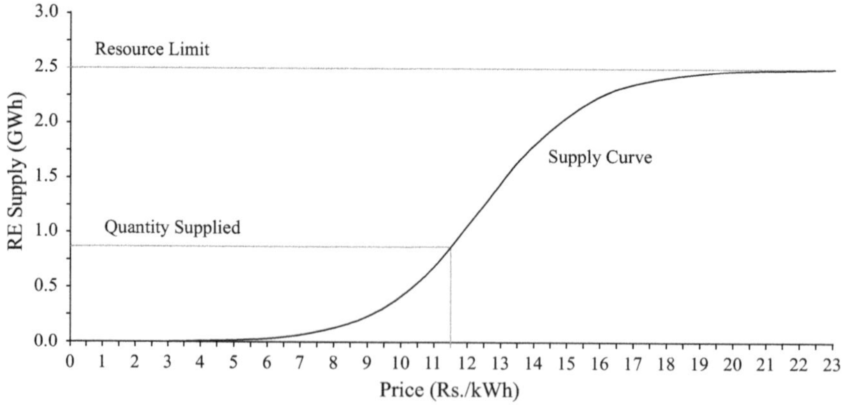

Figure 8.2 Logistic Renewable Energy Supply Curve

up to 10% of the barren land available in the respective state can be used to harness solar energy. Based on this, the estimated electricity generation potential for the four states is given in Table 8.1 and 8.2.

The underlying guiding price for investors is the FiT across the states. We assume that investment decisions are made on the basis of the expected FiT for the future years. Assuming rational expectations of investors, we calculate the levelized tariff for electricity that investors may expect from an RE plant. This is adjusted for price levels to convert it to constant prices. The supply of RE (in MUs) is assumed to be the logistic function of the inflation-adjusted levelized tariff (IALT) calculated per the earlier procedure.

Table 8.1 Renewable Energy Potential of the Four States

	Non-solar (MU)	*Solar (MU)*
State1	5,530	30,540
State2	4,868	104,200
State3	4,584	93,204
State4	11,531	33,096

The projected RE potential is illustrative in purpose. Development in technology and evolving cost of harnessing the RE sources would redefine the potential. For e.g. high efficiency solar cells and efficient power converters expand the potential for solar energy. Similarly, wind energy potential rises with increase in height, and is further supplemented with harnessing of off-shore wind

Table 8.2 Estimated Parameters for Logistic Supply Function

		Non-Solar	
	A	*b*	*c*
State1	5,530	805	2.42
State2	4,868	785	1.55
State3	4,584	780	1.51
State4	11,531	788.5	1.41
		Solar	
	A	*B*	*c*
State1	30,540	18,000	0.51
State2	104,200	18,900	0.52
State3	93,204	20,000	0.50
State4	33,096	19,500	0.44

The state-specific supply functions for solar and non-solar technology are specified as follows:

$$q_{i,k}^{s} = f(p) = \frac{a_{i,k}}{\left(1 + b_{i,k} \times \exp\left(-c_{i,k} \times p_{i,k}\right)\right)}, \tag{8.2}$$

where i = state, k = technology (solar and non-solar).

The parameters b and c are estimated using nonlinear regression (Seber and Wild, 2003). The steps for estimation of the parameters in equation 8.2 include calculation of a Jacobean matrix, which is a rectangular matrix of order equal to the number of data points available and number of parameters to be estimated. This is a least squares estimation method.

$$F(b,c) = \begin{bmatrix} \frac{\partial f_1}{\partial b} & \frac{\partial f_1}{\partial c} \\ \frac{\partial f_2}{\partial b} & \frac{\partial f_2}{\partial c} \\ \frac{\partial f_n}{\partial b} & \frac{\partial f_n}{\partial c} \end{bmatrix}, \tag{8.3}$$

The order of the Jacobean matrix in the current context is ($n \times 2$) as there are two unknown parameters in the original non-linear equation and the 'n' data points each for a set of levelized tariff and capacity added year-wise. (b_0, c_0) are the starting values of (b, c), which are to be estimated, and the initial iteration count is set to be $p = 1$.

$$\delta_{(2\times1)}{}^{1} = \left[F^{0T} \times F^{0}\right]_{(2\times2)}^{-1} \times F^{0T}{}_{(2\times n)} \times \left[Y - f(x)^{0}\right]_{(n\times1)}, \tag{8.4}$$

where δ is the error in the estimate. By using the Gauss-Newton algorithm, the parameters b and c of the non-linear curve (2) are estimated as shown:

For $p = 2$, Therefore,

$$\begin{pmatrix} b \\ c \end{pmatrix} = \begin{pmatrix} b_0 \\ c_0 \end{pmatrix} + \begin{pmatrix} \delta_1^1 \\ \delta_2^1 \end{pmatrix} \tag{8.5}$$

Thus, the value for the next iteration is obtained by adding δ. We use b_0 and c_0 as starting values:

$$\Delta_{p+1} = \sqrt{\left(b_{p+1} - b_p\right)^2 + \left(c_{p+1} - c_p\right)^2}, \tag{8.6}$$

The Euclidean distance is calculated for the parameters between the initial values of the iteration and the final values of the iteration, and the loop is terminated if $\Delta_k < \in$, where $\in$ is a very small user-defined number. We assumed it to be 0.1. If the Euclidean distance is $\Delta_k \geq \in$, then the counter is incremented by 1; that is, $p = p + 1$ and the previously mentioned steps are followed sequentially till $\Delta_k < \in$ satisfies the condition, and then the iteration is stopped. The supply-cost curve parameters are thus estimated for the four states considered. The supply-cost curves are unique for all the states and for both the technologies.

8.4 Models for alternate scenarios

The modeling framework characterizes a market for electricity generated from RESs. In the current version, a modeling framework is set up to characterize resource base and renewable portfolio obligation for four states, namely Andhra Pradesh, Karnataka, Maharashtra and Tamil Nadu. These are called State 1, State 2, State 3 and State 4, respectively. Under the alternate market environment, the buyers and sellers across the four states can interact depending on trade and its mode. The buyer utility need to satisfy their respective obligation for the purchase of RE based electricity or RECs from solar and non-solar sources while minimizing the cost of compliance.

Table 8.3 defines the alternate scenarios with different market structures. In "No-trade" model, obligated entities can meet their obligation from sources located within the respective state only. "RE_Local" is the local market model where the RE trade is possible across states but the price is determined separately in the state-specific RE market. "RE_National" is a common national market where the RE trade is possible across states with a single price being determined at the national level separately for solar and non-solar technology. "REC_National" simulates the REC market mechanism at the national level. The price of the RECs is determined in a common national market separately for solar and non-solar technology. "REC_Local" simulates an REC market mechanism within each state separately, and the price of the certificate is determined by the local market based on the equilibrium reached locally.

8.4.1 No-renewable energy trade (no-trade model)

In this model, no RE trade is possible between states and obligated entities in each state need to buy solar and non-solar energy from plants located

Table 8.3 Renewable Energy Policy Modelling Scenarios

S. No.	*Model Name*	*RE Trade between States*	*REC Trade between States*	*State-Specific/ National Market*
1	No-trade	–	–	NA
2	RE_Local	Y	–	State-specific
3	RE_National	Y	–	National
4	REC_Local	–	Y	State-specific
5	REC_National	–	Y	National

within the state (Figure 8.3). This is similar to the case in India before the introduction of the REC market.[6] The interaction between the supply and demand for two technologies separately determines the equilibrium price of RE generated in each state.

The four blocks in Figure 8.3 represent separate markets for electricity generated from RESs in the four states. The crossed links between the blocks represent that the trade is not possible between states and the RPO is to be met from RESs sources located within the respective state only. Therefore, the price for RE is determined separately for ($p_{i,k}$, where i = 1 to 4 and k = solar and non-solar) all the states as per market equilibrium between the supply and demand function.

8.4.1.1 *Renewable energy supply function*

The following equation is a supply function to be estimated for non-solar and solar technologies (k) for each state (i) using the nonlinear regression technique discussed earlier:

$$q_{i,k}^{s} = f(p) = \frac{a_{i,k}}{\left(1 + b_{i,k} \times \exp\left(-c_{i,k} \times p_{i,k}\right)\right)}. \tag{8.7}$$

8.4.1.2 *Renewable energy demand function*

The following equation represents the vertical demand for non-solar and solar technologies for each of the states considered:

$$q_{i,k}^{d} = x_{i,k} \times e_i. \tag{8.8}$$

8.4.1.3 *The equilibrium condition*

The following condition specifies that the supply of non-solar and solar energy can be less than or equal to the demanded quantity due to resource

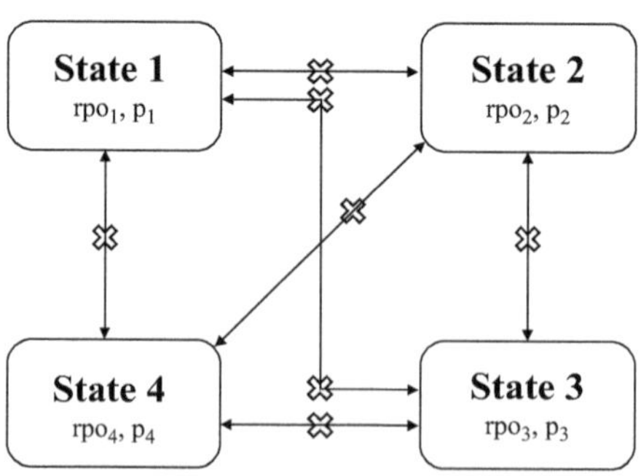

Figure 8.3 "No-Trade" Model

constraints in the respective state. In the case of failing to meet RPO, the obligated entities are penalized:

$$q_{i,k}^{s} \leq q_{i,k}^{d}. \quad (8.9)$$

The following equation defines RE shortage, which may arise because of resource scarcity:

$$q\_shortage_{i,k} = q_{i,k}^{d} - q_{i,k}^{s}. \quad (8.10)$$

The state-specific cost of compliance is given by

$$\text{cost}\_comp_i = \sum_{k} \left(p_{i,k} \times q_{i,k}^{s} + q\_shortage_{i,k} \times penalty_{i,k} \right). \quad (8.11)$$

8.4.1.4 *Objective function*

The objective function minimizes the overall cost of compliance for all states, including any penalty for a shortfall in meeting the RPO:[7] and is subject to all the model constraints defined above.

$$minimize \sum_{i,k} \left(q_{i,k}^{s} \times p_{i,k} \right) + \sum_{i,k} \left(penalty_{i,k} \times \left(q_{i,k}^{d} - q_{i,k}^{s} \right) \right), \quad (8.12)$$

where,

i = state, k = technology;

$a_{i,k}, b_{i,k}$ and $c_{i,k}$ are supply function parameters;

e_i = total electricity consumption in state i;

$x_{i,k}$ = RPO;

$p_{i,k}$ = market clearing price for each state and technology; and

$penalty_{i,k}$ = state - level penalty imposed for each technology for RPO shortfall.

8.4.2 *RE trade with state-specific RE market (RE_Local Model)*

In this model, the obligated entities in a state can purchase electricity generated from RESs located within the host state or, the other states with which adequate transmission capacity is available (Figure 8.4). The obligated entities in each state minimize their cost of compliance by optimizing procurement of RE from indigenous capacity or that available in other states. Due to the possibility of a physical trade of electricity and transmission losses, as well as transmission charges, are appropriately taken into account. Although the model allows for trade of RE, their prices are determined in a market at the respective state level.

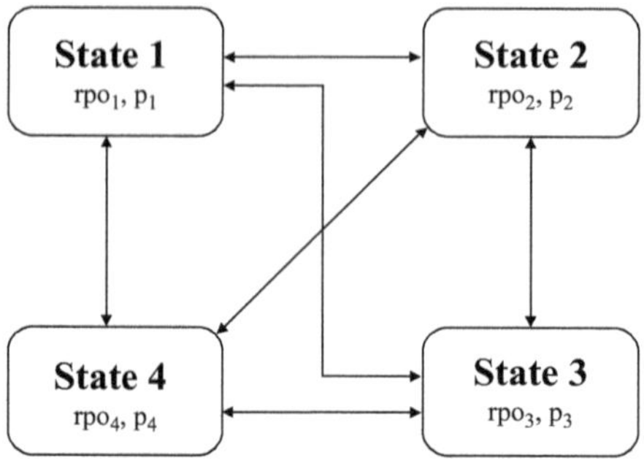

Figure 8.4 Model II: Renewable Energy Trade with State-Specific Renewable Energy Markets

The connections between the states suggest that the states can trade across the transmission links available. In this model, the equilibrium price for the electricity generated from RE is established separately in each state.

8.4.2.1 Renewable energy supply function

The supply-cost function, to be estimated for non-solar and solar technologies for each state using the nonlinear regression technique, is given as follows:

$$q_{i,k}^{s} = \frac{a_{i,k}}{\left(1 + b_{i,k} \times xp\left(-c_{i,k} \times p_{i,k}\right)\right)}. \tag{8.13}$$

The following equation given specifies that the total RE quantity traded, including transmission losses, must be equal to the total quantity generated:

$$q_{i,k}^{s} = \sum_{j} q\_trade_{i,j,k}. \tag{8.14}$$

8.4.2.2 Renewable energy demand function

The vertical demand for non-solar and solar technologies for each state is specified as

$$q_{j,k}^{d} = x_{j,k} \times e_{j}. \tag{8.15}$$

8.4.2.3 The equilibrium condition

The following condition specifies that the total RE quantity demanded can be greater than or equal to the total RE quantity received by a state through trade after adjusting for transmission losses:

$$q_{i,k}^{d} \geq \sum_{i} q\_trade_{i,j,k} \times \left(1 - transloss_{i,j}\right). \tag{8.16}$$

The following equation specifies that the total RE traded between states is limited to the transmission limit or the trade limit allowed between states:

$$\sum_k q\_trade_{i,j,k} \leq translim_{i,j}. \tag{8.17}$$

The RE shortage is defined as a difference between the total quantity demanded and the total traded RE quantity realized at the state, which includes indigenous generation and imports from other states, after adjusting for transmission losses:

$$q\_shortage_{i,k} = q^d_{i,k} - \sum_j q\_trade_{i,j,k} \times (1 - transloss_{i,j}). \tag{8.18}$$

The cost of compliance of each state for non-solar and solar technology together is given by the following equation:

$$\begin{aligned} cost\_comp_j = & \sum_{i,k} \left((p_{i,k} + transcharge_{i,j}) \times q\_trade_{i,j,k} \right) \\ & + \sum_k penalty_{i,k} \times q\_shortage_{i,k}. \end{aligned} \tag{8.19}$$

The first term in the equation represents the cost of compliance due to trade, that is, the cumulative indigenous generation and trade between states, and the second term represents the penalty due to shortfall in meeting the RPO.

8.4.2.4 Objective function

The objective function minimizes the overall cost of compliance for the utilities across all the states subject to the above equations defining this model: The cost of compliance also includes RPO shortfall penalty.

$$\begin{aligned} minimize \sum_{i,j,k} q\_trade_{i,j,k} \times (p_{i,k} + transcharge_{i,j}) \quad + \\ \sum_{j,k} penalty_{j,k} \times \left(q^d_{i,k} - \sum_i q\_trade_{i,j,k} \times (1 - transloss_{i,j}) \right), \end{aligned} \tag{8.20}$$

where

i, j = state, k = technology;
$a_{i,k}, b_{i,k}, c_{i,k}$ are supply function parameters;
e_i = total electricity consumption in state i;
$x_{i,k}$ = RPO;
$p_{i,k}$ = market clearing price for each state and technology;
$transcharge_{i,j}$ = transmission charge for trade between states;
$penalty_{i,k}$ = state - level penalty imposed for each technology for RPO shortfall;
$translim_{i,j}$ = transmission capacity between states; and
$translim_{i,j}$ = transmission capacity between states.

8.4.3 Model III: Renewable energy trade with national renewable energy market

In this model, we simulate a common national model encompassing the four states (Figure 8.5). We determine the aggregate demand and aggregate supply curves for the two technologies separately. The interaction between the two gives a common equilibrium price for RE across the states. The RE trade across states takes into account transmission charge, transmission loss and transmission limit as in the case of the previous model.

8.4.3.1 Renewable energy supply function

The following equation specifies the supply function to be estimated for non-solar and solar technologies for each state using the nonlinear regression technique discussed earlier:

$$q_{i,k}^{s} = \frac{a_{i,k}}{\left(1 + b_{i,k} \times \exp(-c_{i,k} \times p_{i,k}\right)} \tag{8.21}$$

The equivalence of total RE procured is specified as

$$q_{i,k}^{s} = \sum_{j} q\_trade_{i,j,k} \tag{8.22}$$

The total RE generated is equal to the total traded RE quantity which includes the transmission losses.

The following equation is the aggregate supply function for non-solar and solar technology respectively in the national market:

$$q_{k}^{s\_ag} = \sum_{i} \frac{a_{i,k}}{\left(1 + b_{i,k} \times \exp\left(-c_{i,k} \times p_{k}\right)\right)}. \tag{8.23}$$

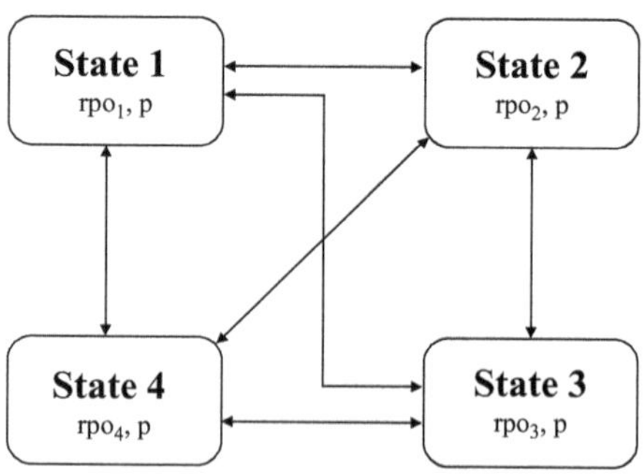

Figure 8.5 Model III—Renewable Energy Trade with National Renewable Energy Market

8.4.3.2 RE demand function

The vertical demand function for non-solar and solar technologies for each state is given by:

$$q_{j,k}^{d} = x_{j,k} \times e_j. \tag{8.24}$$

The aggregate vertical demand for non-solar and solar technologies in the national market can be written as

$$q_k^{d\_ag} = \sum_j x_{j,k} \times e_j. \tag{8.25}$$

The following condition suggests that the total RE quantity demanded, due to RPO, can be greater than or equal to the total supply of RE for each technology in the respective states after adjusting for transmission losses:

$$q_{j,k}^{d} \geq \sum_i q\_trade_{i,j,k} \times \left(1 - transloss_{i,j}\right). \tag{8.26}$$

8.4.3.3 The equilibrium condition

The following equilibrium condition equates aggregate supply with aggregate demand and total transmission loss:

$$q_k^{d\_ag} + \sum_i^j q\_trade_{i,j,k} \times \left(transloss_{i,j}\right) \geq q_k^{s\_ag}. \tag{8.27}$$

The following equations ensure that the total RE quantity, irrespective of technology of generation that can be traded between states, is limited by the transmission capacity between the states:

$$\sum_k q\_trade_{i,j,k} \leq trans\lim_{i,j}; \tag{8.28}$$

$$q\_shortage_{i,k} = q_{i,k}^{d} - \sum_j q\_trade_{i,j,k}\left(1 - transloss_{i,j}\right). \tag{8.29}$$

The RE shortage is defined as the difference between the total quantity demanded as per RPO and the total traded RE procurement realized for a state, including indigenous generation and imports from other states. The cost of compliance for each state together for non-solar and solar technology is given by the following equation:

$$\begin{aligned} \text{cost\_comp}_j = &\sum_{i,k}\left(\left(p_k + \text{transcharge}_{i,j}\right) \times q\_trade_{i,j,k}\right) \\ &+ \sum_k penalty_{i,k} \times q\_shortage_{i,k}. \end{aligned} \tag{8.30}$$

The first term in equation 8.30 gives the cost of compliance due to indigenous generation and trade between states, and the second term gives the cost of compliance due to shortfall in meeting the RPO, if any.

8.4.3.4 *Objective function*

The objective function minimizes the overall cost of compliance across all states, subject to the defined RE trade model (equations 8.21–8.30):

$$minimize \sum_{i,j,k} q\_trade_{i,j,k} \times \left(p_k + \text{transcharge}_{i,j}\right) \quad +$$
$$\sum_{j,k} penalty_{j,k} \times \left(q^d_{i,k} - \sum_i q\_trade_{i,j,k} \times \left(1 - transloss_{i,j}\right) \right), \quad (8.31)$$

where
i, j = state, k = technology,
$a_{i,k}, b_{i,k}, c_{i,k}$ are supply function parameters,
e_i = total electricity consumption in state i,
$x_{i,k}$ = RPO,
p_k = national market clearing price for each state and technology,
$transcharge_{i,j}$ = transmission charge for trade between states,
and
$penalty_{i,k}$ = state-level penalty imposed for each technology for RPO shortfall
$translim_{i,j}$ = transmission capacity between states.

8.4.4 *National REC market (REC_National Model)*

In this model, RPO is met by the procurement of RECs instead of physical RE procurement. We assume that there is no physical RE trade in this case and REC mechanisms in the only avenue available for the obligated entities to meet their RPO (Figure 8.6). The average pool price is declared by the SERCs, and the certificate price is derived from the market equilibrium between national demand and supply function for REC. We also apply the floor and the forbearance price as per the CERC (2010b) regulations. The average power purchase cost (APPC) is estimated and approved by the respective SERCs.

8.4.4.1 *Renewable energy certificate supply function*

The following equation is the supply equation for RECs to be estimated for non-solar and solar technologies for each state using the nonlinear regression technique discussed earlier. It is emphasized here that *qs* represent the quantity of RECs for all the discussion in Sections 8.4.4 and 8.4.5.

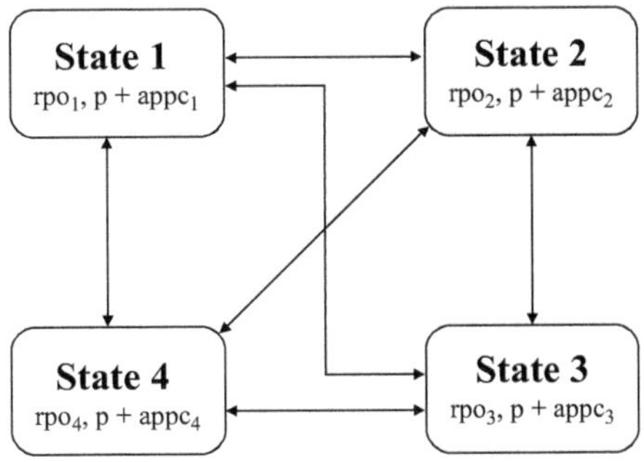

Figure 8.6 Model IV National REC Market

$$q_k^{s\ ag} - \sum_i \frac{a_{i,k}}{\left(1 + a_{i,k} \times \exp\left(-c_{i,k} \times \left(p_k + appc\right)\right)\right)} \tag{8.32}$$

$$q_{i,k}^{s} = \sum_j q\_trade_{i,j,k} \tag{8.33}$$

The total RECs generated is equated to the total RECs traded. Also note that in the case of RECs, no transmission charge and transmission loss is applicable.

8.4.4.2 Renewable energy certificate demand function

The following equation represents a vertical demand function for non-solar and solar RECs for the respective state:

$$q_{j,k}^{d} = x_{j,k} \times e_j. \tag{8.34}$$

The aggregate vertical demand function for non-solar and solar RECs in a national market is specified as:

$$q_k^{ad} = \sum_j x_{j,k} \times e_j. \tag{8.35}$$

8.4.4.3 The equilibrium condition

The following condition represents that the total RECs demanded can be greater than or equal to the total RECs availability, including that procured from within the state or that traded with other states:

$$q_{i,k}^{d} \geq \sum_i q\_trade_{i,j,k}. \tag{8.36}$$

The RECs shortage is measured as a difference between the demand for RECs and that actually met by the obligated entities in each state for the respective technology:

$$q\_shortage_{i,k} = q^{d}_{i,k} - \sum_{j} q\_trade_{i,j,k}. \tag{8.37}$$

The cost of compliance of each state together for non-solar and solar technology is given by the following equation. The first term gives the cost of compliance due to indigenous procurement and trade, while the second term specifies the penalty due to shortfall in meeting the RPOs:

$$cost\_comp_j = \sum_{i,k}\left(\left(p_k + appc_i\right) \times q\_trade_{i,j,k}\right) + \sum_{k} penalty_{i,k} \times q\_shortage_{i,k}, \tag{8.38}$$

where,

i, j = state, k = technology;

$a_{i,k}, b_{i,k}, c_{i,k}$ are supply function parameters;

e_i = total electricity consumption in state i;

$x_{i,k}$ = RPO;

p_k = national market clearing price for each state and technology;

$penalty_{i,k}$ = state-level penalty imposed for each technology for RPO shortfall; and

$appc_i$ = average power purchase cost.

8.4.4.4 *Objective function*

The objective function minimizes the overall cost of compliance for the four states for two technologies taken together:

$$minimize \sum_{k}\sum_{i,j} q\_trade_{i,j,k} \times \left(p_k + appc_i\right) + \sum_{j,k} penalty_{j,k} \times \left(q^{d}_{i,k} - \sum_{i} q\_trade_{i,j,k}\right). \tag{8.39}$$

8.4.5 *State-specific renewable energy certificate market (REC_Local Model)*

In contrast to the previous model, we consider a model wherein each state separately develops a market for RECs (Figure 8.7). The obligated entities can only trade in RECs to meet the RPO, and no physical trade of RE electricity is allowed. Separate REC prices are determined in each state for the two types of technologies being considered.

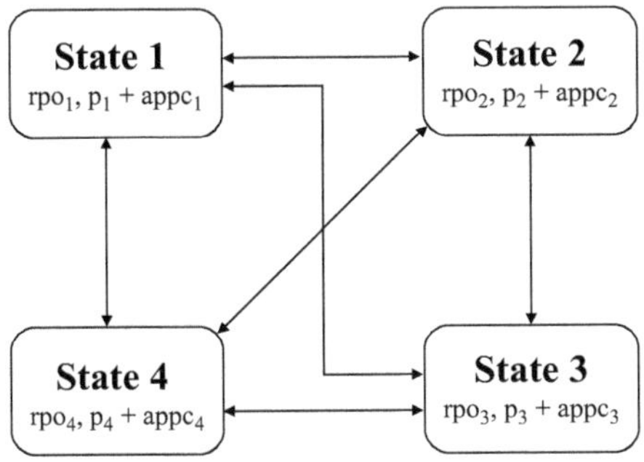

Figure 8.7 Model V—Renewable Energy Certificate Trade with State-Specific Market

The connectors between states show that the REC trade possibility between states.

8.4.5.1 *Renewable energy certificate supply function*

The following equation is a supply-function, in the case of certificate mechanism with the local market, to be estimated for non-solar and solar technologies for each state using the nonlinear regression technique:

$$q_{i,k}^{s} = \frac{a_{i,k}}{\left(1 + b_{i,k} \times \exp(-c_{i,k} \times \left(p_{i,k} + appc_i\right)\right)}. \tag{8.40}$$

The total RECs generated should be equal to the total traded RECs:

$$q_{i,k}^{s} = \sum_{j} q\_trade_{i,j,k}. \tag{8.41}$$

8.4.5.2 *Renewable energy certificate demand function*

The following equation is the vertical demand function for non-solar and solar RECs for each of the state considered:

$$q_{j,k}^{d} = x_{j,k} \times e_j. \tag{8.42}$$

8.4.5.3 *The equilibrium condition*

The following condition specifies that the total RECs demanded can be greater than or equal to the total RECs realized at each state by both indigenous supply and trade between states:

$$q_{i,k}^{d} \geq \sum_{i} q\_trade_{i,j,k}. \tag{8.43}$$

The RECs' shortage may arise if the demand for RECs is greater than its supply in each state for the respective technology:

$$q\_shortage_{i,k} = q^d_{i,k} - \sum_j q\_trade_{i,j,k}. \tag{8.44}$$

The cost of compliance of each state for non-solar and solar together includes that due to trade plus any penalty due to shortfall in meeting the RPOs:

$$\begin{aligned} \text{cost\_}comp_j = &\sum_{i.k}\left(\left(p_{i,k} + appc_{i,j}\right) \times q\_trade_{i,j,k}\right) \\ &+ \sum_k penalty_{i,k} \times q\_shortage_{i,k}. \end{aligned} \tag{8.45}$$

8.4.5.4 *Objective function*

The objective function minimizes the cost of compliance for all the four states for two technologies taken together:

$$\begin{aligned} minimize &\sum_k \sum_{i,j} q\_trade_{i,j,k} \times \left(p_{i,k} + appc_i\right) \\ &+ \sum_{j,k} penalty_{j,k} \times \left(q\_rpo_{i,k} - \sum_i q\_trade_{i,j,k}\right), \end{aligned} \tag{8.46}$$

where

i, j = state, k = technology;

$a_{i,k}, b_{i,k}, c_{i,k}$ are supply function parameters;

e_i = total electricity consumption in state i;

$x_{i,k}$ = RPO;

$p_{i,k}$ = market clearing price for each state and technology;

$penalty_{i,k}$ = state-level penalty imposed for each technology for RPO shortfall; and

$appc_i$ = average power purchase cost.

The preceding five models are evaluated for two cases one with all states having equal RPO and the other with each state having a different RPO. The models are set up in the General Algebraic Modeling System (GAMS) environment, and the results are discussed in Section 8.5. The outcomes are evaluated in terms of cost of compliance and welfare of the society. The latter represents economic efficiency and is measured in terms of social surplusincluding consumer and producer surplus.

8.5 Measure of economic efficiency

Consumer and producer surplus are measures of social welfare. Producer surplus is measured as the difference between the price at which producers are willing and able to supply a good, and the price they actually receive. Given the vertical demand curve, consumer surplus can't be estimated. Hence, we use producer surplus as a welfare measure. The producer surplus is calculated as the area between the supply curve below the equilibrium price and above the price axis. This is expressed as

$$producer\ surplus = \int_{0}^{p^*} \frac{a}{\left(1 + b \times \exp(-p \times c)\right)} \partial p,$$

where
p^* = market clearing price for optimal solution and
q^* = market clearing quantity for optimal solution.
The producer surplus is calculated for the five Models using MATLAB.

8.6 Results and discussion

The five models were set up in a GAMS environment with the estimated supply functions for the respective technology. We evaluate and compare the outcome of the models on the basis of the overall cost of compliance and social surplus.

The following charts present the model outcomes for the five Models considered for the two cases, that is, with equal RPO across states and unequal RPO across states.

Figures 8.8 and 8.9 show that the cost of compliance is low in the case of certificate mechanism with local market compared to certificate mechanism with the national market, and the same is true with the trade-of-electricity case. The local market has a low cost of compliance as compared to national market mechanism as the average per unit cost of compliance is more in the case of latter. This is because in the case of former, the market

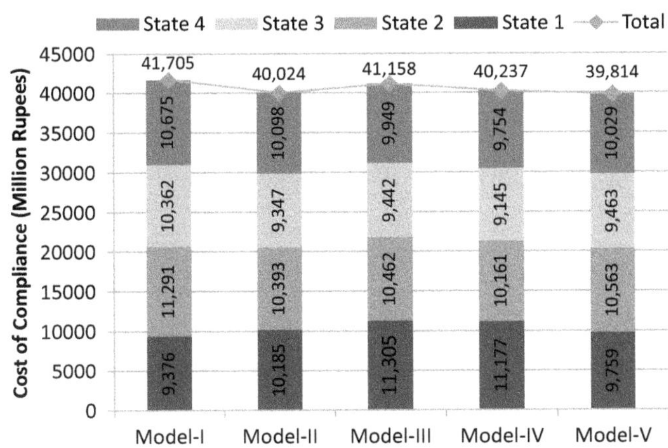

Figure 8.8 Cost of RPO Compliance (equal RPO)

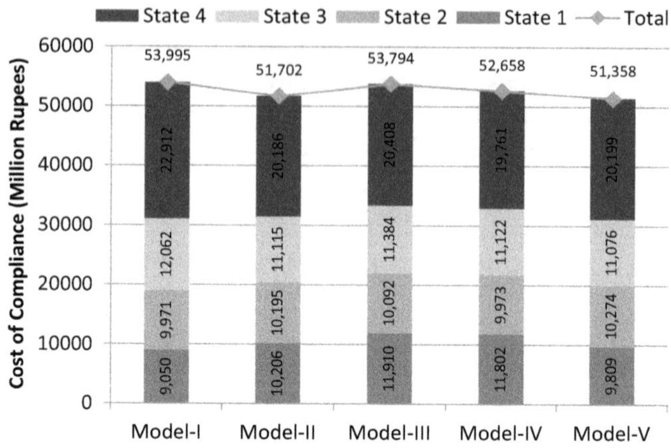

Figure 8.9 Cost of RPO Compliance (unequal RPO)

with cheaper resources is utilized to its maximum extent when compared to the national market mechanism.

The models for REC trade have a lower cost of compliance as compared to those with physical trade of electricity. This is expected as the former avoids both the transmission losses and the transmission charges. An interesting outcome is that amongst the three models of physical trade of electricity, the cost of compliance is found to be the lowest in the case of trade with state-specific markets. This is primarily due to the fact that the model with the national market observes a higher trade of RE amongst states (Figure 8.10), which leads to higher transmission loss and higher transmission

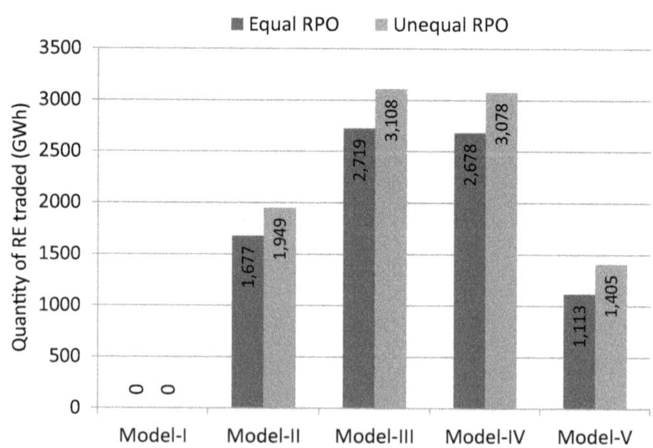

Figure 8.10 Cumulative RE Traded between States (equal and unequal RPO)

charges. Similarly, in the case of REC models (IV and V), the cost of compliance is a bit lower in the case of state-specific model. As discussed in the following, there also seems to be uneven distribution of benefits of trade across states, with some of the states with cheaper resources tend to lose due to trade as cost of compliance rises in such states as compared to a no-trade model.

The state-specific outcomes are a bit different in terms of cost of compliance. We can note from Figure 8.9 that the reason for higher overall cost of compliance in the case of national RE market (Model III) versus a state-specific RE market (Model III) is the distribution of benefits due to the national market. While cost of compliance for states 1, 2 and 3 are higher in the case of a national market, it is lower in the case of state 4 (Figure 8.9). In the case of unequal obligations, only state 2 marginally gains from a national RE market.

In the case of models for RECs, states 2, 3 and 4 gain from a common market while state 1 loses due to higher costs of compliance (Figure 8.9). In the case of unequal obligations, states 1 and 3 find their cost of compliance is higher in a national model while the other two reap the benefits of reduction (Figure 8.10). Due to the high penalty for shortfall in RPO, none of the models witnesses an RPO shortfall in any simulation.

The overall societal benefit across the alternative market design can help prioritize the potential candidates for market design. The final choice of market design would depend on the overall regulatory and policy environment as well as historical policy legacy and institutional setup that may tilt the favor in case of a particular choice.

We can observe from Figure 8.11 that producer surplus, used as a measure of societal welfare, is higher in the case of national markets, both in the case

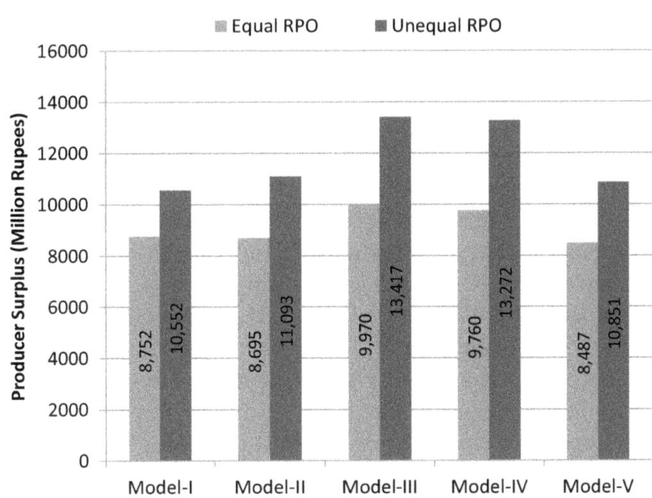

Figure 8.11 Producer Surplus (equal and unequal RPO)

of RE trade and REC trade, highlighting the role of a national-level market for RESs. Overall, we observe that a market for REC benefits the obligated entities as well as the society, however, there are differences between the national and state-specific markets. Given the concurrency of the REC market amidst RE-based power trading arrangements across buyers and sellers across states, the chapter also highlights the need for developing an RE based electricity contract that can be traded competitively on the country's power exchanges. Significant growth in solar energy development, driven by favourable economics, seems to have altered the supply side conditions and may require recalibration of the models for further research. A dynamic version of the model may be developed to evaluate medium- to long-term impact of alternate policy choices given a trajectory of RPOs.

8.7 Conclusion

The economic analysis of the energy policies is useful for the design of appropriate policies for the development of energy resources, including RE. Alternate policy choices, emerging with development of the REC market and competitive procurement of RESs, need evaluation not only in terms of the overall cost of compliance but also social welfare. For the deployment of RE technologies, it is advisable to allow for trade between states, which was absent in the case of the regulations designed by the respective SERC as per the Electricity Act 2003. This forced the obligated entities to procure RE from resources located within the state only, leading to an RPO shortfall in most cases. The result of the modeling simulations highlights the need for developing a competitive market for the trade of RE, which could be designed through RE-based electricity-trading contracts for the power exchanges in the country. This would also allow to capture the time variant value of RE as significant growth in solar energy generation would reduce its value to the power system as compared to wind energy which is generally available during later part of the day. REC mechanism enhances overall economic efficiency by prioritizing development of economic national RESs, reduces the overall cost of compliance and removes physical transmission bottlenecks as compared to the trade of electricity. While we have ignored the administrative costs associated with the REC market, we believe that these are not very significant, at least in comparison to transmission losses and the transmission charges. Further development of the models presented fhere should include more RE rich Indian states. Furthermore, an additional modeling approach can allow coexistence of the physical RE trade and the REC market as prevailing in India. This is an area of research extension in the future. The model is currently static in nature and does not explicitly account for dynamic nature of decisions and their impact across time. The generic modeling framework developed in the chapter can be applied in the context of a country or region and help choose between alternate RE policy choices and market design.

Notes

1 The significant reduction in the cost of harnessing solar energy has accelerated its growth in India. The target to achieve 100,000 MW of solar energy capacity by 2022 is being aggressively followed.
2 Amidst a lack of electricity access, the role of policy, as well as institutional framework, was highlighted to improve the adoption of "uneconomic" RE (Cust et al. (2007)). Furthermore, Singh (2009b) identified a role for new business models in addressing the electricity access gap in the "informal market".
3 The scheme was applicable to solar generators in the capacity range of 100 kW to 2 MW, and those connected to the grid at a voltage below 33 kV.
4 Adoption of competitive bidding, first for solar and now also for wind energy development, led to the lowest solar bid at Rs. 2.36/kWh in 2020 (compared to the previous low of Rs. 2.44/kWh in 2018). However, there is a wide range of site-specific solar bids across states due to differences in resource intensity, local policy and regulatory framework and risk factors influencing these costs.
5 Investment decision for RE projects leads to the creation of capacity and electricity generated from such projects cannot be guaranteed. To keep the modeling simple, we assume the supply function to be defined in terms of electricity generated, given a normative capacity utilization factor (CUF).
6 The state-specific RPO regulations prescribed these to be met only out of the RE electricity procured from within the state. However, the interstate trade of RE was permitted for sale to third-party or for captive consumption.
7 There are no specific penalties for the noncompliance of RPO till recentlky. However, SERCs may also use generic penalty provided for noncompliance with its directions/regulations as per the Electricity Act 2003.

References

CEA, *Executive Summary on Power Sector: March 2020*, Central Electricity Authority, New Delhi, (2020).

CERC. *Regulatory Commission (Terms and Conditions for Recognition and Issuance of Renewable Energy Certificate for Renewable Energy Generation) Regulations, 2010*, Central Electricity Regulatory Commission, New Delhi (2010a).

CERC. *Order on Determination of Forbearance and Floor Price for the REC framework*, Central Electricity Regulatory Commission, New Delhi (2010b).

Cust, J., Anoop Singh & Karsten Neuhoff. "Rural Electrification in India: Economic and Institutional aspects of Renewables," Working Papers EPRG 0730, Energy Policy Research Group, Cambridge Judge Business School, University of Cambridge 2007). Available at SSRN 2760810.

Government of India. *Electricity Act 2003*. Government of India, New Delhi. (2003).

Mendonça, M. *Feed-in Tariffs: Accelerating the Deployment of Renewable Energy*. London: Earthscan (2007).

Morthorst, P. "Capacity Development and Profitability of Wind Turbines", *Energy Policy* 27 (1999), 779–787.

Mozumdar P., and Achla Marathe. "Gains from an Integrated Market for Tradable Renewable Energy Credits." *Ecological Economics* 49 (2004): 259–272.

Nishio K., and H. Asano. "Supply Amount and Marginal Price of Renewable Electricity under the Renewables Portfolio Standard in Japan." *Energy Policy* 34 (2006): 2373–2387.

Olsina F., F. Garces, and H. J. Haubrich. "Modelling of Long-Term Dynamics of Electricity Markets." *Energy Policy* 34 (2006): 1411–1433.

Respective Regulations of Electricity Regulatory Commissions of Andhra Pradesh, Karnataka, Maharashtra and Tamil Nadu. Accessed from their respective website.

Seber, G. A. F., and C. J. Wild. *NonlinearRegression*. NJ: Wiley (2003).

Singh, Anoop. "A Market for Renewable Energy Credits in the Indian Power Sector." *Renewable and Sustainable Energy Reviews* 13 (2009a): 643–652.

Singh, Anoop. "Informal markets for electricity: economics of lighting for hawkers in India"; *International Journal of Energy Sector Management* 3 (2009b); 308–323.

Singh, Anoop (2010) "Economics, Regulation and Implementation Strategy for Renewable Energy Certificates in India," India Infrastructure Report (2010).

Part III

Environment and resources

9 Integrating natural and human factors for sustainable development in Himachal Pradesh

Anjan Sen and Nitin Punit

9.1 Introduction

In the contemporary decade, it has become a very contentious issue to decide the focus of development. Should the emphasis be on infrastructural development translated as welfare and well-being of humans, or is a better quality of environment preferred? Both issues are equally important for sustainable development. Besides, both human development and environment quality have a mutual cause-and-effect relationship. A high level of infrastructural development is generally believed to be the cause of environmental degradation and a declining rate of environmental quality and natural resources. To assess the well-being of a society, three basic points need to be emphasized: well-being of the ecosystem, well-being of the people and a healthy interaction between people and the ecosystem. By considering the three points together, each society can determine, which combination of human and ecosystem well-being is sustainable? Accordingly, they can make their own decisions about how to achieve them. Human well-being is a requirement of sustainability because no rational person would want to continue a low standard of living. Ecosystem well-being is a requirement because the ecosystem supports life and makes possible any standard of living. It may be necessary to turn forests and wetlands into farms and farms into towns. But people will not survive for long unless the ecosystem is healthy, productive and diverse. At the same time, it does not matter how well the ecosystem is if people cannot meet their needs. Ultimately, the well-being of both humans and the ecosystem are equally important, and a sustainable society needs to achieve both together. The research objectives of this chapter are as follow:

1. To prepare an ecosystem well-being index
2. To prepare a human well-being index
3. To formulate a composite well-being index on the basis of human and ecosystem well-being indices

The state of Himachal Pradesh has been selected as the area of study. The state comprises 12 districts—Bilaspur, Chamba, Hamirpur, Kangra, Kinnaur, Kullu, Lahul and Spiti, Mandi, Shimla, Sirmaur, Solan and

Una, which are further subdivided into tehsils and sub-tehsils. Despite its mountainous nature, the state shows diversity not only in terms of human dimensions but also in terms of environmental dimensions. On one hand, there is the Shimla district that can be termed the leading district of Himachal Pradesh because the district contains the state capital. The district manifests itself as the hub of development activities, which generates a gravitational pull to not only attract natural resources but also human resources from the neighboring districts. On the other hand, there is Lahul district and Spiti where the level of development is poor. Therefore, the state represents a wide disparity in levels of development from one district to another.

9.2 Tools for analysis

Since it is impossible to measure ecosystem or human well-being directly, their assessments include selected indicators. It is very essential to know at the start of an assessment of how well a given set of indicators represents a desirable combination of human and environmental conditions. Since indicators require the collection and analysis of often large amounts of data, choosing the wrong ones can be an expensive mistake (Allen, 2001). The selection of variables is a very significant stage that depends on certain considerations. The first consideration is the availability of data as it is very decisive, such that only those variables for which the data are readily and easily available should be selected. Besides, there should be consistent availability of data for the period of study. As both the concepts, environmental and human dimensions, are dynamic in nature, consistent availability of data is very fundamental. Consequently, it was decided first to take apart the concepts of human and environmental well-being to identify the features that need to be measured, and those that are both representative and measurable. Subsequently, two sets of variables were identified—one reflecting human dimensions and the other showing the quality of the environment. A total of 19 variables were selected and grouped into two sets: Quality of Life (QoL) and Quality of Environment (QoE).

The study represents a comprehensive picture of Himachal Pradesh; therefore, it includes a wide variety of data sources. First and foremost is the *Primary Census Abstract*, *Provisional Population Totals*, *Statistical Abstract*, *Compendium of Environmental Statistics*, *Human Development Reports* and various other reports relevant to the study have been referred.

9.2.1 Human indicators (Quality of life)

The indicators selected for assessing the quality of life in each of the 12 districts of the Himachal Pradesh are 10 in all. These 13 indicators are further subdivided into two groups—demographic, infrastructural, and environmental indicators.

9.2.1.1 *Demographic indicators*

1. Literacy Rate
2. Urbanization Rate
3. Unemployment Rate

9.2.1.2 *Infrastructural indicators*

1. Educational Facilities (Primary and Higher Secondary Schools)
2. Health Facilities (Primary and Community Health Centers)
3. Communication Facilities (Telephone Connections)
4. Electricity (Electrified Villages)
5. Banking Facilities (Credit Deposit Ratio)

9.2.2 Environmental indicators (Quality of environment)

To assess the quality of the environment, five indicators are selected. These indicators largely reflect the environmental quality whether upgraded or degraded:

1. Forest Land
2. Culturable Wasteland
3. Barren and Un-Culturable Land
4. Fallow Land
5. Fertilizer Consumption

9.3 Demographic indicators

9.3.1 Literacy rate (2011)

Data reveal that the total literacy rate has improved considerably in the state from 76.50% in 2001 to 83.78% in 2011. The male literacy rate has been recorded as 90.83% as compared to 85.30% in 2001. The data exhibit that females have shown remarkable progress in literacy, that is, an almost 10% increase as the female literacy rate is 76.60% as compared to 67.40% in 2001. At the district level, the literacy rate is very pleasant because all 12 districts show more than 70% of their population as literate. Hamirpur is the leading district, with 89.01% of literate population. The district-level data of literacy on all districts of the state are given in Table 9.1.

9.3.2 Urbanization rate (2011)

According to data, the status of Himachal Pradesh has barely changed in reference to the composition of the urban population to total population. The state comprises only 10.04% of its population as urban, compared

Table 9.1 Human Development Indicators for Composite Index, 2011

Districts	*Literacy (%)*	*Urban Population (%)*	*Vacancies (%)*	*Primary Schools %)*	*High & Secondary Schools (%)*	*Primary Health Centers (%)*	*Community Health Centers (%)*	*Telephone Connections (%)*	*Villages Electrified (%)*	*Credit-Deposit Ratio*
Bilaspur	85.67	6.58	0.19	6.00	6.00	7.00	8.00	5.00	100.00	36.32
Chamba	73.19	6.98	0.15	10.00	8.00	9.00	9.00	3.80	100.00	35.17
Hamirpur	89.01	6.91	0.13	5.00	6.00	6.00	6.00	8.80	100.00	20.63
Kangra	86.49	5.73	0.33	16.00	19.00	17.00	18.00	21.70	100.00	25.24
Kinnaur	80.77	0.00	18.97	2.00	2.00	5.00	5.00	1.90	90.60	41.19
Kullu	80.14	9.43	0.15	7.00	5.00	4.00	8.00	5.90	100.00	46.62
Lahul & Spiti	77.24	0.00	0.02	2.00	2.00	4.00	4.00	0.50	91.29	22.52
Mandi	82.81	6.27	0.15	16.00	16.00	13.00	16.00	13.40	99.68	39.00
Shimla	84.55	24.77	0.57	15.00	15.00	17.00	9.00	16.10	97.18	36.62
Sirmaur	79.98	10.8	2.84	9.00	7.00	8.00	4.00	5.40	100.00	89.52
Solan	85.02	17.7	2.77	7.00	7.00	7.00	6.00	10.30	100.00	76.60
Una	87.23	8.62	1.02	5.00	7.00	4.00	6.00	7.80	75.86	35.27

Source: Authors.

to 9.80% in 2001. Shimla, being the capital of the state, is still leading in terms of urban population, with 24.77% urbanized population. Solan is the second-leading district in 2001, however, the percentage of urban population declined from 18.22% in 2001 to 17.70% in 2011. Sirmaur is the third-largest district in urban population experiencing a very minimal increase from 10.39% in 2001 to 10.80% in 2011. Except for these three districts, all remaining districts show urban population less than 10% of total population of the district. Two districts—Kinnaur and Lahul and Spiti—do not have any urban population, and hence are wholly rural.

9.3.3 Unemployment rate (2011)

According to government statistics, the employment status is miserable in the state of Himachal Pradesh. It reveals that the total notified vacancies during 2010 were 6,732 against 825,764 applicants on the live register, indicating a success rate of 0.82%. At the district level, the situation is so miserable that eight districts of the state have a success rate of less than 1% of notified vacancies to the applicants on the live register. Kinnaur (18.97%), recorded the highest success rate of notified vacancies during the year. Sirmaur, Solan and Una are the other three districts to successfully notify more than 1% vacancies with 2.84%, 2.77% and 1.02%, respectively.

9.4 Infrastructural indicators

9.4.1 Educational facilities (2010–2011)

The total number of schools has increased in the state on an average as compared to 2006–2007. In 2010–2011, the state of Himachal Pradesh has shown an increase in all categories—primary, middle and high senior secondary schools. In 2010–2011, the total number of primary schools grew to 10,767 as compared to 10,607 in 2006–2007. So the state has witnessed an increase of almost 1.5% in five years, with respect to primary schools. On the other hand, the number of high senior secondary schools has increased from 1,850 in 2006–2007 to 2,094 in 2010–2011, which exhibits an increase of more than 13%. The district-level data on educational facilities for all districts are given in Table 9.1. At the district level, the highest percentage of primary schools have been recorded in Kangra and Mandi districts, both with 16%; however, the absolute number of primary schools is higher in Kangra than in Mandi. Kinnaur and Lahul and Spiti are two districts with the least number of primary schools, as both stand with only 2% each of total number of primary schools in the state.

9.4.2 Health facilities (2010–2011)

The total number of medical institutions has increased in the state from 2006–2007 to 2010–2011. The medical institutions—community health centers (CHCs) and primary health centers (PHCs)—those that have been selected for the study, have shown an increasing trend. In 2006–2007, the total number of CHCs was 71 in the state which has increased to 77, showing an increase of more than 8% in just five years. On the other hand, the number of PHCs has also increased, but it is just 2% as the PHCs have increased from 443 in 2006–2007 to 435 in 2010–2011. District-level data are given in Table 9.1. At the district level, the highest numbers of CHCs are found in the Kanga district. Kangra district alone consists 18% of the total CHCs of the state in 2010–2011. Mandi is the second-largest district of the state in terms of CHCs, as almost 16% of the total CHCs in the state are lying in this district. Lahul and Spiti and Sirmaur stand last, as both districts show the lowest figure in terms of CHCs. With respect to PHCs in 2010–2011, Kangra is still leading, with 17% of the total strength of PHCs of the state; however, Shimla is another leading district with the same number of PHCs. Kullu, Lahul and Spiti, and Una show the lowest percentage of PHC.

9.4.3 Communication network (Telephone connections)

The total number of households using the telephone facilities has increased in the state from 204,684 in 2000–2001 to 1,017,292 in 2010–2011. It indicates that the communication facilities have improved, as the figure accounts for an almost fivefold increase in the decade. At the district level, Kangra is the leading district as almost 23% of the total number of households using telephone connections in Himachal Pradesh. Mandi is the second-largest district, with 14% households of the state using telephone facility. The districts of Lahul and Spiti (0.5%) and Kinnaur (1%) are at the bottom as in both the districts very meager shares of households are using telephones.

9.4.4 Electricity

The government statistics on electrified villages reveal that the state has achieved a lot by providing electricity to almost all villages in 2006–2007. The total number of villages electrified in the Himachal Pradesh is 17,169 out of 17,495, which accounts for a remarkable 98.14%. There are six districts in the state where 100% of the inhabited villages were electrified in 2006–2007. These districts are Bilaspur, Hamirpur, Kangra, Kullu, Sirmaur and Solan. Out of the remaining six districts, two are very close to 100%, namely, Mandi (99.58%) and Chamba (98.93%). Una district comes at the bottom, with only 75.33% villages with electricity.

9.4.5 Banking facilities (Credit–deposit ratio)

The banking sector is also considered an important indicator of development. It highlights the saving capacity of the people. As in the previous section, here, too, the credit–deposit ratio has been used to observe the economic status of Himachal Pradesh, in general, and all 12 districts of the state, in particular. The credit–deposit ratio in the state of Himachal Pradesh has jumped from 22.00 in 2000–2001 to 39.86 in 2010. It is an achievement that the credit–deposit ratio in the state has shot up by 18% in just a decade. The district-level numbers (displayed in Table 9.2) indicates that Sirmaur is the leading district, with credit–deposit ratio of 89.52 followed by Solan (76.60) as both districts have interchanged their ranking of 2000–2001. Hamirpur (20.63) is observed as the district with the lowest credit–deposit ratio.

9.5 Environmental indicators

9.5.1 Forest land

The state figure shows that 1,083,100 ha or 23.89% of land in 2007–2008 was devoted to forest. It reveals that the forest cover has declined in the state as compared to 1999–2000. Data given in Table 9.3 reveal that Mandi district still ranks first as it had more than 44% of its geographical area to forest cover in 1999–2000. Kangra is the second-largest district with 40.24%, and Hamirpur is still leading in this category.

9.5.2 Culturable wasteland

In 2007–2008, the culturable wasteland of the state of Himachal Pradesh is 123,200 ha that account for 2.80% of state's geographical area. Una is the largest district in terms of percentage of culturable wasteland of 14.97% in 2007–2008 as compared to 13.23% in 1999–2000. Hamirpur (11.07%) is the second-highest in terms of culturable wasteland in 2007–2008. Lahul and Spiti is the lowest ranked district in terms of culturable wasteland, with only 0.06%.

9.5.3 Barren and un-culturable land

Barren and un-culturable land has decreased in the state of Himachal Pradesh from 856,900 ha in 1999–2000 to 657,500 ha in 2007–2008. This decrease in the extent of barren and un-culturable land has resulted in a decline in percentage of this land from 18.91 in 1999–2000 to 14.98 in 2007–2008. At the district level, Lahul and Spiti still ranks as the largest district, but the percentage has declined from 60% to 45.63% in 2007–2008. The percentage of barren and un-culturable land has also declined in the Kinnaur district from 22.81 to 21.36, but it still ranks as the second-largest district. Hamirpur and Una have interchanged their previous decade ranks, as the percentage of barren and un-culturable land has shown a considerable fall in Hamirpur from 16.98% to 12.34%, allowing Una to become the third-largest district, with 14.33% of barren and un-culturable land.

Table 9.2 Human Indicator Ranking (composite index), 2011

Districts	*Literacy (%)*	*Urban Population (%).*	*Vacancies (%)*	*Primary Schools %)*	*High/Sr. Schools (%)*	*Primary Health Centers (%)*	*Community Health Centers (%)*	*Telephone Connections (%)*	*Villages Electrified (%)*	*Credit-Deposit Ratio*	*Total*	*Total/10*	*Final Ranking (composite index)*
Bilaspur	9	4	4	5	5	3.5	4.5	4	5.17	6	50.17	5.017	5
Chamba	1	6	2.33	9	9	6	6.5	3	5.17	4	52.00	5.200	6
Hamirpur	12	5	2	3	3	3	2.33	8	5.17	1	44.50	4.450	3
Kangra	10	2	5	12	12	7.5	8	12	5.17	3	76.67	7.667	11
Kinnaur	5	0.5	10	1	1	2	2	2	2	9	34.50	3.450	2
Kullu	4	8	2.33	6	6	0.33	4.5	6	5.17	10	52.33	5.233	7
Lahul & Spiti	2	0.5	1	2	2	0.33	0.5	1	3	2	14.33	1.433	1
Mandi	6	3	2.33	11	11	7	7	10	5	8	70.33	7.033	9
Shimla	7	11	6	10	10	7.5	5.5	11	4	7	79.00	7.900	12
Sirmaur	3	9	9	8	8	5	0.5	5	5.17	12	64.67	6.467	8
Solan	8	10	8	7	7	3.5	2.33	9	5.17	11	71.00	7.100	10
Una	11	7	7	4	4	0.33	2.33	7	1	5	48.66	4.866	4

Source: Authors.

Table 9.3 Environmental Indicators for Composite Index, 2011

Districts	*Forest Land (%) out of Total Land*	*Culturable Wasteland (%) out of Total Land*	*Barren and Un-Culturable Land (%) out of Total Land*	*Current Fallow Land (%) out of Total Land*	*Fertilizer Consumption (%) of Whole State*
Bilaspur	12.52	5.54	3.94	5.59	4.15
Chamba	39.28	0.92	0.77	5.28	2.39
Hamirpur	16.60	11.07	12.34	13.69	4.70
Kangra	40.24	5.03	2.53	8.55	16.22
Kinnaur	6.00	0.56	21.36	21.52	0.46
Kullu	NA	4.38	2.19	8.60	9.16
Lahul & Spiti	14.85	0.06	45.63	NA	0.74
Mandi	44.04	1.13	5.28	11.34	12.05
Shimla	25.62	4.16	3.09	19.41	22.63
Sirmaur	21.48	5.56	3.78	12.84	6.47
Solan	1.89	6.63	9.00	15.15	7.37
Una	5.29	14.97	14.33	9.46	13.65

Source: Authors.

9.5.4 *Fallow land*

Fallow land in the state of Himachal Pradesh has increased from 71,900 ha in 1999–2000 to 72,500 ha in 2007–2008. In 1999–2000, it was 1.58% of total geographical area, which in 2007–2008 increased to 1.65%, showing an increase of 0.7%. The district-level data of the state show a variation similar to what it did in 1999–2000; however, the values have changed. In 2007–2008, Kullu district is the largest under the fallow land category, with 5.85%. Hamirpur is the second-largest district with 5.80%, showing a decline from 7.17% of 1999–2000. Kangra with 1.93% of fallow land is followed by Chamba (0.43%), Kinnaur (0.28%), and Lahul and Spiti (0.01%), the three low-category districts in terms of fallow land.

9.5.5 *Fertilizer consumption (2010–2011)*

The total fertilizer consumption has increased in the state from 35,552 metric tonnes in 2000–2001 to 55,133 metric tonnes in 2010–2011, which accounts for an increase of more than 55% in fertilizer consumption from the previous decade. At the district level, Shimla shows the largest percentage consumption of fertilizers as it consumed 22.63% of total fertilizers consumed in the state. Kangra is the second-largest district following Shimla. Kinnaur and Lahul and Spiti are still termed the two lowest ranked districts with respect to the percentage of fertilizer consumption (Table 9.4).

Table 9.4 Environmental Indicators Ranking (composite index), 2011

Districts	*Forest Land (%) out of Total y*	*Culturable Wasteland (%) out of Total Land*	*Barren and Un-Culturable Land (%) out of Total Land*	*Current Fallow Land (%) out of Total Land*	*Fertilizer Consumption (%) of Whole State*	*Total*	*Total/5*	*Final Ranking (composite index)*
Bilaspur	5	8	7	2	9	31	6.20	3.5
Chamba	10	3	12	1	10	36	7.20	8.0
Hamirpur	7	11	4	8	8	38	7.60	9.0
Kangra	11	7	10	3	2	33	6.60	6.0
Kinnaur	4	2	2	11	12	31	6.20	3.5
Kullu	1	6	11	4	5	27	5.40	3.0
Lahul & Spiti	6	1	1		11	19	4.75	1.0
Mandi	12	4	6	6	4	32	6.40	4.5
Shimla	9	5	9	10	1	34	6.80	7.0
Sirmaur	8	9	8	7	7	39	7.80	10.0
Solan	2	10	5	9	6	32	6.40	4.5
Una	3	12	3	5	3	26	5.20	2.0

Source: Authors.

9.6 Composite index

The statistical technique that is used to analyze the data is the composite index method (Ramachandran, 2000). The data of human development and the environment quality used in construction are applied for the composite index. The technique is very simple, based on the rank allotted to each district with reference to the ranks obtained by the other districts in reference to each indicator. The total number of districts was 12; therefore, the ranks range from 1 to 12.

The district that scored the lowest is ranked 1. If two or more districts show the same value, the same rank is given to each of them then averaged by dividing by the number of districts having the same rank. The rank allotted to the next district in order of value then is given on the basis of the number of ranks already allotted. The same method of ranking is applied to each indicator, following which a composite index is obtained by adding the ranks scored by each district for all the indicators.

9.7 Results and discussion

The analysis incorporates critical and strategic indicators of human development, for which the data are available consistently. Finally, all relevant variables were selected to represent the level of human development and the quality of the environment. Subsequently, the composite index for both, human development as well as the environmental quality, was determined using the composite index method. On the basis of this technique, ranks were assigned to all the 12 districts of Himachal Pradesh in reference to human indicators as well as environmental indicators (Table 9.5). Finally, the bivariate analysis of both indicators resulted in 9 different typologies to classify 12 districts of Himachal Pradesh into 9 different

Table 9.5 Composite Index Ranking of Himachal Districts for 2010–2011

Districts	*Human Index*	*Environmental Index*
Bilaspur	5.0	3.5
Chamba	6.0	8.0
Hamirpur	3.0	9.0
Kangra	11.0	6.0
Kinnaur	2.0	3.5
Kullu	7.0	3.0
Lahul and Spiti	1.0	1.0
Mandi	9.0	4.5
Shimla	12.0	7.0
Sirmaur	8.0	10.0
Solan	10.0	4.5
Una	4.0	2.0

Source: Authors.

combinations of levels of human development and environmental degradation (Table 9.6).

Shimla represents the leading district with high human development and a high level of environmental degradation. The districts of Kangra, Mandi and Solan also fall in the category of high human development and a medium level of environmental degradation. There is no district in the category of high human development and low level of environmental degradation. The category of medium human development and a high level of environmental degradation is represented by three districts, namely, Sirmaur, Kullu and Bilaspur. Chamba scored medium in terms of both dimensions—human development and environmental degradation. Una, Kinnaur, Lahul and Spiti ranked under the category of low human development and a high level of environmental degradation. Finally, Hamirpur is the only district with low human development and a low level of environmental degradation.

The debate between the proponents of infrastructural development, on one hand, and the supporters of preserving the environment, on the other, is very long and extensive. Both issues are equally important for sustainable and holistic development. Besides, both human development and environmental quality have a mutual cause-and-effect relationship. A high level of human or infrastructural development is generally believed to be the cause for equally high level of environmental degradation and a declining rate of environmental quality and natural resources. This work attempts to contribute to this debate with an empirical analysis of district-level data of Himachal Pradesh. However, this is a simple pilot study, based on basic secondary data. A more detailed quantitative research based on current and primary data is highly warranted to bring out more details.

Table 9.6 Typology of Districts Based on Composite Index

2010–2011		*Environmental Degradation*		
		High	*Medium*	*Low*
Human Development	High	Shimla	Kangra Mandi Solan	X
	Medium	Sirmaur Kullu Bilaspur	Chamba	X
	Low	Una Kinnaur Lahul & Spiti	X	Hamirpur

Source: Authors.

References

Allen, R. P. *The Wellbeing of Nations: A Country by Country Index of Quality of Life and The Environment*, Washington, DC: Island Press (2001).

Ramachandran, N. *Monitoring Sustainability: Indices and Techniques of Analysis*, New Delhi: Concept Publishing Company (2000).

10 An input–output approach to study environmental impact

Aparna Mishra and Kaushal Kumar Saxena

10.1 Introduction

Last few decades have witnessed a quantum jump in the growth of the industries that has led to high economic growth as well as created severe environmental issues. As a result, the stress on natural resources has been on the rise, and the quality of air and water are seriously affected. The water-polluting industries always required special attention and have been a cause of concern. Most of the water pollution is generated in the form of industrial effluents that contains solid wastes, as well as other hazardous material. Unfortunately, the industrial waste is either left untreated or not treated fully, and consequently, the level of toxins has increased over time. Surface water is the only medium for industrial water disposal. From time to time, there have been multiple studies on the impact of rising water pollution on the environment. A close examination of the literature suggests that a majority of these studies have taken into account only the direct pollution effect that is generated in the process of production. The indirect effect generated during the process of production has been utterly ignored in the literature of the environmental policy. This chapter explains how total pollution effects are decomposed into direct pollution and indirect pollution effects using input–output (I-O) tables.

In a particular industry sector, direct pollution indicates the generation of pollutants per unit of output whereby production takes place directly and indirect effect implies for the generation of pollutants whose output has been used as an input in the production process of a particular sector. For example, if we increase the paper and pulp production, then the level of pollution is affected by the increased production of paper and pulp. The pollution is also affected indirectly due to the industry sectors whose inputs have been used in producing the paper and pulp. There is an interdependence found among the producing sectors. In the preceding example, to produce paper and pulp several other inputs are required from the agriculture sector, coal, electricity, engineering products and so on. To produce these inputs, further inputs are required from other sectors and the process continues.

In summary, if production in one sector is increased due to which pollution increases, the level of output of other sectors is also increased, and that further adds to the pollution in an indirect manner. Since pollution is directly related with the output, the overall magnitude of pollution, no longer remains the same as that of direct pollution.

The entire process can be explained in a simple terminology. One industry depends on other industries for its intermediate product requirements. There is an inevitable flow of resources in the form of goods and services where one sector supplies to another sector of the economy. An interdependence is found among different producing and consuming units. It is not possible to change the output of one sector without affecting the output of another sector. Environmental pollution is also generated in a similar manner where changing output in one sector impacts environment in a direct and indirect manner. To understand the right magnitude of the pollution, it is important that total effects are decomposed and calculated from direct plus indirect effects. The environmental quality therefore is not just the direct effect of the production but also the indirect effect, and both together are called "total effect".

In an Indian context, highly polluting industries are segregated together where sectors having a high discharge of pollutants are treated uniformly. The indirect pollution generated and caused due to interlinkages of the industries is completely ignored and left unaccounted for. A careful review of the literature suggests that the significant portion of the Indian studies have been at an aggregate level that results in only high-level insights. There have been minimal attempts to calculate the total pollution impacts, due to interlinkages of the industries, that could be decomposed into direct plus indirect impact.

The present study elaborates an approach to identify the interlinkages of the sectors and calculate the water pollution intensity. For this purpose, I-O tables published by planning commission of the government of India, have been used, and the open static I-O model, pioneered by Leontief, has been applied.

The interdependence among the producing sectors of the Indian economy is captured in the form of a simplified model of production. The production in any sector is contingent on two factors—first, the demand that is sourced due to consumers and households and, second, the input requirement generated from different industry verticals that use the output of that sector as an input. Pollution is nothing but a by-product that is generated in the process of production and meant for satisfying the demand of consumers and different industrial sectors. Therefore, the level of pollution is directly proportional to the industrial production and output. A change in the output level of pollutants can be attributed to the change in the level of final demand from consumers and to the change in the technological structure of industrial sectors where demand for raw material is generated. In summary, a combination of the change in final demand and industry demand leads to a dynamic situation where the level of industrial pollution changes with the production. An I-O model of generalized nature is a unique technique through which extra rows and columns are added to represent the generation and abatement of pollution.

The I-O model introduced by Leontief (1970) and Leontief and Ford (1972) is the kind of model that has been used for demonstrating the environmental impact in the Indian industry context.

10.2 The input-output approach

The I-O approach is a unique approach that can capture intersectoral relationships and can identify the interdependence of different producing and consuming units in an economy. These models are originated from Tableau Economique of Quensey (1758), providing diagrammatic depiction whereby expenditures are traced in a systematic manner for an economy. Although the idea was given, but the systematic modern approach was developed by Leon Walras in 1877. The Walrasian system introduced, a system of equations to explain the interdependence among productive sectors of the economy. The intent was to highlight the relationships for consumer income and expenditure, production cost in each sector, total demand/supply of commodities and factors of production (Kuenne, 1954). The idea behind the mathematical formulation was to explain the existence of a determinate solution for the quantities and prices in the system under the assumption of maximizing behavior of economic agents.

Walrasian model, on one hand, had very few economic realities to offer and, on the other hand, was too complicated for empirical verification. As a result, the gap between non-operational theory and inadequate empirical research became wider. The Leontief (1936) approach to I-O analysis proposed to eliminate this gap (Leontief, 1986). The conception of I-O analysis led to the practical solution of the problem created by the general equilibrium theorists. In Leontief's conception, statistical data were used in hypothetical production and consumption equations in the form of explicit multisectoral producing and consuming units. The systematic I-O approach was applied on the American economy after the Great Depression of the 1930s.

Leontief's I-O technique offers a quantitative approach to analyze the interdependence among various producing as well as final consuming sectors of the economy. The information for the development of I-O model is acquired from the interindustry table, which is a table of I-O transactions for various sectors. In the next section, such tables are discussed in detail.

10.3 Input–output tables and the fundamental relationships

I-O tables in an economy are constructed for an accounting period to explain the flow of goods and services from one industrial productive sector to another. For that purpose, the whole economy is segregated into finer sectors where each sector has a dual role. The first role for individual sectors is defined as a "producer of output" that relates to the generation of actual production in an economy. The second role is defined as a "user of inputs" that calls for the usage of different inputs (from other sectors) and its utilization in the process of production.

The I-O table formulation provides the system of interdependence of various sectors of the economy by means of a two-way table. The row elements of the table signify the output distribution among producing and final demand sectors. The column entries indicate the composition of intermediate inputs (depicted by the values in the row vector) and primary inputs. Primary inputs such as land, labor and capital, among others, are primary in the sense that these are determined by factors outside the system and supplied by the value addition sectors. These factors are utilized by the industries in the process of production. Originally, the I-O table was constructed in physical terms, but to avoid the anomalies associated with the different measuring units of the sectors, it is expressed in value terms. The representative structure of an I-O table is given in Table 10.1.

As already mentioned, that the output of a sector is either consumed as an intermediate use in the process of production, or it is consumed by the final demand category. Therefore, the output distribution among intermediary sectors and final demand sectors and distribution between produced and primary inputs lead to a transaction system that could be divided into four quadrants. These quadrants are marked in Table 10.1.

The first part indicates transactions for final demand where the final use of products and services takes place. The total final use of a product can be decomposed into various categories such as investment, private consumption, government consumption and exports. A major portion of the gross national product is absorbed by the final demand quadrant. The sectoral deliveries to final demand are shown as the output of sector i delivered to the kth category of final expenditures, denoted as Y_{ik}.

Table 10.1 Input–Output Table Structure

	Purchasing Sectors	
	Intermediate Use	*Final Demand*
	Sectors *i…j…n*	
1. Producing sectors i.n	QUADRANT II (X_{ij})	QUADRANT I (Y_{ik})
Primary Inputs (value added) Total production	QUADRANT III	QUADRANT IV

Source: Chenery and Clark (1965).

The second category is the most important part of the interindustry transactions where each sector in the transaction table appears twice, as a producing and as a purchasing sector. Thus, the resulting matrix would be a square matrix. Each element X_{ij} of this matrix denotes the quantity of product *i* used by sector *j* at constant prices. The total intermediate demand for product *i* can be shown by W_i and U_j records.

The third part of the table indicates the value addition in different sectors. These are the primary inputs in the sense that they are determined outside the system such as land, labor, and capital. The payment to the primary sector can be calculated by subtracting the input cost with the value of output, incurred in a sector. In this sense, it is produced outside the system. V_j denotes the value addition generated in sector *j*.

Finally, the fourth part contains the primary factors put into the final use under the categories of government employment and domestic service. Although it does not enter as a regular transaction in inter-industry modeling, it is important to maintain the consistency in regular I-O table. The last column comprises the total supply of a product equal to the domestic production of sector *i* plus the imports of commodity *i*.

The previously mentioned I-O formulation leads to two balance equations. First, the rows of the I-O tables state that the total supply and total demand are always equal where demand can be split into intermediate demand and final demand. In this way, the sum total of sector i output represented by $X_{i,}$ is the summation of all deliveries (sector *i*) to all producing sectors and final users. Thus,

$$Zi = Mi + Xi = \sum\nolimits_j Xij + Yi = Wi + Yi, \tag{10.1}$$

where Z_i in equation (10.1) is the total supply of commodity *i*, M_i is the import of commodity *i* and Y_i, the final demand for commodity *i*. The second equation is derived from the column entries of the transaction matrix and describes the input structure of various sectors. The total production for a sector is composed of purchases equivalent to the value of inputs from other sectors plus generated value addition within that sector:

$$Xj = \Sigma Xij + Vj = Uj + Vj\left(j = 1,\ldots,n\right), \tag{10.2}$$

where V_j, in equation (10.2) denotes primary inputs total usage in sector *j* and U_j is the total use by sector *j* of inputs purchased from other industries ($\sum X_{ij}$).

In this way, the final demand vector appears to be a difference between the total production and intermediate use of the product, and value addition becomes the difference between the total value and payments for inputs purchased from the rest of the sectors. We can write

$$\sum Xij + Yi = \sum Xij + Vj. \tag{10.3}$$

This relation also holds for the sum total of all intermediate products, considering all sectors at a time.

$$\sum_{i=1}^{n}\sum_{j=1}^{n} Xij + \sum_{i=1}^{n}\sum_{j=1}^{n} Yik = \sum_{j=1}^{n}\sum_{i=1}^{n} Xij + \sum_{j=1}^{n} Vj. \tag{10.4}$$

It follows from equation (10.3) that the sum total of value addition is equivalent to the sum total of all deliveries to final users. That is,

$$\sum_{j=1}^{n} Vj = \sum_{i=1}^{n}\sum_{j=1}^{n} Yik \tag{10.5}$$

This is the most important result of the I-O technique. With the preceding formulation, the relationship between I-O accounts and national income aggregates can also be established.

10.3.1 Basic input–output model

The I-O model is intended to provide a basic framework to analyze the production process of different sectors of the economy. These production processes, in turn, are governed by the technological characteristics of the economy. Under a certain set of assumptions, it is possible to write a technical coefficient matrix. This matrix would exhibit the technology of the production system. The individual elements in the technical matrix illustrate the amount of input requirement from each sector in the process of production.

Now it is possible to write an equation for demand X_{ij} of each industry j, for each commodity i as a function of its own level of output X_j:

$$Xij = \bar{X} + aijXj, \tag{10.6}$$

where a_{ij} in equation (10.6) is the input coefficient. The first term indicates any fixed cost element, which does not vary with the level of production. If the first term is zero, then only second term appears in the demand function. If it is not zero, then it becomes the part of final demand and only second term appears. To produce a unit of *j*th product, the input requirement for the *i*th product must be a fixed amount denoted by a_{ij}. Thus,

$$Aij = \frac{Xij}{Xj}. \tag{10.7}$$

The relationship in equation (10.7) illustrates that for producing each unit of the *j*th product, an a_{ij} amount of the first product and a_{nj} of the nth product is required. By the same definition, the level of technology can be defined for an "n-sector" economy. The input coefficients compose of matrix $A = [a_{ij}]$, where each column entry signifies the input requirement for producing a unit of output for an industry sector. Now the technical coefficients matrix $[A]$, can be written as shown in Table 10.2.

The total output for an industry sector is split into two parts, intermediate demand and final demand. By intermediate demand, we mean the use of the commodity to produce other commodities, and final demand is the amount that is used for nonproductive purposes, such as consumption, stock building, export and so forth. Now the demand function for n industries can be written as

$$
\begin{aligned}
X_1 &= X_{11} + X_{12} + \ldots + X_{1n} + F_1 \\
X_2 &= X_{21} + X_{22} + \ldots + X_{2n} + F_2 \\
X_n &= X_{n1} + X_{n2} + \ldots + X_{mn} + F_n.
\end{aligned} \tag{10.8}
$$

From equation (10.7) we can substitute the corresponding values of $a_{ij}X_j$ in equation (10.8). Thus,

$$
\begin{aligned}
X_1 &= a_{11}X_{11} + a_{12}X_{12} + \ldots + a_{1n}X_{1n} \\
X_2 &= a_{21}X_{11} + a_{22}X_{22} + \ldots + a_{2n}X_{2n} \\
X_n &= a_{n1}X_{n1} + a_{n2}X_{n2} + \ldots + a_{mn}X_n;
\end{aligned} \tag{10.9}
$$

$$X = AX + F, \tag{10.10}$$

where $X = (n \times 1)$ vector of gross output, $A = (n \times n)$ matrix of technical coefficients and $F = (n \times 1)$ vector of final demand.

Table 10.2 Mathematical Representation of Input–Output Table

INPUT	*OUTPUT*			
	1 2 3……n			
1	a_{11}	a_{12}	$a_{13}\cdots a_{1n}$	
2	a_{21}	a_{21}	$a_{21}\cdots a_{2n}$	
3	a_{31}	a_{31}	$a_{31}\cdots a_{3n}$	
	.	.	.	.
	.	.	.	.
	.	.	.	..
	.	.	.	.
⋮	.	.	.	.
	.	.	.	.
n	a_{n1}	a_{n2}	a_{n3}	a_{nn}

Source: Leontief (1936, 1974).

By solving equation (10.10), we get

$$X = (1 - A)^{-1} \tag{10.11}$$

or

$$X = BF, \tag{10.12}$$

where $B = (1 - A)^{-1}$.

Equation (10.12) establishes the relationship between the final demand and total output. This equation is used to determine the level of output for meeting the final demand of goods and services, and it also captures a change that is required to support all the producing activities involved in providing these final goods. The inverse matrix $(1 - A)^{-1}$ of equation (10.11), popularly known as "Leontief inverse", calculates the direct plus indirect demand of industry i generated by a unit of final demand of industry j. Any change in final demand brings a simultaneous change in all the producing sectors of the economy. The Leontief inverse attempts to calculate all such changes in an economy where there is a change in demand. This inverse can also be used to calculate the effect on primary inputs, such as labor and employment. Recently, it has been applied in the diversified area for the analysis of total change in an economy.

10.4 Environmental studies under input–output framework

The interrelations among economic units and sectors lead to a multidirectional flow of inputs and outputs. The environment provides all the resources for economic activities, and in return, the residual wastes are released back to the environment (see Muller, 1979). Therefore, a two-way physical flow of materials between environment and economy is found. The generation of waste is a normal and inherent part of the production and consumption processes. Ayres and Kneese (1969) consider them as externalities that provide disservices to the producers as well as the consumers. Various environmental media, such as air and water, have been used for the discharge of these unwanted residuals.

The flow of material between producing as well as final consuming units can very well be traced both in terms of value and physical units. However, the material flow from and back to the environment is seldom reflected completely, neither in value nor in physical terms. The inter-industry flow models provide a very convincing tool for studying these interactions and for projecting future residual production.

In recent years, considerable research has been performed on extending the Leontief I-O table. Such extensions mainly capture the factors associated with inter-industry transactions in the economy. With the environment being a focal point of all activities, it has been very well studied under these inter-industry transactions.

The systematic approach to the environment under an I-O framework is due to two seminal articles by Leontief and Ford (1970, 1974) and Leontief and Ford (1972). Later, Miller and Blair (1985) classified the environmental I-O models into three categories:

1. The generalized models of I-O whereby the regular I-O table is extended in the form of extra rows and columns to represent the generation and abatement of pollutants, respectively (Leontief 1970; Leontief and Ford 1972)
2. Economic-ecologic models to integrate ecological commodities or residuals from production and consumption processes, as an inputs. The augmentation to basic I-O framework is provided on certain key performance indicators of ecosystem sub-matrices that flows between economic and ecological system rather than the flows in an interregional I-O model (Daly, 1968; Isard et al., 1968; James et al., 1978).
3. Commodity by industry models, whereby additional rows and columns are introduced to represent the ecological inputs and output, respectively (Victor, 1972)

Because of the exhaustive data requirement involved with ecological models and commodity-by-industry models, most empiricists apply the generalized I-O models of the Leontief type (Kohn, 1975; Cumberland and Stram, 1976; Herendeen, 1978). Flick (1974), Steenge (1978), Lowe (1979), Lee (1982), Rhee and Miranowski (1984), Qayum (1991), Arrous (1994), Luptacik and Böhm (1994) and many others followed the Leontief framework and reformulated the model in different ways. For empirical analysis and research, the Leontief's I-O model has a wider scope.

10.4.1 Leontief's model of environmental repercussions

In an economy, the I-O table traces the intermediate demand from different industry sectors as well as final demand from consumers and households. Pollution generation is a normal and inherent part of production and consumption processes. The level of pollution depends on the technological structure of the economy. Technological characteristics are explained by the structural coefficients of the conventional sectors. To determine the environmental repercussions these structural coefficients have been incorporated in the regular I-O table of the economy.

Leontief contributed to the basic environmental I-O model in two ways. First, he explicitly introduced the involuntary by-products (pollutants) of the regular economic activities and combined them with the output coefficients of the conventional sector of the economy. Second, antipollution sectors were combined with the conventional sectors, and the pollutants of the conventional sectors were supposed to be treated in this antipollution sector. In this way, the amount of emitted pollutants was endogenized. The basic I-O model with the aforesaid features has the following form:

$$\begin{matrix} A_{11} & A_{12} \\ A_{21} & A_{22} \\ v_1 v_2 \ldots . v_m & v_{m+1} \ldots . v_n \end{matrix} = \begin{bmatrix} a_{11}a_{12} & \cdots & a_{1m} & a_{1m+1} & \cdots & a_{1n} \\ a_{21}a_{22} & \cdots & a_{2m} & a_{2m+1} & \cdots & a_{2n} \\ \vdots & \cdots & \vdots & \vdots & \cdots & \vdots \\ a_{m1}a_{m2} & \cdots & a_{mm} & a_{mm+1} & \cdots & a_{mn} \\ a_{m+11}a_{m+12} & \cdots & a_{m+1m} & a_{1m+1} & \cdots & a_{m+1n} \\ a_{m+21} & \cdots & \vdots & \vdots & \cdots & \vdots \\ \vdots & \cdots & \vdots & \vdots & \cdots & \vdots \\ a_{n1}a_{n2} & \cdots & a_{nm} & a_{nm+1} & \cdots & a_{nn} \\ v_1 v_2 & \cdots & v_m & v_{m+1} & \cdots & v_n \end{bmatrix}$$

In the preceding equation, A_{11}, A_{12}, A_{21} and A_{22} are the key partitioned matrices. The right side of the equation expands and explains the contents of the key elements and matrices. Now, taking the $(1 - A)^{-1}$ of the partitioned matrix, we get the following:

$$\begin{pmatrix} I - A_{11} & -A_{12} \\ -A_{21} & I - A_{22} \end{pmatrix} \begin{pmatrix} x_1 \\ x_2 \end{pmatrix} = \begin{pmatrix} y_1 \\ y_2 \end{pmatrix} \tag{10.13}$$

The n-dimensional vector of gross industrial outputs is depicted by x_1; x_2 depicts the k-dimensional vector of antipollution activity levels; A_{11} an $(n \times n)$ matrix comprising inter-industry a_{ij} coefficients, which shows inputs for commodity i for per unit of output of commodity j (produced by sector j); A_{12} denotes $(n \times k)$ matrix of antipollution activities comprising the input structure coefficients with a_{ig} to show the input of good i per unit of the eliminated pollutant g (eliminated by sector g); A_{21} a $(k \times n)$ matrix shows direct pollution output coefficients for antipollution activities a_{gk} with the output of pollutant g per unit of the eliminated pollutant k (eliminated by sector k); I is the identity matrix; y_1 is the n-dimensional vector of final demand; and y_2 is the k-dimensional vector of the net pollutant effect that remains untreated even after abatement activity. Unlike the final demand of the conventional sectors, the untreated pollutants are not demanded; rather, they are tolerated. This tolerated level is fixed at a standard level. It is therefore entered with a negative sign, and the gth element of this vector is used to show the pollution standard of pollutant g as an indicator of its permitted level.

Equation (10.13) can be solved for the level of final demand as y_1 and y_2 (given pollution standards), through the inversion of coefficients, such that

$$\begin{pmatrix} x_1 \\ x_2 \end{pmatrix} = \begin{pmatrix} I - A_{11} & -A_{12} \\ -A_{21} & I - A_{22} \end{pmatrix}^{-1} \begin{pmatrix} y_1 \\ y_2 \end{pmatrix}. \tag{10.14}$$

As mentioned, y_2 value is entered with a negative sign because the level of unwanted pollution is not determined rather it is tolerated. We know the relevance of the $(I - A)^{-1}$ matrix: that it describes the total effects (direct plus indirect) of an incremental change in final demand on the total output of all the industries. Accordingly, the change in final demand leading to the change in output is expected to change the level of pollution. The following matrix shows the change behavior:

$$A_{21}(I - A_{11})^{-1} \tag{10.15}$$

Where A_{11} is the inter-industry coefficient matrix and A_{21} matrix captures the direct pollution output coefficients. The result in equation (10.15) provides only the incremental change effect of direct and indirect pollution that is an impact depicted by a rupee worth increase in final demand. The total impact of final demand on pollution output generation is obtained by multiplying the direct–indirect coefficients with that of final demand. The following relationship explains this relationship:

$$A_{21}(I - A_{11})^{-1} Y_k, \tag{10.16}$$

where Y_k is the column vector of deliveries to final consumers of one kind.

10.5 Industrial structure, technical change and effluent generation

Technology is an important determinant responsible for industrial pollution. The technological structure in I-O is depicted by the structural coefficients of the conventional sectors. These structural coefficients are expressed in the form of production function (see Camerson 1952). This implies that for each sector there is a specified production function which is relatively fixed for the year in which I-O table is being constructed. The change in technology only indicates the change in the relationship among different sectors and the change in primary input requirements. Technological change under I-O modeling framework therefore is measured by keeping the final demand vector constant and computing the difference between the I-O tables of the two different periods. Then the output is calculated with different technological matrices (period "0" and period "t") but with the same final demand (say, period "0") vector. If more output is produced in time period t, it means improved technology has bought more output. A fall in output implies deterioration in technology. In this way, changes in output become a resultant factor of the change in technology between periods.

We know that the changing output level leads to change in pollution level. Technology is an important factor that causes the change in output as well as the change in pollution generation. Thus, there is an interdependence between the undesirable pollutants and the production level. The effect of technology on pollution generation is explained by the structural coefficients' matrix, and previously mentioned approach is followed to capture the effect of the same.

The basic methodology of technological change is adopted from Carter (1967), Leontief and Ford (1972) and Forsell (1988). In simple terminology, the model can be explained. The quality of residuals can be estimated as a function of the industry outputs. These can be used for the estimation of direct residuals output or pollution output coefficients. In equation form, we have

$$R = EX, \tag{10.17}$$

where R having r_{kj} amount of residual k generated by industry j in preceding equation is a matrix of residuals generated by industry outputs, E is a direct pollution coefficient matrix and each element e_{kj} is the quality of residual k generated by industry j and X is the diagonal matrix of industry outputs measured in monetary units.

Equation (10.17) determines the direct pollution effect generated by each industry in its operation. However, change in final demand will have repercussions on the output of different other sectors of the economy, and the same will be ultimately reflected on residual flows. The residual flows generated in different stages of the interaction process can be expressed to be dependent on final demand. From the basic I-O model, we have

$$X = (I - A)^{-1} F, \tag{10.18}$$

where F denotes the final demand vector, X is the output vector equilibrium and $(I - A)^{-1}$ is the Leontief inverse matrix, all expressed in monetary units. From equation (10.17), the total quantity of each residual generated by all industries can be found by summing the rows of R, and we have

$$Z = Ri^1, \tag{10.19}$$

$$Z = EX, \tag{10.20}$$

where Z is a vector of total residuals by type and i is a sum vector. Now equation (10.18) can be subsided in equation (10.20) to obtain the relationship between the final demand and residual flow. Now

$$Z = E(I - A)^{-1} F. \tag{10.21}$$

If the same equation is to be used to compare the two I-O tables for the base period 0 and current period t, then the total pollution generated can be specified by the product $A_{21,}$ which is represented by E and F, is replaced by Y_k:

$$A_{21}(I - A_{11})^{-1} Y_k. \tag{10.22}$$

On the same argument, technological change can be defined as the difference in the total pollution generated between year 0 and year *t*. The following relationship explains the technological effect on pollution generation:

$$X_{0-t} = A_{21}\left(I - A_{11}^{t}\right)^{-1} Y_k^0 - A_{21}\left(I - A_{11}^{0}\right)^{-1} Y_k^0 \tag{10.23}$$

Equation (10.23) is a unique way of capturing the technological effect on the pollution-generation matrix comprising all sectors of the I-O table. A_{11} is the usual structural coefficient matrix. Subscripts 0 and t indicate the period. Unlike the case of technological analysis of the conventional economic sectors, here the positive values of pollution output (in period t) does not imply improved technology. In environmental analysis, the pollution output is expected to fall when there is a change in technology from period 0 to period *t*. If the relationship in equation (10.23) results into a positive value of pollution output, then that is an indication of deterioration in the environment. It shows that over a period, technology has not been efficient enough in improving the environmental quality by reducing pollution.

For the present study, "open, static Leontief" type model has been considered as a basic analytical framework. The model is defined under the following mentioned set of simplified assumptions where the technological impact on pollution generation has been studied:

a. Only technological matrix changes over time
b. Final demand is constant
c. No change in pollution characteristics of industry's technology takes place

This has been done for two reasons: First, as per central pollution control board (CPCB) documents, the abatement technology for effluent treatment has not changed much in India. Several measures to improve the technology have been taken, but overall, the technology to abate pollution has remained constant over a period. Second, no industrial pollution database exists in our country, and therefore, the preparation of the environmental pollution matrix for different periods has been a herculean task. Thus, it has been assumed that environmental technology does not change with time.

Empirically, the reference period has been from 1983–1984 to 1993–1994, which is depicted in the following equation:

$$X_{83-93} = A_{21}\left(I - A_{11}^{93}\right)^{-1} Y_k^{83} - A_{21}\left(I - A_{11}^{83}\right)^{-1} Y_k^{83}, \tag{10.24}$$

where X_{83-93} is the change in pollution between two periods. The associated subscript represents the year under consideration. Thus, change in pollution attributed to technology between years 1983–1984 to 1993–1994 is explained in equation (10.24). A_{21} and A_{11} matrices have predefined meanings.

By using the relationship explained in equations (10.19) and (10.21), we attempt to calculate the effect of technology on the generation of industrial water pollutants in the economy, over the period 1983–1984 to 1993–1994. This has been done by examining the pollution generation pattern at the aggregated as well as at the disaggregated level.

10.6 Empirical analysis using I-O tables

The I-O tables of the Indian economy are the open, static Leontief-type tables. For the present purpose of the study, 1983–1984 (CSO, 1992), 1989–1990 (CSO, 1997a) and 1993–1994 (CSO, 2000) I-O tables have been selected. The selection of these years has been subject to the availability of I-O tables with the CSO.

All the I-O tables under the reference years are prepared for the common 60-sector classification. However, detailed classification of 1968–1969 I-O table is prepared for 230 sectors, while other I-O tables for the later periods, that is, 1973–1974, 1978–1979, 1983–1984, 1989–1990 and 1993–1994 have been prepared for 115 sectors.

Point to remember, all I-O tables are constructed at the current prices. Before we use these tables for the environmental analysis, it should be converted to the common base. Hence, the first adjustment made to them was the conversion of all the tables to a constant base.

For the present study, the 1983–1984 and 1989–1990 tables have been converted to the 1993–1994 factor prices as that has been the most recent year in terms of availability of these tables. Moreover, the analysis in terms of the most recent year is more meaningful, and the price deflators required for the conversion of the base were available for most of the sectors in that year.

The second adjustment was made in terms of aggregation of sectors. For the study, the original 115 sectors I-O table has been collapsed into 56 meaningful sectors. These redefined sectors were based on environmental implications as well as economic rationale. All newly formed 56 aggregated sectors classification is given in Appendix 10.1.

10.7 Water pollution data

Present study introduces 36 different type of pollutants. Thus, the resulting environment pollution flow matrix which would be of (36 × 56) size. To complete the matrix of this size is not a simple task as there is no environmental database exists in India that covers industrial pollution for all the sectors. Thus, this matrix is formed by collecting the data from multiple secondary sources. The main sources of the data for the purpose have been CPCB and Uttar Pradesh state pollution control board (SPCB). In addition to this, some of the data gaps have been filled by other publications (Manivasakam, 1997; Nemerow, 1978; Goel and Sharma, 1996).

In the augmented model of environmental repercussions, pollutants are entered in physical terms under the linearity assumption. This implies that pollutants vary linearly with output. Thus, the additional rows created for different pollutants represent the pollution output of the corresponding industry that varies with output. In this way (36 × 56), an environment pollution flow matrix is formed. Clearly, not all the pollutants are discharged by all the industries; thus, the resulting matrix will have both zero and nonzero characters. Most of the data requirement for this matrix is met by CPCB and SPCB. The CPCB has industry specific environmental accounting documents of different industrial units, which clearly mentions the pollution characteristics, load factors and abatement cost alternatives. This accounting is done for most of the highly polluting industries. In addition to this, CPCB has some documents that provide guidelines to different industries regarding pollution discharge, treatment and disposal. Besides, some of the database has been prepared by the SPCB survey of industries. Uttar Pradesh (SPCB) has conducted a survey of highly polluting industries. This provides detailed information regarding effluent characteristics, quality and abatement status of large and medium-sized 450 industrial units.

For the present purpose of the study, the data set for livestock products, hydrogenated and other edible oil, food products, synthetic textile, jute paper, leather, rubber, petroleum products, heavy chemicals, fertilizers, pesticide, paints, drugs, soaps, cosmetics and glycerin, other nonmetallic mineral products, iron and steel, nonferrous basic metals, batteries, machine tools, other non-electrical machinery and electricity have been taken from the CPCB data source. The data for dairy products, tobacco, metal products, hand tools, transport equipment, watch and clocks are taken from SPCB data source.

Apart from this, there are some polluting industries left that could not be covered by either of the aforesaid sources or in certain cases, the information provided by the CPCB and the SPCB was incomplete. Data for such industries have been taken from Manivasakam (1997) and Nemerow (1978). These industries include coal and lignite, beverages, cotton textile, woolen textile, plastic, fertilizer and synthetic fibers. Still, there are some industrial categories left for those where no data source was available. Such industries include metallic mineral, nonmetallic minerals, silk textile, structural clay products, cement, electrical and electronic equipment and transport services, the data for these industrial categories has been substituted from the general standard figures (CPCB, 19; Goel and Sharma, 1996).

Finally, there are some industrial categories that do not emit any liquid effluent or it is negligible. A zero pollution value has been taken for such industries, which include agricultural products, wooden products, printing and publishing, construction, gas and water supply and other services.

Initially, pollution output data was obtained as a function of physical output of the different sectors. In this case, units of the different sectors were different. To make them consistent with each other, it was necessary to convert them on a common scale. Thus, the environment matrix was to be expressed in value terms. Moreover, I-O tables were available in value terms. It was done in such a way that the pollution output became the function of the value of output instead of physical output.

To solve the preceding problem, absolute prices have been used for different sectors of the input-output table. These processes are derived from the 'Annual Survey of Industries' (CSO, ASI 1997–98). The ASI provides data for products and by-products, in value as well as in quantitative terms. In deriving such prices, we did not face any problem as information was available at three- to four-digit industrial classification. In this way, prices for all the sectors are easily calculated for all the manufacturing sectors. For some service sectors such as transport services, prices were calculated from the information given in the statistical abstract (CSO, 1997b).

As mentioned earlier, all analysis has been done at 1993–1994 prices. Therefore, the environmental matrix has also been converted at 1993–1994 prices. This was done by multiplying the physical output with the monetary value of output. Then, it was multiplied by 1 lakh because in the present analysis, intermediate and final demand categories are also expressed in terms of 1 lakh rupees.

Finally, the pollution output flow matrix is constructed both for before and after abatement of pollution. These two matrices remain the same for all the years. This has been done under the assumption that environment technology remains the same and does not change frequently. Within a 10-year time frame, it has remained mostly unaltered. A survey of most of the industries also reveals the same. It has been verified from some of the highly polluting sectors that the effluent treatment technology has not changed much over past several years. Thus, the after-abatement pollution characteristics have remained more or less the same through the last several years. In India, most of the effluent treatment plants were installed when industry came into existence. The changes in environment technology have been present in very few cases. In some of the industrial units, there have been updates but because of the unavailability of data, we have resorted to this assumption. The list of pollutants is provided in Appendix 10.2.

10.8 Results and discussion

The entire period was divided into two relevant sub-periods, 1993–1994 to 1989–1990 and 1989–1990 to 1993–1994. However, for convenience, we will be discussing results only for the period covering 1983–1984 to 1993–1994. The same 56 sectors aggregation has been followed for 36 different pollutants while studying the impact before as well as after abatement. The contribution of technological change in the form of changing a_{ij}s has been estimated in aggregate pollution as well as for individual industrial categories. Also, the contribution of highly polluting industries has been examined by taking the pollution from other sectors to be zero. Thus, the effect of technology has been examined in two aspects: first, the change in aggregate pollution and, second, the change in the level of pollution at an individual industry level. This analysis has been done for all sector categories as well as highly polluting industrial categories. Results for the entire period (1983–1984 to 1993–1994) are shown in Appendix 10.2.

10.8.1 Changes in aggregate pollution due to change in technology

The change in pollution attributed to input technology, between 1983–1984 and 1993–1994, is calculated for all sectors combined and for highly polluting industrial sectors. Figure 10.1 displays the comparison of top 10 pollutants at before and after treatment stages.

The detailed results depicted in Appendix 10.2 show how pollution discharged by the various industries would change if we keep the final demand vector constant and change the "A" matrix between base year and current year. The first column represents the total pollution generated by all the sectors of the economy, without any abatement measures. The second column illustrates the total pollution after abatement. In columns 3 and 4, similar information is given for highly polluting industries.

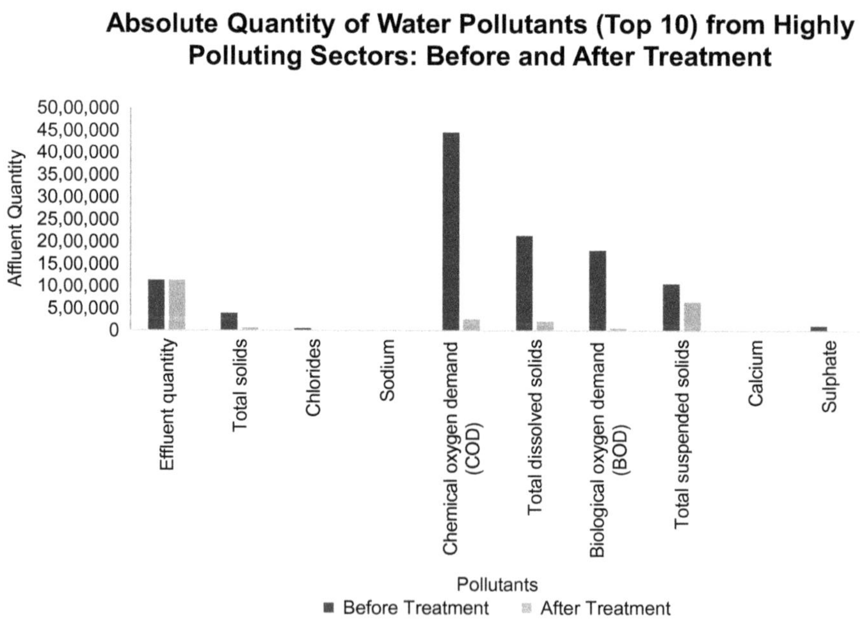

Figure 10.1 Absolute Quantity of Water Pollutants (Top 10) from Highly Polluting Sectors: Before and After Treatment

The analysis shows that the overall situation has worsened, which is apparent from the positive entries of all pollutants. The same trend has been observed in all sectors. Highly polluting sectors have again appeared with high pollution growth. All major pollutants have shown a high, positive contribution for both categories.

The negative trend is shown by only TKN (11), which is very insignificant in comparison to the high growth of other pollutants. The study suggests that over a period of time, use of inputs in industries has generated more pollutants. Overall, the state of the environment cannot be considered satisfactory on account of technological change.

10.9 Summary and conclusion

The chapter demonstrated the use of open, static Leontief-type I-O model as a general framework to study the impact of industrial water pollution on the environment. We have seen that interdependencies among producing sectors has direct as well as indirect impact on the economy which has been studied with the help of a simplified model of production. The total output of a sector is determined by the final demand from consumers and the intermediate demand from other sectors of the economy. The level of output changes when there is a change in either final demand or intermediate demand. On the same analogy, the level of pollution cannot be considered independent from the level of output, and therefore, it changes when the production output for a sector changes. Ultimately, there is a change in pollution level when the level of final demand or intermediate demand changes. The entire process is considered as an "I-O model of environmental repercussions".

The role of technology under I-O framework has been studied with the help of structural coefficients. Change in technology indicates the change in intersectoral relationships and its impact on environment can be analyzed by taking the difference in pollution level when final demand is kept constant. Pollution is an unwanted element that is tolerated and not demanded. Therefore, while keeping the final demand constant, if the level of pollution increases, then there is an indication of deterioration in the environment as a change in technology doesn't reflect an impact on bringing down the level of pollution.

Empirically, I-O models have not been leveraged much for studying the impact of industrial water pollution. The empirical literature mostly suggests the calculation and use of energy-based emissions. The studies of Leontief and Ford (1972) for the US, Miernyk and Sears (1974) for West Virginia in the US, James et al. (1978) for the Netherlands and Førsund and Strom (1975) for Norway are some of the great examples which have been mostly conducted on air pollutants. On the level of water pollution assessment, there has not been any such study. The present analysis is an attempt in that direction that has aimed to assess industrial water pollution in India by using the I-O framework.

References

Arrous, Jean. "The Leontief Pollution Model: A Systematic Formulation." *Economic Systems Research* 6, no. 1 (1994): 105–107.

Ayres, Robert U., and Allen V. Kneese. "Production, Consumption, and Externalities." *The American Economic Review* 59, no. 3 (1969): 282–297.

Camerson, B. "The production functions in Leontief Models", *Review of Economic Studies* 20 (1952): pp.62–69.

Central Statistical Organization, *Input-Output Transaction Tables, 1983-84*, New Delhi: Department of Statistics, Minitry of Planning and Program Implementation. Govt. of India (1992).

Central Statistical Organization. *Input-Output Transaction Tables, 1989-90*, New Delhi: Department of Statistics, Minitry of Planning and Program Implementation. Govt. of India (1997a).

Central Statistical Organization. *Input-Output Transaction Tables*, 1993-94, New Delhi: Department of Statistics, Minitry of Planning and Program Implementation. Govt. of India (2000).

Central Statistical Organization. Annual Survey of industries 1993-94, Calcutta: Minitstry of Planning and Program Implementation, Govt. of India (1997–98).

Central Statistical Organization. Statistical Abstract, New Delhi: Ministry of Pnanning and Program Implementation, Govt. of India (1997b).

Carter, Anne P. "Changes in the Structure of the American Economy, 1947 to 1958 and 1962." *The Review of Economics and Statistics* 49 (1967): 209–224.

Chenery, H. B., and P. G. Clark. *Interindustry Economics*. New York: John Wiley and Sons (1965).

Cumberland, J. H., and B. N. Stram. "Empirical Application of Input-Output Models to Environmental Problems, K. Polenske and JV Skolka." *Advances in Input-Output Analysis* 22, no. 3 (1976): 365–382.

Daly, Herman E. "On Economics as a Life Science." *The Journal of Political Economy* 76 (1968): 392–406.

Flick, Warren A. "Environmental Repercussions and the Economic Structure: An Input-Output Approach: A Comment." *The Review of Economics and Statistics* 56 (1974): 107–109.

Forsell, Osmo. "Growth and Changes in the Structure of the Finnish Economy in the 1960s and 1970s" in Ciaschini, M (ed.) *Input-Output Analysis: Current Developments*, New York: Chapman and Hall (1988), 287–302.

Førsund, Finn R., and Steinar Strøm. (1975) "Industrial Structure, Growth and Residual Flows,University of Oslo, Institute of Economics, Reprint no. 139, pp.21-69

Goel, P. K., and K. P. Sharma. *Environmental Guidelines and Standards in India*. Jaipur: Technoscience Publications (1996).

Herendeen, Robert A. "Input-Output Techniques and Energy Cost of Commodities." *Energy Policy* 6, no. 2 (1978): 162–165.

Isard, Walter, Kenneth Bassett, Charles Choguill, John Furtado, Ronald Izumita, John Kissin, Eliahu Romanoff, Richard Seyfarth, and Richard Tatlock. "On the Linkage of Socio-Economic and Ecologic Systems." *Papers of the Regional Science Association* 21, no. 1 (1968) 79–99.

James, David Edward, Huib M. A. Jansen, and Johannes Baptist Opschoor. *Economic Approaches to Environmental Problems*. Amsterdam: Elsevier (1978).

Kohn, Robert. "Input-Output Analysis and Air Pollution Control." In Mills, E.S. (Ed.) *Economic Analysis of Environmental Problems*. Cambridge, Massachusetts: NBER (1975), 239–274. https://www.nber.org/system/files/chapters/c2837/c2837.pdf

Kuenne, Robert E. "Walras, Leontief, and the Interdependence of Economic Activities." *The Quarterly Journal of Economics* 68, no. 3 (1954): 323–354.

Lee, Kwang-Soo. "A Generalized Input-Output Model of an Economy with Environmental Protection." *The Review of Economics and Statistics* 64 (1982): 466–473.

Leontief, Wassily W. "Quantitative Input and Output Relations in the Economic Systems of the United States." *The Review of Economic Statistics* 18 (1936): 105–125.

Leontief, Wassily W. "Environmental repercussions and the economic structure: an input-output approach", *The Review of Economics and Statistics* 52 (1970): pp. 262–271.

Leontief, Wassily W. "Structure of the World Economy—Outline of a Simple Input-Output Formulation." *American Economic Review* 64 (1974): 823–834.

Leontief, Wassily (ed.). *Input-Output Economics*. New York: Oxford University Press (1986).

Leontief, Wassily, and D. Ford. "Environmental Repercussions and the Economic Structure: An Input-Output Approach." *Review of Economics and Statistics* 52, no. 3 (1970): 262–271.

Leontief, W., and D. Ford. "Air Pollution and Economic Structure: Empirical Results of Input-Output Computations in Input-Output Techniques." In A. Brody and A. P. Carter. Amsterdam: North Holland Publishing Company (1972), 9–30.

Lowe, Peter D. "Pricing Problems in an Input-Output Approach to Environment Protection." *The Review of Economics and Statistics* 61 (1979): 110–117.

Luptacik, Mikulas, and Bernhard Böhm. "Reconsideration of Non-Negative Solutions for the Augmented Leontief Model." *Economic Systems Research* 6, no. 2 (1994): 167–170.

Manivasakam, N. *Industrial Effluents*. Coimbatore: Sakthi Publications (1997).

Miernyk, William H., and John Thomas Sears. "*Air Pollution Abatement and Regional Economic Development*." Lexington Books, 1974.

Miller, Ronald E., and Peter D. Blair. *Input-Output Analysis: Foundations and Extensions*. Englewood Cliffs, NJ: Prentice Hall (1985).

Muller, Frederik. *Energy and Environment in Interregional Input-Output Models*. Boston: Martinus (1979).

Nemerow, Nelson L.. *Industrial Water Pollution*. Massachusetts: Addison Wesley, (1978).

Qayum, Abdul. "A Reformulation of the Leontief Pollution Model." *Economic Systems Research* 3, no. 4 (1991): 428–430.

Rhee, Jeong J., and John A. Miranowski. "Determination of Income, Production, and Employment under Pollution Control: An Input-Output Approach." *The Review of Economics and Statistics* (1984): 146–150.

Steenge, Albert E. "Environmental Repercussions and the Economic Structure: Further Comments." *The Review of Economics and Statistics* 60 (1978): 482–486.

Victor, Peter A. *Pollution: Economy and Environment*. London: Allen and Unwin (1972).

Appendix 10A

Table 10.A1 List of Aggregated Sectors for the Purpose of Analysis

1. Agriculture products	29. Fertilizers
2. Dairy products and animal services	30. Pesticides
3. Other livestock products	31. Paint, varnishes and lacquers
4. Forestry, logging and fishing	32. Drugs and medicines
5. Coal and lignite	33. Other chemicals
6. Crude petroleum and natural gas	34. Synthetic fibers and resins
7. Metallic minerals	35. Structural clay products
8. Nonmetallic mineral	36. Cement
9. Sugar	37. Other nonmetallic mineral product
10. Hydrogenated and edible oil	38. Iron-steel industries and foundries
11. Food products	39. Non-ferrous basic metals
12. Beverages	40. Metal products including hand tools
13. Tobacco products	41. Tractors and agricultural implements
14. Cotton and other textile products	42. Industrial machinery
15. Woolen textiles	43. Machine tool and other non-electrical machinery
16. Silk textiles	44. Electrical and electronic equipment
17. Art silk and synthetic fiber textiles	45. Batteries
18. Jute, Hemp Mesta textile	46. Ships and boats
19. Wood and wooden products	47. Rail equipment
20. Paper and paper-based products	48. Motor vehicles and scooters
21. Printing and publishing	49. Cycle, rickshaw and other transport equipment
22. Leather footwear	50. Watches and clocks
23. Leather and leather products	51. Miscellaneous manufacturing
24. Rubber products	52. Construction
25. Plastic products	53. Electricity
26. Petroleum products	54. Gas and water supply
27. Coal tar products	55. Transport services
28. Organic–inorganic heavy chemicals	56. Other services

Note: A 56-sector aggregation from 115 original sectors in Input-Output Transaction Table of Central Statistical Organization.

Table 10.A2 List of Water Pollutants and Increase in pollution (Effluent quantity in ,000 cu m and other pollutants in tons)

Pollutants	*Table 10.2 Absolute Increase in Pollution*			
	All sectors		*Highly polluting sectors*	
	BT	*AT*	*HIBT*	*HIAT*
Effluent quantity	1,817,307	1,817,092	1,137,068	1,136,854
Total suspended solids	1,062,264	638,511	1,058,048	634,533
Total dissolved solids	2,138,963	210,972	2,137,564	209,512
Total solids	40,577,106	8,106,969	383,197	68,188
Dissolved fixed solids	23,322	-1,573	23,322	-1,573
Today volatile solids	139,372	23,333	139,372	23,333
Biological oxygen demand (BOD)	1,810,622	54,635	1,809,355	53,439
Chemical oxygen demand (COD)	4,459,466	258,031	4,449,423	248,127
Oil and greases	60,094	14,265	59,708	13,878
Ammonical nitrogen	47,464	7,258	45,531	5,325
Total Kjedahl nitrogen (TKN)	3,865	3,854	-12	-1
Total nitrogen	183	8	183	8
Organic nitrogen	11,036	1,391	11,036	1391
Nitrate nitrogen	7,033	1,082	7,033	1,082
Iron	12,518	4,139	4,788	2,501
Chlorides	20,625,356	4,144,733	44,845	1,876
Zinc	1,896	925	1,728	757
Sulphate	6,32,413	148,871	108,962	17,427
Chromium	1,410	338	1,343	271
Phenolic compounds	750	122	716	89
Lead	131	63	128	60
Cyanides	101	12	94	5
Phosphate	1,946	405	1,778	237
Mercury	107	55	107	54
Hexavalent chromium	27	6	23	3
Fluorides	220,949	27,900	220,882	27,833

(Continued)

Table 10.A2 (Continued)

Pollutants	*Table 10.2 Absolute Increase in Pollution*			
	All sectors		*Highly polluting sectors*	
	BT	*AT*	*HIBT*	*HIAT*
Calcium	2,158,101	430,685	12,111	1,517
Copper	145	123	44	23
Cadmium	76	72	9	5
Sodium	10,502,623	2,100,539	1,651	315
Nickle	125	113	24	12
Magnesium	699,578	139,916	2	1
Potassium	14,508	203	14,508	203
Aluminum	9,358	2,988	3,475	1,811
Total acids	161,178	32,249	–98	–6
Dissolved phosphate	292	30	292	30

Note: In the analysis, effluent quantity (1) will be expressed in thousand cu. m. and other pollutants (2–36) in tons. The immediate description will be given wherever any other unit will be used (e.g., percentage). BT = Before abatement values (all sectors); AT = After abatement values (all sectors); HIBT = Before abatement values (highly polluting sectors); HIAT = After abatement values (highly polluting sectors).

11 Investigating the existence of environmental Kuznets curve hypothesis for the South Asian region

Mohd Irfan, Sarani Saha and Sanjay Kumar Singh

11.1 Introduction

Over the past few decades, globalization, rapid industrialization, and structural changes have contributed significantly to the economic growth of the South Asian region. In recent years, many countries of this region have liberalized their economic policies to integrate with the world economy. The integration has helped countries to achieve unprecedented growth in the past years with significant improvement in economic development. Together, South Asian countries accounted for about 1.5% of the world's total gross domestic product (GDP; at constant 2005 US$) in 1990 and 3.35% in 2014 (World Bank, 2017). In the past 20 years, the region has enjoyed an average growth rate of more than 5%. However, approximately 40% of people still live below the poverty line. Therefore, the current economic policies of this region are emphasized towards alternate development strategies such as reducing inequalities, development of the human resource, employment generation, and inclusive growth. South Asian countries continue to face the challenge of environmental pollution, which arises due to an increase in income, improvement in the living standard, population growth, trade openness and urbanization. In the last two decades, environmental pollution has increased manifold in all the economies of this region. More specifically, in the overall environmental pollution, the contribution of carbon emissions was approximately around 60%, followed by deforestation, which was around 10% in this region (Raghuvanshi et al., 2006). At the regional level, the level of carbon dioxide (CO_2) emissions in 2011 was 1.4 metric tons per capita as compared to 1.1 metric tons per capita in 2006 (World Bank, 2017). The plausible reasons for such an increase are mainly associated with an increase in income, growing population, rampant poverty, urbanization and changes in consumption patterns.

The rising environmental pollution in the South Asian region is a serious cause of concern because countries of this region have strong linkages between growth, development, population growth, resource depletion and environmental damages. Furthermore, such linkages are not confined to one individual country. Although the South Asian region is biodiversified, it has a unified ecosystem that mainly relies on the Hindu Kush Himalayan

(HKH) region for environmental resources such as freshwater, rivers, monsoon and rainfall. The HKH feeds four major rivers (the Ganges, Sutlej, Indus and Brahmaputra) of South Asia, which are significant sources of freshwater and hydro-energy generation for downstream countries such as India, Pakistan, Nepal and Bangladesh. South Asian countries are agrarian economies where agriculture contributes around one-fourth of total GDP and supports the livelihood of over 50% of the population. Moreover, the critical dependence on HKH for monsoon, rainfall and agricultural yields poses challenges for food security in this region. Since the environmental stress and greenhouse gas (GHG) emissions have increased due to economic development in the region, the HKH ecosystem is under serious threat of global warming and climate change. As a result, this also influences the entire region's (South Asian) ecosystem as a whole.

For instance, growth and development in India or any other country of South Asia create pressure on environmental resources in the own country through changes in land use and pattern, loss of biodiversity, climate change and environmental pollution such as water and air pollution. Furthermore, as the region's ecosystem is interlinked, all of these changes in one country make the whole region vulnerable to global warming and climate change and poses a threat to water, food and energy security for all. Specifically, expansion of economic activities and growing energy needs in Nepal may result in the construction of new hydro-energy generation plants in its rivers, which may further alter the whole region's ecosystem and obstruct the availability of freshwater resources for downstream countries like India and Pakistan. To overcome such challenges, economists and environmentalists have agreed that the agenda of sustainable development ought to be given priority in the South Asian region. Following the agenda of sustainable development, South Asian countries established South Asia Environmental Cooperation Programme in 1982 with an emphasis on regional cooperation on environmental issues, climate change and the sustainable management of environmental resources of the region. Recently, the Dhaka Climate Change Action Plan has been formulated to increase the cooperation between South Asian countries to mitigate and manage the impacts of climate change. It is now recognized that in South Asian countries, there is an urgent need for a reduction in GHG emissions, and therefore, a clear understanding of the role of an increase in income on carbon emissions is inevitable. Such understanding would eventually be useful in providing some framework for the formulation of integrated environmental policies and support of sustainable development for the region.

As suggested in the literature, the environmental Kuznets curve (EKC) hypothesis defines the impact of income on carbon emissions to be an inverted-U-shape-type relationship (Grossman and Krueger, 1995). The EKC hypothesis suggests that emissions and income are linked through a non-linear relationship. The relationship implies that there is an increase in emissions with a rise in income but after attaining a certain level of income (which is the turning point), a further increase in income leads to lower emissions. This is possible because, in the initial stages of industrialization, countries start exploiting their natural resources, which gives carbon emissions as a by-product. Nonetheless, countries also emphasize on clean, renewable and energy-saving technologies after achieving the threshold to reduce the emissions for the sustainability of environmental resources (Stern, 2004; Dinda, 2004, 2005). Soon after the formulation of the EKC hypothesis, there have been ample theoretical and empirical works conducted to validate the EKC hypothesis in several countries. In the past studies, different econometric methodologies have been utilized to investigate the emissions-income nexus for individual countries and regions. A detailed review of recent literature in this field is provided in the study of Al-Mulali et al. (2015). Although broad empirical support exists for the EKC, one may argue that the findings of existing empirical studies on the existence of the EKC hypothesis are mixed. More specifically, results vary across countries (Richmond and Kaufmann, 2006) and regions (Dijkgraaf and Vollebergh, 2005) as well as findings are not uniform across the type of data (time series or panel) used.

Despite the validity of the EKC hypothesis in some individual South Asian countries, to the best of our knowledge, there have been no studies that empirically confirmed the existence of EKC for the South Asian region as a whole. In other words, the validity of EKC for South Asia is elusive in nature, and therefore, conducting a study of EKC at the regional context may elicit a better understanding of the impact of income on carbon emissions. Countries of this region have a strongly interlinked ecosystem, where any changes in biodiversity and ecosystem of one country influence the whole region. It is, therefore, necessary to discover the true shape of EKC for this region using cross-country panel data. Such kind of analysis will allow a researcher to integrate the effects of South Asian countries on each other in determining the extent of the relationship between emissions and income.

Furthermore, India is home to 1.2 billion people with just 2.4% of the world's land area is industrially diversified across states/provinces. In some states such as Maharashtra and Haryana, industrial development has been successful in promoting growth, reducing inequalities and fostering structural changes, but other states like Bihar and Uttar Pradesh are lagging behind their development targets with significant contribution to state-wise emissions. The existence of widespread disparities in income, emissions and other regional characteristics across Indian states may possibly influence the relationship between emissions and income. Moreover, it is also widely visible that India relative to other South Asian countries, had taken significant steps towards the reduction of emissions through environmental policy reforms and emphasized the use of green energy sources to meet the country's growing energy demands. Over the last two decades, the progress in economic development and the adoption of national mitigation measures to reduce emissions may provide experience and lessons for other South Asian countries. Thus, accommodating widespread economic disparities and regional differences across states is necessarily needed to estimate the true shape of EKC for India. Therefore, a state-level analysis is required to fulfill this gap and could also corroborate the findings of regional analysis in the South Asian context. The analyses presented in this study rely on panel data econometric techniques for accommodating heterogeneity across South Asian countries and the Indian states to model the relationship between emissions and its underlying factors.

The remainder of this chapter is organized as follows: The literature on EKC studies in the South Asian context is briefly reviewed in Section 11.2. The data used and methodology adopted to estimate the EKC for the South Asian region as a whole and for India are discussed in Section 11.3. The findings of a regional analysis utilizing the cross-country panel of South Asian countries are presented in Section 11.4. Section 11.5 presents the estimation results of the state-level analysis for India. Finally, the last section presents the concluding remarks.

11.2 Theoretical background and review of past studies

The impact of income on environmental pollution is studied in the seminal work of Grossman and Krueger (1995). Their empirical findings show that as income level increases, there is an increase in environmental pollution at first, and then it starts declining after reaching a threshold. This empirical relationship is termed as the EKC hypothesis, which suggests an inverted-U-shape-type impact of income on environmental pollution. After their seminal work, several studies have examined the effect of income level on environmental pollution in an empirical setting. These studies have used different countries for analysis with different measures of environmental pollution. Evidence from these studies suggests that environmental degradation is associated with the level of income for most of the countries; this relationship has been best described as an inverted U–shape or N-shape. However, considering the existing empirical studies, one may argue that the existence of EKC is country-specific with a high level of dependence on measures of environmental pollution.

Apart from income level, the other reasons for rising environmental pollution are energy use, population, international trade and urbanization (Ahmed and Long, 2012; Shahbaz et al., 2012; Al-Mulali et al., 2015). In most of the developing countries, it is well established that a rise in income level leads to higher energy demands, which are mainly met through coal consumption either for electricity generation or steel production. This has increased the CO_2 emissions and raises the concern for clean and renewable energy for sustainable development. Another issue of great practical importance is the population growth in developing countries. A higher population puts more stress on the resources and ultimately fuels the demand for more natural resources, including land, water, coal, minerals, in a short span of time that has led to rampant deforestation in developing countries. Another critical factor that is responsible for environmental pollution is trade openness. After the establishment of the World Trade Organization, developing countries have participated significantly in international trade with little consideration over environmental issues. Hence, export-oriented polluting industries have brought environmental degradation in the form of pollution and deforestation. In most of the studies, it was found that trade openness has increased environmental pollution (Suri and Chapman, 1998; Abler et al., 1999; Beghin et al., 1999). Urbanization is another reason behind environmental pollution because it encroaches on farmland, destroys natural habitat and diminishes biodiversity.

Since all the countries in the South Asian region are in their initial stages of economic development, it is obvious that countries may find themselves at different points on the EKC. Most empirical studies using annual time-series data for individual South Asian countries have found the EKC hypothesis is evident. For instance, Managi and Jena (2008) use panel data of states for the period from 1991 to 2003 to examine the EKC between environmental productivity and income for India. Their study concluded that environmental productivities decline with income in high-income states relative to low-income states. In a study for 46 developing countries, Narayan and Narayan (2010) investigate the validity of the EKC hypothesis. Their study utilized panel co-integration techniques to investigate the presence of EKC for five different panels. The results suggest the validity of EKC for India and Pakistan only. Kanjilal and Ghosh (2013) investigated the existence of the EKC-type relationship between income and carbon emissions for India during the study period from 1971 to 2008. Their study using co-integration technique suggests that CO_2 emissions are highly responsive to changes in per capita income in India. A study by Tiwari et al. (2013) on the causality between coal consumption and environmental pollution, based on time series data, found a dynamic long-run relationship between these two variables. Ahmed and Long (2012) in their study on Pakistan using the Auto-Regressive Distributed Lag approach, showed that economic development impacts CO_2 emissions and it is the form of an inverted-U-shape-type relationship. In a similar approach, Nasir and Rehman (2011), using the Johansen co-integration technique, validated the EKC for Pakistan and suggest that the major factors responsible for environmental pollution are energy consumption and economic development. On the contrary, Shahbaz et al. (2012) found opposite evidence of the EKC hypothesis for Pakistan. Their findings show that in the long run, a decrease in carbon emissions is a result of an increase in trade openness. In a study for Bangladesh, Miah et al. (2011) examined the shape of EKC, and it was found that there is no specific shape of EKC for Bangladesh. Moreover, the shape of EKC varies with the specification of the econometric model. Similarly, Ghosh et al. (2014) in their study observes that economic growth can be attained without polluting the environment for Bangladesh. Aspergis and Ozturk (2015) in a recent study tested the relevance of EKC in 14 Asian countries, including India, Pakistan and Bangladesh. Their study revealed that environmental pollution increases with per capita income; however, after achieving a certain level, it decreases with per capita income. The review of existing studies suggests that the existence of EKC hypothesis for South Asian countries or region as a whole is ambiguous in nature and could possibly be associated to econometric issues lying beneath the quality of data and varying measures of environmental pollution (Stern, 2004; Itkonen, 2012; Kaika and Zervas, 2013).

11.3 Materials and methods

11.3.1 Data and sources

The sample utilized in this study is cross-country data for Bangladesh, India, Nepal, Pakistan and Sri Lanka, covering the study period from 1978 to 2011. The level of environmental pollution in each country defined in terms of carbon dioxide emissions per capita and the income or level of economic development is measured by GDP per capita. The other control variables which could possibly affect the level of carbon emissions are energy use, population density, international trade and urbanization. The data series of variables of interest were extracted from the World Bank development indicators database. Table 11.1 shows the descriptive statistics of variables used in this study, and Table 11.2 describes the variable construction details. Since carbon emissions and GDP data of Afghanistan, Bhutan and the Maldives are not available for all the years of our study period, these countries have been dropped from the sample. The choice of panel data for regional analysis is supported by the fact that in this region, countries have some kind of homogeneity in the characteristics such as common structural changes in the economy, trade as their engine of growth and rapid urbanization in the past few decades. Thus, an empirical study based on cross-country panel data could allow us to examine the EKC hypothesis for South Asian countries where strong linkages in ecosystems exist amongst the countries.

Table 11.1 Details of Variable Construction

Variables	*Details*
CO_2 emissions per capita (CO_2pc)	Carbon dioxide emissions in metric tons per capita
GDP per capita (*gdppc*)	Gross domestic product (GDP) per capita in 2005 U.S. dollars
(GDP per capita) ^{2} ($gdppc^2$)	Square term of GDP per capita in 2005 U.S. dollars
energy consumption per capita (*ec*)	Energy use expressed in kgs of oil equivalent to per capita
population density (*pd*)	People per sq. km of land area in thousands
trade openness (*op*)	Total trade (sum of exports and imports) as a percentage of GDP
urbanization (*urb*)	Urban population as a percentage of total population

Table 11.2 Summary Statistics of All Variables in Cross-Country Panel

Variable	*Obs.*	*Mean*	*Std. Dev.*	*Min*	*Max*	*Variable*	*Obs.*	*Mean*	*Std. Dev.*	*Min*	*Max*
India						Sri Lanka					
co_2pc	34	0.969	0.356	0.480	1.700	co_2pc	34	0.404	0.168	0.210	0.730
gdppc	34	530.734	228.191	279.780	1,086.050	*gdppc*	34	927.077	340.112	501.530	1,724.850
$gdppc^2$	34	332,218.8	295,399.3	78,276.85	1,179,505	$gdppc^2$	34	971,746.1	723,380.5	251,532.3	2,975,108
ec	34	407.146	90.993	286.150	613.720	*ec*	34	374.887	63.185	304.490	499.340
pd	34	318.192	57.107	224.450	410.720	*pd*	34	283.072	31.250	226.280	332.790
op	34	24.899	13.190	12.010	55.020	*op*	34	71.679	9.297	49.150	88.640
urb	34	26.633	2.515	22.380	31.280	*urb*	34	18.530	0.166	18.310	19.060
Bangladesh						Pakistan					
co_2pc	34	0.192	0.091	0.080	0.370	co_2pc	34	0.666	0.184	0.350	0.970
gdppc	34	389.471	106.246	280.650	649.370	*gdppc*	34	564.707	114.523	370.320	755.410
$gdppc^2$	34	162,643.6	94,512.03	78,764.42	421,681.4	$gdppc^2$	34	331,623.1	131,703.8	137,136.9	570,644.3
ec	34	135.193	32.536	97.120	204.720	*ec*	34	408.181	63.265	298.640	509.600
pd	34	903.445	181.162	599.310	1,174.330	*pd*	34	162.494	40.468	97.120	228.530
op	34	26.389	8.356	16.690	47.420	*op*	34	34.117	2.741	27.720	38.910
urb	34	21.795	4.962	12.630	31.230	*urb*	34	31.887	2.761	27.370	37.010
Nepal						Panel					
co_2pc	34	0.086	0.044	0.020	0.160	co_2pc	170	0.463412	0.377911	0.02	1.700
gdppc	34	265.936	58.698	185.130	384.770	*gdppc*	170	535.5849	296.8614	185.13	1,724.85
$gdppc^2$	34	74,065.82	32,579.54	34,273.12	148,047.9	$gdppc^2$	170	374,459.5	473,537.1	34,273.12	2,975,108
ec	34	335.381	21.736	312.430	382.640	*ec*	170	332.1576	118.09	97.12	613.72
pd	34	142.854	30.200	96.160	189.440	*pd*	170	362.0112	293.2954	96.16	1,174.33
op	34	42.118	10.507	26.040	64.040	*op*	170	39.84029	19.50377	12.01	88.64
urb	34	11.055	3.735	5.550	17.170	*urb*	170	21.98006	7.795887	5.55	37.01

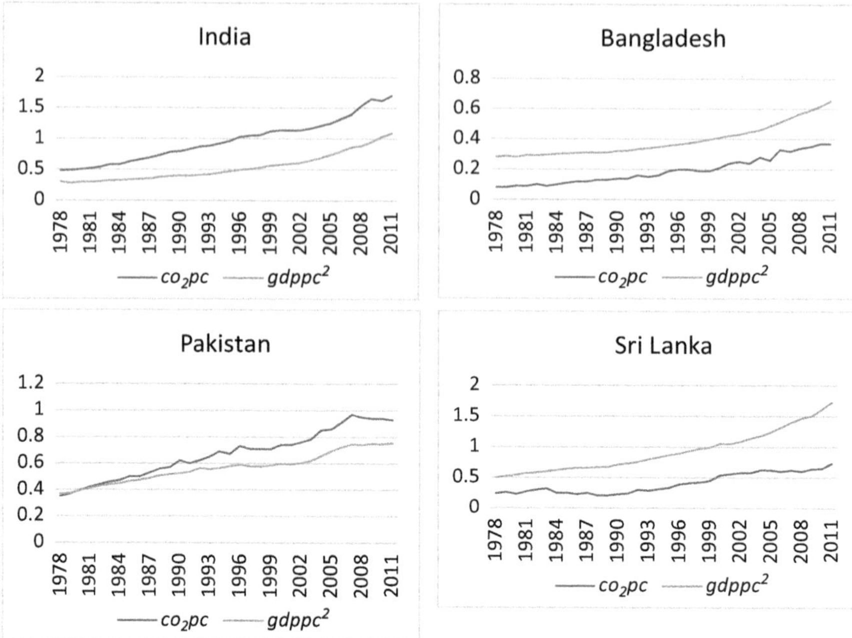

Figure 11.1 Country-wise time series plots (in natural logarithmic value) of gross domestic product per capita/US$1,000 at constant 2005 prices ($gdppc^2$) and carbon dioxide emissions per capita in metric tons (co_2pc)

Figure 11.1 shows the evolution of GDP per capita (in $1,000) and carbon dioxide emissions per capita of India, Pakistan, Bangladesh, Nepal and Sri Lanka. The values of GDP per capita (*gdppc*) and carbon dioxide emissions per capita (co_2pc) are presented in light and dark lines, respectively. In Figure 11.1, there is an increase in *gdppc* and co_2pc for all countries in our sample, but the pattern in gaps between the series is different for some countries. In India, Sri Lanka and Pakistan, the gaps are broadening while the gaps are following no definitive pattern in Bangladesh and Nepal, but the emissions intensity increases with an increase in *gdppc* in these countries.

The sample for state-level analysis of EKC for India includes 19 major states, and the study period spans 39 years from 1980 to 2008. The major source of data for state-wise per capita CO_2 emissions is Ghoshal and Bhattacharyya (2012). Their study estimated the state-wise per capita CO_2 emissions using a methodology provided by the Oak Ridge National Laboratory (ORNL) for the period from 1980 to 2008. The details of the methodology are discussed in Ghoshal and Bhattacharyya (2008). The other sources of data include Reserve Bank of India (RBI) Handbook of Statistics on the Indian economy for state-wise per capita net state domestic product (NSDP) at factor cost (at current prices) during the study period. We deflated the values of per capita NSDP at current prices using GDP deflator for each year, which is available in the World Bank development indicators database. The dependent variable is carbon dioxide emissions per capita (co_2pc) for each state, and the independent variables are per capita NSDP (*nsdppc*) and its square term. Due to the lack of state-wise data for other explanatory variables such as energy consumption, urbanization, population growth and so on during the study period, these were ignored in the state-level analysis. The summary statistics of all the variables of interest are presented in Table 11.3. As can be seen, on average, Madhya Pradesh is the largest contributor to CO_2 emissions per capita, followed by Uttar Pradesh, whereas Jammu and Kashmir had the lowest CO_2 emissions per capita during the study period. On the other side, per capita NSDP is lowest in Bihar, followed by Uttar Pradesh.

11.3.2 *Econometric strategy*

The econometric model in this study is an extension of the model suggested by Stern (2004). In the extended model, energy use is added as an explanatory variable along with other control variables such as population, trade and urbanization. Specifically, the functional relationship between the dependent and independent variables is specified as follows:

$$co_2pc = f\left(gdppc, gdppc^2, ec, pd, op, urb\right). \tag{11.1}$$

In the panel data regression models (fixed and random effects) form, the specification is written as

$$\begin{aligned} co_2pc_{it} &= \alpha + v_i + \beta_1 gdppc_{it} + \beta_2 gdppc_{2it} + \beta_3 ec_{it} \\ &\quad + \beta_4 pd_{it} + \beta_5 op_{it} + \beta_6 urb_{it} + u_{it}, \end{aligned} \tag{11.2}$$

$$\begin{aligned} co_2pc_{it} &= \alpha + v_{it} + \beta_1 gdppc_{it} + \beta_2 gdppc_{2it} + \beta_3 ec_{it} \\ &\quad + \beta_4 pd_{it} + \beta_5 op_{it} + \beta_6 urb_{it} + u_{it}, \end{aligned} \tag{11.3}$$

where subscript $i = 1, 2, \ldots, 5$ stands for country and $t = 1, 2, \ldots, 34$ stands for the time period. The environmental pollution is expressed in terms of carbon dioxide emissions per capita (co_2pc), income level is expressed in terms of GDP per capita (*gdppc*), square term of the income level is $gdppc^2$, energy use is expressed in terms of energy consumption per capita (*ec*), *pd* is the population density, *op* is trade openness ratio, *urb* is level of urbanization. v_i is an unobserved effect that is country-specific and fixed over time, v_{it} is the random component or between country error and follows a normal distribution with $N(0, \sigma^2)$, u_{it} is the idiosyncratic error, and it is uncorrelated with explanatory variables. The theory regarding EKC proposes that the sign of the coefficient β_1 is expected to be positive while the sign of the coefficient β_2 is expected to negative. With respect to the other explanatory variables, the sign of the coefficients β_3 and β_4 are expected to be positive. However, the sign of the coefficients β_5 and β_6 can be either positive or negative.

Table 11.3 Summary Statistics of All Variables in the Cross-State Panel of Indian States

Variable	*Obs.*	*Mean*	*Std. Dev.*	*Min*	*Max*	*Variable*	*Obs.*	*Mean*	*Std. Dev.*	*Min*	*Max*
Jammu and Kashmir						Rajasthan					
co_2pc	29	461.0579	190.6814	168.29	788.58	co_2pc	29	7,394.492	4,554.184	1578.61	16379.91
nsdppc	29	223.0834	76.75344	143.0699	379.0508	*nsdppc*	29	206.068	80.99709	110.9232	392.4378
$nsdppc^2$	29	55,454.16	39,423.26	20,469	143,679.5	$nsdppc^2$	29	48,798.33	38,043.01	12,303.96	154,007.5
Himachal Pradesh						Delhi					
co_2pc	29	497.5841	350.0401	88.75	1128.7	co_2pc	29	5,177.615	1297.01	2791.54	7821.64
nsdppc	29	307.3963	158.979	145.4546	626.1014	*nsdppc*	29	645.3613	296.3002	354.7216	1,402.132
$nsdppc^2$	29	118,895.3	116,709.4	21,157.05	392,002.9	$nsdppc^2$	29	501,257.7	488,476	125,827.4	1,965,974
Punjab						Bihar					
co_2pc	29	8,343.664	3,423.126	2,377.12	14,032.85	co_2pc	29	21,443.33	3,344.196	16,785.77	31,017.39
nsdppc	29	403.6139	128.4482	242.724	694.0023	*nsdppc*	29	106.1314	24.00223	74.07732	184.67
$nsdppc^2$	29	178,834.2	115,318.8	58,914.92	481,639.2	$nsdppc^2$	29	11,820.12	6,053.316	5,487.45	34,103
Haryana						Orissa					
co_2pc	29	5,094.342	2,853.019	1258.36	12,422.89	co_2pc	29	13,829.68	10,609.31	3,568.06	41,454.85
nsdppc	29	394.4846	179.536	215.1293	845.5875	*nsdppc*	29	182.3493	77.43958	113.73	394.1567
$nsdppc^2$	29	186739.7	176355.8	46280.62	715,018.3	$nsdppc^2$	29	39,041.38	37,442.54	12,934.51	155,359.5
Uttar Pradesh						West Bengal					
co_2pc	29	31,896.72	13,925.86	9343.07	54,350.83	co_2pc	29	19,884.33	6,992.617	11,499.33	35,042.07
nsdppc	29	160.2968	40.42057	109.5871	256.2219	*nsdppc*	29	249.7238	89.02563	158.0741	445.2329
$nsdppc^2$	29	27,272.54	14,221.75	12,009.34	65,649.67	$nsdppc^2$	29	70,014.23	51,059.11	24,987.42	198,232.4

(Continued)

Table 11.3 (Continued)

Variable	*Obs.*	*Mean*	*Std. Dev.*	*Min*	*Max*	*Variable*	*Obs.*	*Mean*	*Std. Dev.*	*Min*	*Max*
Assam						Andhra Pradesh					
co_2pc	29	1,049.335	227.6479	589.84	1,470.4	co_2pc	29	6,923.444	3,819.549	1,586.15	14,084.27
nsdppc	29	193.6046	53.75995	116.5511	302.3549	*nsdppc*	29	608.192	1,216.841	140.5524	5,152.724
$nsdppc^2$	29	40,273.23	22,870.05	13,584.15	91,418.5	$nsdppc^2$	29	1,799,541	6,379,988	19,754.98	26,600,000
Gujarat						Karnataka					
co_2pc	29	15,865.03	3,981.289	7,501.63	24,822.56	co_2pc	29	6,923.444	3,819.549	1,586.15	14,084.27
nsdppc	29	336.0636	150.4584	176.0974	690.9034	*nsdppc*	29	274.7794	133.9528	137.9732	603.2795
$nsdppc^2$	29	134,795.9	126,504.6	31,010.3	477,347.5	$nsdppc^2$	29	92,828.33	93,725.21	19,036.61	363,946.2
Maharashtra						Kerala					
co_2pc	29	27,745.65	10,642.7	10,564.23	44,576.08	co_2pc	29	2,270.298	837.689	946.09	4149.56
nsdppc	29	387.479	166.9736	213.4426	783.5708	*nsdppc*	29	295.236	163.1417	129.0802	665.5346
$nsdppc^2$	29	177,058.8	160,694	45,557.75	613,983.2	$nsdppc^2$	29	112,861.8	120,172.2	16,661.7	442,936.3
Goa						Tamil Nadu					
co_2pc	29	529.2576	157.0329	234.43	812.97	co_2pc	29	12,254.94	4,635.285	4,540.14	18,625.57
nsdppc	29	654.427	407.3652	275.9335	1,705.879	*nsdppc*	29	304.3311	158.9402	134.6424	679.2603
$nsdppc^2$	29	588,498.8	726,135	76,139.3	2,910,025	$nsdppc^2$	29	117,008.3	122,260.9	18,128.57	46,1394.6
Madhya Pradesh						Panel					
co_2pc	29	32,682.55	16,797.31	9,890.58	76,159.89	co_2pc	551	12,404.09	12,705.71	88.75	76,159.89
nsdppc	29	186.5077	56.3668	117.6956	315.8548	*nsdppc*	551	329.5982	376.8077	74.07732	5,152.724
$nsdppc^2$	29	37,852.78	22,634.53	13,852.25	99,764.25	$nsdppc^2$	551	250,361.3	1,552,150	5,487.45	26,600,000

Note: *co2pc* is per capita co_2 emissions; *nsdppc* is per capita net state domestic product; *nsdppc2* is the square term of *nsdppc*.

Although region-specific concern for sustainable management of environmental resources is apparent, the strategies for the region's sustainable development rely on the extent of trade-off among carbon emissions and economic development. Nevertheless, the dominance of such trade-offs is higher in India, where food, energy and water security are altogether important. The Indian economy, which is the largest economy of this region with the largest number of inhabitants, has a serious concern over maintaining higher economic growth with minimal adverse impacts on environmental resources. In recent years, the share of the agricultural sector in India's GDP had fallen significantly, but at the same time, the territory sector had economic dominance. Moreover, the economic development also came with substantial demand for energy sources, and a similar shift is observed in other countries of South Asia as well. Therefore, India's policies to reduce the emissions may create a significant impact on the region's sustainable development goals and lessons for other South Asian countries, which are on a similar path. In India, the CO_2 emissions are mainly linked with solid, liquid, and gaseous fuel consumption, whereby solid fuel consumption is the largest contributor to emissions. In addition, CO_2 emissions have no state/local boundaries—there is a mix of local factors and their relationship with the level of development in that state/province, which could vary the degree of emissions in each state. Thus, at this juncture, it is necessary to have a clear understanding of the link between CO_2 emissions and income while considering the disparities and development dynamics across states. Therefore, to investigate the existence of EKC for India, we use a cross-state panel of 19 Indian states, and the econometric model includes two explanatory variables, which are per capita NSDP and its square term. The econometric specification (fixed and random effects models) for panel data of states is as follows:

$$co_2pc_{it} = \alpha + v_i + \beta_1 nsdppc_{it} + \beta_2 nsdppc^2{}_{it} + u_{it}, \tag{11.4}$$

$$co_2pc_{it} = \alpha + v_{it} + \beta_1 nsdppc_{it} + \beta_2 nsdppc^2{}_{it} + u_{it}, \tag{11.5}$$

where $i = 1, 2, \ldots, 19$ states, $t = 1, 2, \ldots, 29$ time periods, co_2pc is the state-specific carbon dioxide emissions per capita, $nsdppc$ and $nsdppc^2$ are per capita NSDP and its square term, respectively. v_i is an unobserved component that is state-specific and fixed over time, v_{it} is the random component or between states error and follows a normal distribution with $N(0, \sigma^2)$, u_{it} is the idiosyncratic error, and it is uncorrelated with explanatory variables.

11.4 Results using country-level data for the South Asian region

The estimation results of fixed effects (FE) and random effects (RE) models are summarized in Table 11.4. We also employ the Hausman (1978)

Table 11.4 FE, RE and FETF Results for CO_2 Emissions in Cross-country Panel

FE			*RE*			*FETF*		
Variable	*Coef.*	*Std. Err.*	*Variable*	*Coef.*	*Std. Err.*	*Variable*	*Coef.*	*Std. Err.*
gdppc	2.65×10^{-05}	1.52×10^{-04}	*gdppc*	1.17×10^{-03}***	2.68×10^{-04}	*gdppc*	4.50×10^{-04}*	2.58×10^{-04}
gdppc²	-7.38×10^{-08}	6.21×10^{-08}	*gdppc2*	-5.59×10^{-07}***	1.32×10^{-07}	*gdppc2*	-1.60×10^{-07}*	8.22×10^{-08}
ec	3.59×10^{-03}***	1.81×10^{-04}	*ec*	2.58×10^{-03}***	2.03×10^{-04}	*ec*	3.50×10^{-03}***	2.09×10^{-04}
pd	1.08×10^{-04}	7.37×10^{-05}	*pd*	3.32×10^{-04}***	6.72×10^{-05}	*pd*	-6.90×10^{-05}	1.13×10^{-04}
op	2.39×10^{-03}***	4.79×10^{-04}	*op*	-7.67×10^{-03}***	1.03×10^{-03}	*op*	1.38×10^{-03}*	7.69×10^{-04}
urb	-1.05×10^{-02}***	2.34×10^{-03}	*urb*	4.79×10^{-03}*	2.83×10^{-03}	*urb*	6.63×10^{-03}	7.45×10^{-03}
			constant	-7.32×10^{-01}	5.82×10^{-02}			
			sigma_u	4.44×10^{-05}				
			sigma_e	0.001509				
			rho	0.2929		Turning Point (in US$) = 1418		
R^2	0.96346		R^2	0.8703		R^2	0.96853	
Model fit	F(6,159) = 698.783		Model fit	F(6,163) = 182.289		Model fit	F(39,126) = 99.424	
	Prob > F = 0.0001			Prob > F= 0.0001			Prob > F = 0.0001	
Obs.	170		Obs.	170		Obs.	170	
BG test:	$Chi^2(34)$ = 101.34 Prob > Chi^2 = 0.0001							
BP test:	$Chi^2(10)$ = 41.077 Prob > Chi^2 = 0.0001							
Hausman-test:	$Chi^2(6)$ = 4462.5†/2611.5‡ Prob > Chi^2 = 0.0001							

Note: ***Significant at 1 % level of significance,

* Significant at 10 % level of significance.

† For FE and RE models.

‡ For FETF and FE models.

specification test to choose between FE and RE models. We find that although the coefficient of variables *gdppc and gdppc*^{2} are statistically insignificant in the results of FE model, the Hausman test favors the FE model over the RE model in our sample. However, the results of the FE model could also be influenced by time-specific unobserved variables, which are common to all countries as well as the presence of heteroskedasticity and serial correlation in the sample. The Breusch–Pagan Lagrange Multiplier test and Breusch–Godfrey test for panel data were conducted to examine the heteroscedasticity and autocorrelation related assumptions in the sample, respectively. The results show that there are heteroscedasticity and autocorrelation issues in our sample, and therefore, an appropriate method should be applied to get the true estimates—Arellano's method for robust covariance matrix estimation.

To get a more convincing model, we estimated the following fixed effects model, which includes time fixed effects (γ_t) and the estimated coefficients are robust to heteroscedasticity and serial correlation:

$$\begin{aligned} co_2pc_{it} &= \alpha + v_i + \gamma_t + \beta_1 gdppc_{it} + \beta_2 gdppc^2{}_{it} + \beta_3 ec_{it} \\ &\quad + \beta_4 pd_{it} + \beta_5 op_{it} + \beta_6 urb_{it} + u_{it}. \end{aligned} \tag{11.6}$$

The results of the model in equation 11.6 are shown in the last two columns of Table 11.4. The Hausman specification test results of the fixed effects model with time fixed effects (FETF) and FE models suggest that the former is the best model. Accordingly, interpretations of coefficients are based on the findings of the FETF model. As evident from Table 11.4, the coefficient of variable *gdppc* is positive and statistically significant at the 10% level, and the coefficient of the variable *gdppc*^{2} (square term of GDP per capita) is negative and statistically significant at the 10% level. This finding suggests that the EKC exists for South Asian countries; in particular, an increase in carbon dioxide emissions is associated with higher economic growth in the initial stages. However, there is no monotonic increase in carbon emissions. That is, after reaching the highest level of emissions (turning point), higher growth in income leads to lesser carbon dioxide emissions. This result is supported by the fact that since South Asian countries are in their initial development path along with a low per capita income, environmental pollution worsens as the level of development increases in these countries. However, reaching a higher level of development with the accessibility of energy-efficient technologies and the use of renewable sources for energy generation reverses this relationship. More specifically, the impact of income on carbon emissions is an inverted U–shape type. The turning point for this region as a whole is estimated to be US$1418 and hence, proposes that the region may experience a fall in emissions after achieving this level of per capita GDP. This finding of an inverted-U-shape-type EKC between income and emissions in line with the findings of some existing studies (Managi and Jena, 2008; Tiwari et al., 2013; Al-Mulali et al., 2015).

Another important result is that the estimated coefficient of variable *ec* is positive and statistically significant at the 1% level. This result suggests that any increase in energy use is associated with increases in carbon emissions in the countries of this region. This is true because countries in the region are developing countries and since their reliance on fossil fuels for energy generation is approximately 67% (World Bank, 2017), fossil fuels are the established cause of changes in climate and global warming. There are several studies in this regard, such as Raghuvanshi et al. (2006) and Aspergis and Ozturk (2015) that reached the same conclusion. The estimated coefficient for variable *op* is statistically significant at the 10% level, and it has a positive sign. This relationship suggests that trade openness could cause more emissions in the region. This result coincides with the study of Antweiler et al. (2001), which suggests that the import of environmentally friendly technology in developing countries is still out of reach because of their high costs (Ozturk and Acaravci, 2010). The estimated coefficients for the *pd* and *urb* are both found to be insignificant.

From the preceding analysis, it is evident that the impact of income on carbon emissions is found to be an inverted U–shape type. This implies that in this region, higher economic development in the initial stages is accommodated with more emissions, but this will not continue forever. The concerns of environmental protection increase after achieving a certain level of income, and thus, necessary measures are undertaken to reduce the emissions and move towards sustainable development. Empirical results also clearly indicate that energy consumption and trade openness play important roles in determining the level of carbon emissions in the set of countries. Specifically, an increase in any of these variables could bring about an increase in carbon emissions. To focus on the implications of our results, it is suggested that the countries should integrate their environmental policies within the region and prioritize alternative sources of energy such as hydro, solar and wind to increase the role of non-fossil-fuel-based (clean and renewable) energy in mitigating the environmental damages occurring due to rapid development and energy demands. Moreover, it is important for countries in the region to focus on trade-related measures (actions and strategies) to reduce the environmental damages as there is a positive impact of trade on carbon emissions. In addition, it is important for policy makers and city planners in the region to create public awareness for clean technologies and generate support for environmental legislation. For the purpose, awareness towards clean technologies and demand for a better environment shall be promoted to reduce the environmental damages caused through the expansion of economic activities in the region.

11.5 Results using state-level data for India

The estimated results of two model specifications, FE and RE models using cross-state panel data for India are shown in Table 11.5. The Hausman test was used to identify the superior model, which is FE or RE. Although

Table 11.5 FE, RE, and FETF Results for CO_2 Emissions in the Cross-State Panel of Indian States

FE			*RE*			*FETF*		
Variable	*Coef.*	*Std. Err.*	*Variable*	*Coef.*	*Std. Err.*	*Variable*	*Coef.*	*Std. Err.*
nsdppc	21.531***	2.1148	*nsdppc*	20.935***	2.116	*nsdppc*	14.599***	2.5915
$nsdppc^2$	−0.00463***	0.000479	$nsdppc^2$	0.00452***	0.00048	$nsdppc^2$	−0.0043***	0.0005
			constant	6.6361×10^{03}**	2.4351×10^{03}			
R^2	0.16435		R^2	0.15235		R^2	0.52048	
			sigma_u	103,300,334				
			sigma_e	43,462,666		Turning point (in US$) = 1697		
			rho	0.8804				
Model fit	F(2,530)= 52.1193 Prob > F = 0.0001		Model fit	F(2,548)= 49.2467 Prob > F = 0.0001		Model fit	F(30,502) =18.1629 Prob > F = 0.0001	
Obs.	551		Obs.	551		Obs.	551	

BG test: $Chi^2(29)$= 382.28 Prob > Chi^2 = 0.0001
BP test: $Chi^2(20)$ = 707.38 Prob > Chi^2=0.0001
Hausman test: $Chi^2(2)$ = 64.712†/574.96‡ Prob > Chi2 =0.0001

**** Note: Significant at 1% level of significance,
** Significant at 5% level of significance.
† For FE and RE models.
‡ For FETF and FE models

the results of the Hausman test suggest that FE model provides better fit than RE model, the estimates from FE model could be plagued by common time-specific effects along with the presence of heteroskedasticity and autocorrelation in the sample. The Breusch–Pagan Lagrange Multiplier test and Breusch–Godfrey test results indicate that our sample has an incidence of heteroscedasticity and autocorrelation, and a necessary correction should be done to get robust estimates. Therefore, we estimated the following fixed effects model with time fixed effects and Arrellano's method was employed to eliminate heteroscedasticity and serial correlation:

$$co_2pc_{it} = \alpha + v_i + \gamma_t + \beta_1 nsdppc_{it} + \beta_2 nsdppc^2_{it} + u_{it}. \qquad (11.7)$$

The last two columns of Table 11.5 present the results of FEFT model shown in equation 11.7 with robust standard errors. In this case, the Hausman test results of FETF and FE models suggest that FETF is the superior model. The results from FEFT model indicate that the sign of estimated coefficients of variables *nsddpc* and $nsdppc^2$ are positive and negative, respectively, and both coefficients are statistically significant at the 1% level. Hence, this finding from state-level panel data corroborates the results from country-level panel data analysis in this study. The turning point in the EKC curve for India while considering the state-level disparities is estimated to be at US$1697 of per capita NSDP.

11.6 Concluding remarks

South Asian region is inhabited by almost 1.3 billion people with the highest concentration of people living below the poverty line. The region has been developing with the unprecedented growth of about 5% per year with significant changes in consumption patterns, land use, energy consumption and exploitation of environmental resources. All countries of South Asia have a unique dependence on each other in terms of their biodiversity and ecosystem interlinkages, which rely on the Himalayan Hindu Kush region for the availability of various environmental resources like freshwater, rivers and monsoon. The climate change and global warming challenges which have mainly emerged from India, Pakistan and Bangladesh have now created a serious threat to the survival of the human race for the whole region. Thus, there is an urgent need to address the issue of relationship, more specifically, the trade-off between emissions and income for the region as a whole. This understanding will be helpful in formulating region-specific policies related to environmental protection and mitigation measures for the sustainable development of the region. Although countries of this region are diversified, there are some common characteristics such as the nature of the economy being typically agrarian, trade dependence, reliance on fossil fuels for energy generation and population growth, which could encourage regional cooperation among countries to face the challenges of environmental degradation.

To examine the role of income in determining the level of carbon emissions for South Asian countries, the validity of the shape of the EKC is examined on two different dimensions—regional and state levels by using cross-country data of five South Asian countries during the study period from 1978 to 2011 and cross-state panel of Indian states during the study period from 1980 to 2008, respectively. The estimation procedure relies on panel data econometric models. The analyses conducted in this study asserted that the inverted-U-shape-type pattern exists for the impact of income on carbon emissions; in other words, the EKC is valid for South Asian countries. The emissions follow a monotonic increasing pattern with a rise in income; however, after achieving a threshold, emissions fall with a further rise in income. The study also attempts to investigate the impact of energy use, international trade, population and urbanization on carbon emissions for India, which are considered to be important determinants of emissions. The results suggest that energy consumption and trade openness positively influence the level of emissions. From the state-level analysis, one can find an inverted-U-shape-type impact of income on carbon emissions for India.

References

Abler, David G., Adrian G. Rodriguez, and James S. Shortle. "Trade Liberalization and the Environment in Costa Rica." *Environment and Development Economics* 4, no. 3 (1999): 357–373.

Ahmed, Khalid, and Wei Long. "Environmental Kuznets Curve and Pakistan: An Empirical Analysis." *Procedia Economics and Finance* 1 (2012): 4–13.

Al-Mulali, Usama, Behnaz Saboori, and Ilhan Ozturk. "Investigating the Environmental Kuznets Curve Hypothesis in Vietnam." *Energy Policy* 76 (2015): 123–131.

Antweiler, Werner, Brian R. Copeland, and M. Scott Taylor. "Is Free Trade Good for the Environment?" *American Economic Review* 91, no. 4 (2001): 877–908.

Aspergis, Nicholas, and Ilhan Ozturk. "Testing Environmental Kuznets Curve Hypothesis in Asian Countries." *Ecological Indicators* 52 (2015): 16–22.

Beghin, John C., Bradley J. Bowland, Sebastien Dessus, David Roland-Holst, and Dominique Van der Mensbrugghe. "Trade Integration, Environmental Degradation, and Public Health in Chile: Assessing the Linkages." *Environment and Development Economics* 7, no. 7 (1999): 241–267.

Dijkgraaf, Elbert, and Herman R. J. Vollebergh. "A Test for Parameter Homogeneity in CO_2 Panel EKC Estimations." *Environmental and Resource Economics* 32, no. 2 (2005): 229–239.

Dinda, Soumyananda. "Environmental Kuznets Curve Hypothesis: A Survey." *Ecological Economics* 49, no. 4 (2004): 431–455.

Dinda, Soumyananda. "A Theoretical Basis for the Environmental Kuznets Curve." *Ecological Economics* 53, no. 3 (2005): 403–413.

Ghosh, Bikash Chandra, Khandakar Jahangir Alam and Md. Ataul Gani Osmani. "Economic Growth, CO_2 Emissions and Energy Consumption: The Case of Bangladesh." *International Journal of Business and Economics Research* 3 no. 6 (2014): 220–227.

Ghoshal, Tapas, and Ranajoy Bhattacharyya. "State Level Carbon Dioxide Emissions of India: 1980–2000." *Contemporary Issues and Ideas in Social Sciences* 4 no 1 47 (2008).

Ghoshal, Tapas, and Ranajoy Bhattacharyya. "Carbon Dioxide Emissions of Indian States: An Update." (2012). Available at SSRN 2166900.

Grossman, Gene M., and Alan B. Krueger. "Economic Growth and the Environment." *The Quarterly Journal of Economics* 110, no. 2 (1995): 353–377.

Hausman, Jerry A. "Specification Tests in Econometrics." *Econometrica: Journal of the Econometric Society* 46, no. 6 (1978): 1251–1271.

Itkonen, Juha V. A. "Problems Estimating the Carbon Kuznets Curve." *Energy* 39, no. 1 (2012): 274–280.

Kaika, Dimitra, and Efthimios Zervas. "The Environmental Kuznets Curve (EKC) Theory—Part A: Concept, Causes and the CO_2 Emissions Case." *Energy Policy* 62 (2013): 1392–1402.

Kanjilal, Kakali, and Sajal Ghosh. "Environmental Kuznet's Curve for India: Evidence from tests for Cointegration with Unknown Structural Breaks." *Energy Policy* 56 (2013): 509–515.

Managi, Shunsuke, and Pradyot Ranjan Jena. "Environmental Productivity and Kuznets Curve in India." *Ecological Economics* 65, no. 2 (2008): 432–440.

Miah, Md Danesh, Md Farhad Hossain Masum, Masao Koike, Shalina Akther, and Nur Muhammed. "Environmental Kuznets Curve: The Case of Bangladesh for Waste Emission and Suspended Particulate Matter." *The Environmentalist* 31, no. 1 (2011): 59–66.

Narayan, Paresh Kumar, and Seema Narayan. "Carbon Dioxide Emissions and Economic Growth: Panel Data Evidence from Developing Countries." *Energy Policy* 38, no. 1 (2010): 661–666.

Nasir, Muhammad, and Faiz Ur Rehman. "Environmental Kuznets Curve for Carbon Emissions in Pakistan: An Empirical Investigation." *Energy Policy* 39, no. 3 (2011): 1857–1864.

Ozturk, Ilhan, and Ali Acaravci. "CO_2 Emissions, Energy Consumption and Economic Growth in Turkey." *Renewable and Sustainable Energy Reviews* 14, no. 9 (2010): 3220–3225.

Raghuvanshi, Shiv Pratap, Avinash Chandra, and Ashok Kumar Raghav. "Carbon Dioxide Emissions from Coal-Based Power Generation in India." *Energy Conversion and Management* 47, no. 4 (2006): 427–441.

Richmond, Amy K., and Robert K. Kaufmann. "Is There a Turning Point in the Relationship between Income and Energy Use and/or Carbon Emissions?" *Ecological Economics* 56, no. 2 (2006): 176–189.

Shahbaz, Muhammad, Hooi Hooi Lean, and Muhammad Shahbaz Shabbir. "Environmental Kuznets Curve Hypothesis in Pakistan: Cointegration and Granger Causality." *Renewable and Sustainable Energy Reviews* 16, no. 5 (2012): 2947–2953.

Stern, David I. "The Rise and Fall of the Environmental Kuznets Curve." *World Development* 32, no. 8 (2004): 1419–1439.

Suri, Vivek, and Duane Chapman. "Economic Growth, Trade and Energy: Implications for the Environmental Kuznets Curve." *Ecological Economics* 25, no. 2 (1998): 195–208.

Tiwari, Aviral Kumar, Muhammad Shahbaz, and Qazi Muhammad Adnan Hye. "The Environmental Kuznets Curve and the Role of Coal Consumption in India: Cointegration and Causality Analysis in an Open Economy." *Renewable and Sustainable Energy Reviews* 18 (2013): 519–527.

World Bank. Development Indicators for South Asia. Accessed from: https://data.worldbank.org/country/8S. (2017)

12 Coping with changing climate

The case of water conservation structures in Eastern India

Bhagirath Behera, Pulak Mishra and Dil Bahadur Rahut

This chapter is based on the research project titled "Coping with Changing Climate: The Role of Traditional Water Harvesting System in Odisha and West Bengal", funded by the Indian Council of Social Science Research, New Delhi (Reference Number: 02/84/2014-15/RPR). The authors acknowledge the generous support by the funding agency. The authors are also thankful to Mr. Debananda Bhindani for his excellent research support throughout the period. This chapter was presented in the Sustainability and Development Conference held at the University of Michigan from November 9–11, 2018. The chapter has improved immensely with the constructive suggestions received from participants in the conference.

12.1 Introduction

Water is essential for the survival of life on Earth. Like other living things on Earth, human beings depend on water for biological needs and food to survive. Water is a renewable natural resource governed by the hydrologic cycle of the planet. Only 2.5% of water (around 1.4 billion km^3) is freshwater, out of which only 2,00,000 km^3 or less than 1% is used by human beings and the ecology (Gleick, 1993; Gleick and Howe, 1993). Water is available mainly from the surface (from different water bodies) and underground (from aquifers). Although groundwater is generally recharged through rainwater harvesting and melted snow, in some cases it cannot be renewed due to locational constraints (Tietenberg, 2006). Importantly, groundwater accounts for around 90% of freshwater, but only 2.5% of it is renewable basis. This is important given that more than 1.5 billion world population use groundwater for drinking (UNEP, 2002). Furthermore, agriculture uses 67% freshwater for irrigation and related activities (UNESCO, 2001).

Thus, food security and sustainability require rainwater harvesting and groundwater recharge, especially in emerging economies like India. This is particularly so because, with increasing variability in climatic conditions in recent years and frequent occurrences of droughts, untimely and heavy rains and floods, the agriculture sector and related livelihood activities in

India are facing serious challenges. The problem is more critical in the rain-fed regions of the country that house the largest proportion of the country's total poor.[1] Furthermore, rain-fed agriculture has also immense importance for providing food grains and poverty alleviation when productivity is stagnant in the green revolution regions. Hence, the identification of the right coping strategies is imperative to facilitate conservation of water and generating sustainable livelihoods. Notably, a part of groundwater gets renewed by percolation of rainwater and hence its storage through traditional rainwater-harvesting structures such as tanks/ponds. Such water-harvesting structures can serve as important links between the surface and groundwater sources. Accordingly, reviving the traditional water-harvesting systems such as tanks and ponds is seen as an alternative strategy for efficient storage and sustainable use of rainwater. The revival of tanks and ponds can also help in improving land productivity by reducing land degradation and runoff of rainwater and improving soil moisture and the groundwater table.

One may expect that greater access to irrigation through tanks would improve resilience to climatic shocks. In addition, the traditional water-harvesting system would help in integrating agriculture and allied activities resulting in greater vertical linkages and diversification of economic activities. Furthermore, the adoption of small and medium-sized irrigation systems through the traditional water-harvesting structures at a large scale has the potential to avoid adverse consequences of input-intensive farming practices. It can also reduce inequality in access to irrigation facilities and help in maintaining ecological balance and agricultural sustainability. Historically, Indian farming communities have developed water-harvesting structures/systems to store rainwater and subsequently use the same for cultivation, fishing and other livelihood activities.

Given this backdrop, this chapter identifies the factors that affect households' perceptions about and adaptation to rapidly changing climatic conditions in the eastern Indian states of Odisha and West Bengal. These two states suffered from the problem of prolonged agricultural stagnation in respect of both area under cultivation and production till the end of the 1970s. Farmers in rain-fed districts of West Bengal and Odisha faced repeated droughts, which adversely affected their livelihoods and put survival at stake. Hence, Odisha and West Bengal need greater emphasis on the development of necessary irrigation facilities for facilitating sustained growth of agriculture, reduction in inequality access to irrigation facilities and ecological balance. It is expected that the restoration of traditional harvesting structures would play a crucial role in this regard.

12.2 Analytical framework, data and methodology

12.2.1 Analytical framework

Based on the review of related literature, experiences from several rounds of visits to the study areas and interactions with various stakeholders, the conceptual/analytical framework used in this chapter is presented in Figure 12.1. It is observed that tanks improve the biophysical and hydrological conditions of the region, which, in turn, generates a positive influence on overall ecological conditions in terms of improvement of groundwater recharge and the improvement in biodiversity and ecology. Better ecological conditions would improve the resilience of local people to deal with climate change.

As shown in Figure 12.1, households' resilience increases with the increased ability of households in terms of their socioeconomic conditions captured in five capitals. There are four aspects of resilience that are important in the present context. These include (a) the propensity to retain functions and organizations, (b) change in the system without change in the state (IPCC, 2001), (c) ability to recover from adverse effects of extreme load (UKCIP, 2003) and (d) conditions to recover after shocks.[2]

Overall, resilience enables the households to cope with the changes in climatic conditions in terms of both magnitude and intensity, although the underlying reasons may differ. For example, rain-fed areas often experience drought. In such situation, resilience signals the number of droughts

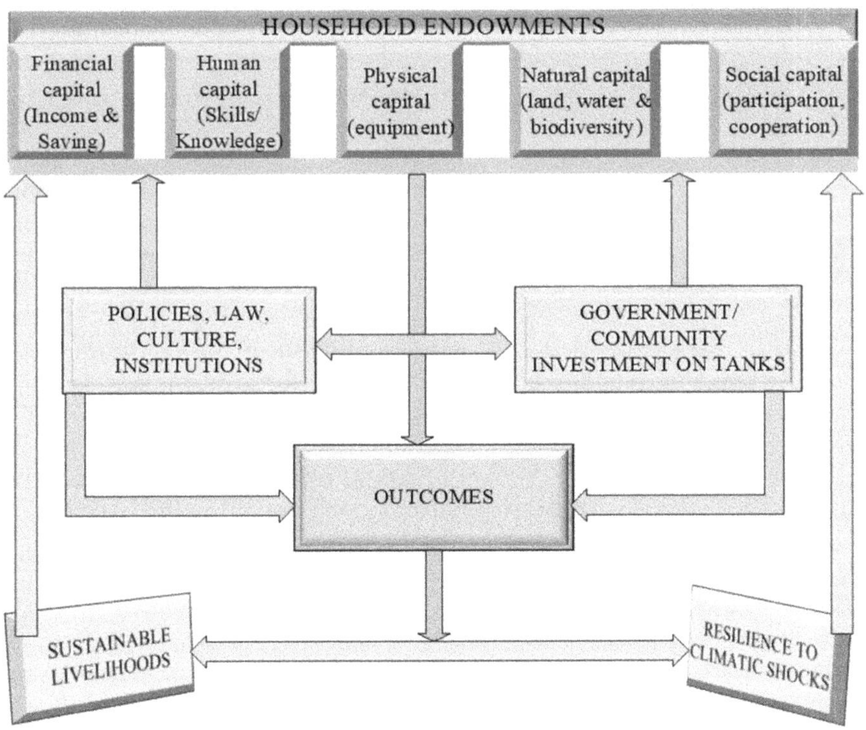

Figure 12.1 Conceptual Framework for Tanks and Socioeconomic Resilience: Sustainable Livelihood Approach

survived. A higher survival rate signals greater resilience. Here, the capacity of a household depends on its access to five capitals—physical, natural, financial, human and social capitals (Figure 12.1). Households' perceptions about these capitals are crucial as they provide a deeper understanding of assets and capabilities for adapting to the changing climate. It also recognizes that focusing only on income is not enough to design adaptation and livelihood strategies. The framework can also evaluate role of these capitals in a dynamic context and at different levels.

Figure 12.1 presents the Sustainable Livelihood Approach (SLA) and its applicability to tanks and the resilience of the households. Researchers have emphasized on the importance of asset creation at the household level that would provide the capacity to withstand any shock and uncertainty. Such an approach focuses on assets that help the households to enhance capabilities, encounter stresses and shocks and enable them to overcome poverty on sustained basis in the long run. However, the focus may vary from simply financial capital to a wider set like physical, social and natural assets.

The SLA framework has a holistic and asset-based structure and contributes immensely to deeper understanding of poverty- and livelihood-related complex issues. It is based on the five capitals model of the Department for International Development (DFID). The framework is suitable at different levels as a conceptual framework or a tool to design programs and evaluate strategies (DFID, 1999). According to the DFID,

> [a] livelihood comprises the capabilities, assets (including both material and social resources) and activities required for a means of living. A livelihood is sustainable when it can cope with and recover from stresses and shocks and maintain or enhance its capabilities and assets both now and in the future.

Figure 12.1 depicts the details of various components of SLA framework for asset creation in individual land under the forestry projects. In total five assets/capitals have been identified which can be applied to assess the impact of land assets developed under the forestry projects. They are (1) Financial Capital (Assets), (2) Social Capital (Assets), (3) Human Capital (Assets), (4) Physical Capital (Assets), and (5) Personal Capital (Assets).

12.2.1.1 Data

The chapter uses primary and secondary data. While secondary data are gathered from the Census of India (2011) and other government records, the primary data were collected from 110 sample tanks located in different parts of Odisha (60 tanks) and West Bengal (50 tanks). The data and information gathered from the secondary sources are used to identify the sample tanks and their beneficiaries. In addition to quantitative data collected using a structured questionnaire and from various secondary sources, the study also uses qualitative data gathered from numerous field visits including focus group discussions and interactions with households from diverse socioeconomic and demographic backgrounds and other stakeholders. A total number of 1,100 households of selected tanks were surveyed using a structured questionnaire for the collection of primary data. Such a large sample was required to capture diversities in size and patterns of landholding and livelihoods within a command area of tanks. It is also expected that the increase in sample size would enhance the efficiency of the findings, particularly of the econometric models. The sample size varies across the selected tanks depending on the population in the command areas to make it representative. Binary logistic regression models and multivariate probit econometric models have been applied to address the research objectives. A detailed exposition of econometric model specifications is presented in the following.

12.2.2 Econometric model specifications

12.2.2.1 Binary logit

Generally, multiple linear regression models are estimated to examine relationships between a continuous (or interval scale) dependent variable and a set of explanatory variables. However, since the dependent variable is categorical in the present case, the factors influencing households' perceptions about and adaptation to climate change are identified by estimating binary logistic regression models. They allow the estimation of probability that an event occurs given the explanatory variables. Here, the natural logarithm of the odds ratio is assumed to depend linearly on the explanatory variables; that is,

$$L_i = \ln\left(\frac{P_i}{1-P_i}\right) = \alpha + \sum_{j=1}^{k} \beta_j X_{ij} + u_i \tag{12.1}$$

Here, the probability that the dependent takes a value 1, P_i follows logistic distribution; that is,

$$P_i = \frac{1}{1+e^{-Z_i}} = \frac{e^{Z_i}}{1+e^{Z_i}}; Z_i = \alpha + \sum_{i=1}^{k} \beta_j X_{ij} + u_i \tag{12.2}$$

Here, P_i is nonlinearly related to Z_i. Importantly, it is nonlinear in variable as well as parameters. Hence, the method of ordinary least squares (OLS) cannot be used to estimate parameters. Furthermore, as the dependent variables, in the present context, are binary in nature, and we have data on the individual household level, the logit becomes $L_i = \ln\left(\frac{1}{0}\right)$ of $L_i = \ln\left(\frac{0}{1}\right)$ depending on if the dependent variable is true and has no meaning. Hence, the model is estimated by applying the maximum likelihood method. Under usual conditions, it gives consistent, asymptotically normal, and asymptotically efficient estimators. Hence, the estimators of the logit model are expected to satisfy these properties. However, there may be a problem of omitted-variable bias. The coefficient of an included variable can be inconsistent even when it is uncorrelated with the omitted variable. Hence, one needs to be careful about the omitted variables.

12.2.2.2 Multivariate probit model

In the multivariate probit model, the binary dependent variable is extended to an arbitrary number of discrete responses of either nominal or ordinal type. The multivariate probit models estimate the parameters of q probit models jointly to examine the dependence of q polytomous variables on a set of independent variables taking into account the association among q categorical variables. Such models parameterize the joint probabilities through a set of parameters known as the marginal parameters, and another set of parameters known as the association parameters. The marginal parameters are those of the probit models for the marginal density functions, whereas the association parameters indicate that the ith order interactions follow log-linear relations. The multivariate probit model assumes the random disturbance term to follow normal distribution.

This chapter estimates multivariate probit models to analyze what influence households' adaptation strategies. Here, the dependent variable is household choices of adaptation strategies: change of planting dates, more irrigation, more pesticides and change in crop variety. The explanatory variables are various socioeconomic, demographic and institutional characteristics.

As there are more than two dependent variables, the univariate probit is not appropriate. Multinomial and multivariate probit models are suitable when there are three or more discrete dependent variables. The multinomial probit model is estimated when the dependent variables are mutually exclusive. On contrary, multivariate probit models are estimated when the dependent variables are not mutually exclusive, that is, when choice of one influences the other(s). To understand what influence households' adaptation strategies for climate risks, multivariate probit models are estimated as they capture strategies applied concurrently to mitigate climate risks.

Since the respondents here use four types of adaptation strategies, we specify models with four categorical dependent variables, $y_1 \ldots y_4$ as suggested by Lin et al. (2005):

$$y_{im} = 1 \, if \; \beta_m' X_{im} + \varepsilon_{im} > 0 \tag{12.3}$$

and

$$y_{im} = 1 \; if \; y_{im}^* > 0, \; and \; 0 \; otherwise; \; m = 1,2,\ldots,4. \tag{12.4}$$

Here, x includes the independent variables, β_1, β_2, β_3 and β_4 are the vectors of coefficients and ε_1, ε_2, ε_3, *and* ε_4 are the random disturbance terms that follow multivariate normal distributions with zero mean and unitary variance.

Furthermore, the dependent variable takes a value 1 when there is *change of plantation date* and 0 otherwise. Similarly, it assumes the value 1 if *more irrigation* is used and 0 otherwise. It also takes the value 1 when *more pesticides* are applied and 0 otherwise and assumes the value 1 when there are *changes in crop variety.* Here, m indicates the number of equations. In the present case, m = 4 because of four alternative strategies to apply.

As mentioned in the introductory section that rainwater harvest and conservation through tanks play an important role in rapidly changing climatic conditions. Rainwater stored in tanks is largely used as irrigation water during summer crops. However, tanks are primarily constructed for providing water for both irrigation and domestic uses during droughts and in periods of extreme water scarcity. Hence, one needs to examine the perceptions and adaptation strategies. Based on the literature and our own experience during the field study, several important factors were identified. The hypotheses are set accordingly. The dependent variables in the models are households' perceptions about and adaptation to changes in climate, while the independent variables include household characteristics (age, sex, wealth and education of household head), institutional factors (agricultural extension services, institutional credit, weather information and land tenure) and so on. The list of these variables and their summary statistics are given in Table 12.1.

12.2.2.3 Independent variables and their hypothesized effects

Perceptions about climate change are likely to depend on personal experiences relating to water, land, forest and so on. Experienced farmers possess more information and knowledge of climatic variations and management of crops and livestock (Nhemachena & Hassan, 2007). However, experience or knowledge may not necessarily improve with age. There is evidence of the negative impact of age on soil conservation (Shiferaw & Holden, 1998). We hypothesize that experience raises the likelihood of adaptation to change in climatic conditions. Similarly, experience in farming can cause positive influence on households' perceptions and adaptation behavior. On contrary, the literature is not conclusive about the relationship between gender and adaptation strategies. Evidence suggest that limited information limits efforts towards soil and water conservation by women-headed households, whereas male-headed households have more information and undertake risky measures (Deressa et al., 2008; Tenge et al., 2004). However, Nhemachena and Hassan (2007) showed a more active response to climate change by female-headed households. As regards education, it provides greater information on modern technologies (Lin, 1991) and better management (Norris and Batie, 1987; Nhemachena & Hassan, 2007).

Table 12.1 Descriptions of Variables Used in Econometric Analysis

Variable	*Definition*	*Expected Effects*	*Mean*	*Std. Dev.*
Perception of climate change	Household's perception of climate change: Dummy variable, = 1 if the household perceive climate change, 0 otherwise	Dependent variable	0.83	0.38
Adaptation to climate change	Household's adaption to climate change: Dummy variable = 1 if the household has adopted at least 3 or more strategies to climate change, 0 otherwise	Dependent variable	0.78	0.42
Age of household head	Age of the household head (years)	Positive	49.90	10.68
Farming experience	Number of years the household head has farming experience (years)	Positive	23.47	12.39
Gender of the household head	Gender of the household head: Dummy variable = 1 if the household head is male, 0 otherwise	Positive/ Negative	0.97	0.17
Gender ratio	The ratio of female to male members in the household	Positive/ Negative	1.17	1.11
Family size	Total number of family members of the household	Positive	5.10	1.85
Workforce	Ratio of family members who are able to work (more than 14 years old) to the total family members	Positive	0.67	0.19
Education of household head	Years of education of the household head (years)	Positive	6.44	3.94
Log of farm income	Log of household income from farming	Positive	2.90	2.00
Log of livestock income	Log of household income from livestock rearing	Negative	4.01	1.32
Land tenure	Dummy variable, = 1 if the household own the land, 0 otherwise	Positive	0.93	0.26
Landholding size	Household's landholding size (acre)	Positive	4.27	6.50
Irrigation	Dummy variable =1 if the watering is done either manually, or using a pump or tank water (other than direct rainwater), 0 otherwise	Positive	0.73	0.44

(Continued)

Table 12.1 (Continued)

Variable	*Definition*	*Expected Effects*	*Mean*	*Std. Dev.*
Location of land (head reach)	Dummy variable, = 1 if the household's land is located in head reach of the tank, 0 otherwise	Positive	0.68	0.47
Ownership of television	Dummy variable, = 1 if the household owns a TV, 0 otherwise	Positive	0.69	0.46
House type	Dummy variable, = 1 if the house is concrete, 0 otherwise	Positive	0.34	0.52
Weather information	Dummy variable, = 1 if the household received timely information about weather from extension service centre, 0 otherwise	Positive	0.35	0.48
Institutional credit	Dummy variable, = 1 if the household received institutional credit, 0 otherwise	Positive	0.13	0.33

Access to credit has been considered by numerous studies as an important determinant of adaptation (Pattanayak et al., 2003). Credit availability enables farmers to apply modern technologies. However, dependence on moneylenders often lowers adaptive capacity (Kelkar et al., 2008). Family size and workforce are another two important demographic characteristics of households, which influence households' perception and adaptation to climate. One can assume a positive impact from the size of family or workforce on perceptions and adaptation measures. Larger households have greater chance to use modern technologies intensively (Croppenstedt et al., 2003). We hypothesize that size of family has a direct influence on adaptation.

Irrigation reduces vulnerability vis-à-vis rain-fed farming (Mendelsohn, 2008; Mendelsohn & Dinar, 2009). Therefore, irrigation is considered as crucial for drought-related risk mitigation (Wehbe et al., 2006). Threats of climate change force to enhance irrigation efficiency as a part of adaptation process (World Bank, 2008). One may expect that more wealth would increase risk-bearing capacity (Gbetibouo, 2009) and hence long-term planning by the farmers. Evidence show that the farm income/wealth has a positive impact on soil conservation (Deressa et al., 2008). We hypothesize that household income, as well as off-farm employment, positively contributes to adaptation. In this study ownership of television, house type and off-farm income are used as a proxy for household wealth. Land ownership often helps in adaptation through the use of new technologies more frequently (Lutz et al., 1994; Shultz et al., 1997). They are thus more likely to make more investment the in land. The location of the land can also influence perceptions and adaptation measures. Households with head-reach or tail-reach land can have different perceptions and hence adaptation measures, as the tail-reach farmers may face frequent water shortages.

12.3 Results and discussions: Perceptions about and adaptation to climate change

12.3.1 Household perception of climate change

Table 12.2 reports the results of the logistic regression model on households' perceptions about climate change. The estimated models are significant. Regarding the individual coefficients in the combined (pooled) model, family size is negatively associated with perception implying that smaller households have larger chance to perceive the changes.

Furthermore, farm income has a negative impact on perceptions about changing climate in all the three models, which means that households with less farm income would perceive the change. Land tenure has a positive impact on the same in Odisha. This implies that landowners would perceive more vis-à-vis the tenants. This could be because of the fact that landowners are intimately associated with cultivation process than their tenant counterparts and hence have greater chances to perceive changes in climatic conditions. Surprisingly, the coefficient of land tenure is not significant in the case of West Bengal as well as in the pooled regression model, although the sign is consistent.

The location of land (head-reach) is negatively associated with perception in all the three models. This indicates that households with tail-end land are more likely to perceive climate change unlike those located in head-reach of the tank. This may be because tail-end farmers may face acute problems of water shortage compared to head-reach farmers and hence perceive climate change. Interestingly, landholding size has turned out to be positive, implying that larger landholding raises the chances to perceive climate change.

The result also shows a positive relation between farmers' perception of and provision of an irrigation facility. It is observed that irrigation facilities raise the chances to perceive climate change. Possibly, a lack of irrigation facilities as a result of water shortage increases farmers' vulnerability to scant rainfall, and as such, they are more likely to be affected (Ndambiri et al., 2012). It is also seen that information on weather raises the chance to perceive the changes in climatic conditions. This is because access to information broadens the information base of farmers regarding climate variability (Ndambiri et al., 2012). A farmer with access to necessary information has more chance to perceive the changes. Finally, ownership of television has a positive impact on perception, indicating that richer households have a greater chance to perceive the changes.

Table 12.2 Determinants of Perception of Climate Change

	Odisha	*West Bengal*	*Odisha and West Bengal (Pooled Regression)*
Variable	Coefficient	Coefficient	Coefficient
Age of household head	0.018 (0.019)	0.003 (0.020)	0.008 (0.014)
Farming experience of the household head	0.009 (0.017)	–	0.001 (0.012)
Gender ratio	−0.153 (0.136)	–	−0.052 (0.082)
Gender of the household head	–	0.695 (0.705)	–
Family size	–	–	−0.146** (0.061)
Workforce	3.601 (3.452)	−0.246 (0.936)	0.673 (0.821)
Education of household head	0.081** (0.038)	−0.027 (0.048)	0.02 (0.025)
Log of farm income	−0.185* (0.102)	−0.245** (0.100)	−0.109** (0.057)
Land tenure	2.301** (0.704)	0.516 (0.777)	0.254 (0.472)
Landholding size	0.01 (0.016)	−0.001 (0.093)	0.046* (0.028)
Location of land (head reach)	−1.532** (0.338)	−0.879** (0.411)	−0.98*** (0.232)
Irrigation	1.915*** (0.270)	1.346*** (0.402)	1.31*** (0.218)
Ownership of television	1.025*** (0.272)	1.251*** (0.414)	1.024*** (0.214)
House type	0.086 (0.378)	0.908 (0.813)	0.153 (0.280)
Weather information	0.896 (0.454)	–	2.005*** (0.482)
Institutional credit	–	−3.109*** (0.641)	–
Constant	−4.728* (2.738)	0.576 (1.236)	0.199 (0.992)
Number of observations	603	500	1103
Wald chi^2 (13)	108	135	183
Prob > chi^2	0.000	0.000	0.000
Pseudo-R^2	0.219	0.403	0.250
Log pseudo likelihood	−229	−129	−384

Note: Robust standard errors are given in the parentheses.
*p < 0.10, **p < 0.05, ***p < 0.01.

12.3.2 Household adaptation to climate change

Table 12.3 reports the regression results on the determinants of adaptation to climate change. The overall models are significant. Regarding individual variables in the pooled adaptation model, access to weather information and institutional credit are two new variables that have turned out to be significantly influencing household adaptation to climate change. Similarly, access to irrigation facilitates adaptation, as the variable irrigation is positively associated with adaptation. Irrigation provides technical supports to mitigate drought-related risks (Wehbe et al., 2006) and increases resilience (Gbetibouo, 2009).

Location of land (head-reach) has a negative association with adaptation, implying that households with tail-end land have larger possibility of adaptation vis-à-vis their head-reach counterparts. This is so possibly because the tail-reach farmers often encounter acute scarcity of water and hence may compel them to take adaptation measures. Finally, ownership of television is positively associated with adaptation suggesting that richer households are more likely to apply adaptation strategies than the poor are. More wealth is expected to enhance risk-bearing capacity (Gbetibouo, 2009) and help the farmers in long-term planning.

12.4 Results and discussion: Determinants of household choice of adaptation strategies

Several studies on adaptation observed that adaptation could potentially counter the adversities. For instance, Downing (1991) showed that adaptation could lower food shortage from 50 to 20% in Africa (Nhemachena and Hassan, 2007). However, the forms of adaptation may vary from one region to the other. (Bradshaw et al., 2004) found the diversification of crops in Canadian prairie agriculture has helped farmers in adaptation (Nhemachena and Hassan, 2007). According to Kapetanaki and Rosenzweig (1997), changing planting dates has increased agricultural yield in Greece by 10%. In another study, Iglesias and Minguez (1997) found that, with combinations of adaptation strategies, yield of wheat and maize increased despite higher temperatures. In India, Kumar and Parikh (1998) observed that by adjusting cropping patterns and input use by farmers, losses due to climate change can be minimized significantly. In Africa, farmers choose crops according to local conditions (Kurukulasuriya and Mendelsohn, 2008), although migration influences adaptation strategies significantly in Zimbabwe (Mutekwa, 2009). Similarly, Hassan and Nhemachena (2008) found that shift from farming to non-farming activities as an effective way of adaptation amongst other methods in African agriculture.

Based on related empirical literature, few common adaptation methods in agriculture are identified. These are soil conservation; crop switching; irrigation; changes in plantation dates; shift to non-farm works; diversification;

Table 12.3 Determinants of Household Adaptation to Climate Change

Variable	*Odisha*	*West Bengal*	*Odisha and West Bengal (Pooled Regression)*
	Coefficient	*Coefficient*	*Coefficient*
Age of household head	0.021 (0.013)	0.019 (0.024)	0.009 (0.012)
Farming experience	–	0.013	0.008 (0.010)
Gender ratio	−0.116 (0.120)	–	0.099 (0.084)
Gender of the household head (dummy)	–	1.081* (0.589)	
Family size	–	–	-0.12** (0.058)
Workforce	2.812 (2.962)	−0.681 (0.873)	1.017 (0.697)
Education of household head	0.053* (0.029)	0.034 (0.046)	0.033 (0.022)
Log farm income	−0.145** (0.072)	0.192** (0.097)	0.005 (0.047)
Log non-farm income	0.055 (0.079)	–	–
Land tenure	1.285** (0.597)	−0.116 (0.641)	−0.155 (0.422)
Landholding size	0.022 (0.017)	0.03 (0.073)	0.035 (0.022)
Irrigation (dummy)	1.74*** (0.231)	−0.87** (0.395)	1.002*** (0.196)
Location of land (head reach) (dummy)	−0.795*** (0.269)	1.978*** (0.460)	−0.571*** (0.211)
Ownership of television (dummy)	0.694*** (0.224)	1.47*** (0.385)	0.693*** (0.194)
House type (dummy)	–	-0.315 (0.501)	0.13 (0.219)
Weather information (dummy)	0.167 (0.330)	–	1.528*** (0.363)
Institutional credit (dummy)	–	−2.884*** (0.604)	−0.626** (0.256)
Constant	−4.044* (2.350)	−1.161 (1.237)	-0.631 (0.876)
Number of observations	603	500	1103
Wald $chi^2(12)$	90.72	150.18	202.4
Prob > chi-	0.000	0.000	0.000
Pseudo-R^2	0.141	0.4107	0.2109
Log pseudo likelihood	−291	−143	−461

Note: Robust standard errors are given in the parentheses.
*p < 0.10, **p < 0.05, ***p < 0.01.

livestock management; the use of more pesticides, insecticides and fertilizers; and migration, among others (Waha et al., 2013). Suitability of these adaptation methods in a region largely depends on local socioeconomic conditions (Acosta-Michlik et al., 2008). A particular approach may be successful in one locality, but it may not result favorably elsewhere. In Philippines, the location of markets and irrigation are important for vulnerability as compared to socioeconomic attributes (Acosta-Michlik and Espaldon, 2008), whereas it is opposite in China (Liu et al., 2008). Therefore, set of appropriate adaptation strategies is location-specific and requires detailed investigation.

Generally, multinomial or multivariate logit or probit models are estimated to understand adaptation decisions when multiple options are available (Hausman and Wise, 1978; Wu and Babcock, 1998; Hassan and Nhemachena, 2008). However, the exact choice between depends on the interdependence of alternatives and the distribution of the random disturbance term. Since the alternatives are not mutually exclusive here, we estimate multivariate probit models to analyze adaptation behavior. The response variable indicates alternative strategies including change of plantation dates, more irrigation, more pesticides and change in crop variety. The independent variables control different social, economic and institutional aspects. This is based on the premise that resource constraints along with household-level factors and poor infrastructure limit farmers' ability to take appropriate adaptation measures (Kandlikar and Risbey, 2000).

Tables 12.4 and 12.5 reports the results of multivariate probit models indicating the nature (positive or negative) of the relationship the factors have with adaptation strategies. The result supports that of similar studies conducted across the globe. Education, wealth, location of the land (role of the tank) and timely provision of weather information are found to influence most of the adaptation choices.

While wealth has a direct relation to changing planting dates, crop switching, the use of pesticides and insecticides, conservation of soil and increase in irrigation and non-farm works, it has inverse relation with migration. The possession of greater wealth helps farmers adjust the planting of seeds as and when rain sets in and switch crops to adapt to climate change. Similarly, more wealth enables a farmer to use more pesticides and insecticides, irrigation, soil conservation technique and shift to non-farm works. Farmers who lack wealth are found to migrate more as they have very limited options to adaptation.

Education has been found to directly affect all the choices of adaptations. Even educated farmers are found to migrate more. The farmers who are less wealthy but educated are often found to migrate may be because they understand their situation better and often consider migration as the best option to adapt. Weather information proved to be an important determinant of choice of adaptation. Better weather information helps in adjusting planting dates, switching crop, using greater irrigation, pesticides and insecticides and soil conservation. One who has better access to weather

Table 12.4 Determinants of Household Adaptation Strategies in Odisha and Bengal

	Odisha				*West Bengal*			
Variable	*Change of Plantation Date*	*More Irrigation*	*More Pesticides*	*Change in Crop Variety*	*Change of Plantation Date*	*More Irrigation*	*More Pesticides*	*Change in Crop Variety*
Age of household head	0.026 (0.022)	0.010 (0.022)	0.004 (0.011)	0.010 (0.010)	0.032*** (0.012)	–0.011 (0.009)	–0.006 (0.009)	0.001 (0.023)
Farming experience	–0.054** (0.023)	–0.016 (0.015)	0.009 (0.009)	0.003 (0.008)	–0.021** (0.010)	0.007 (0.008)	–0.005 (0.008)	–0.011 (0.018)
Gender ratio	0.039 (0.220)	0.222* (0.132)	–0.069 (0.087)	–0.064 (0.075)	–0.169** (0.077)	0.043 (0.070)	0.166** (0.078)	–0.174 (0.260)
Family size	–0.406*** (0.104)	–0.040 (0.122)	–0.002 (0.060)	–0.014 (0.055)	0.034 (0.078)	0.022 (0.060)	–0.067 (0.066)	–0.936*** (0.233)
Workforce	4.058 (4.962)	–12.69* (6.523)	2.054 (2.363)	0.490 (2.095)	–1.570** (0.638)	0.796* (0.483)	1.373*** (0.495)	2.314* (1.273)
Education of household head	0.021 (0.052)	–0.006 (0.045)	0.041* (0.023)	0.030 (0.020)	0.019 (0.023)	–0.038* (0.020)	–0.011 (0.020)	–0.129** (0.060)
Log of farm income	–0.251* (0.132)	–0.006 (0.108)	–0.124** (0.059)	–0.096* (0.052)	–0.032 (0.058)	–0.051 (0.048)	0.054 (0.049)	–0.034 (0.133)
Log of non-farm income	–0.671 (0.453)	–0.077 (0.104)	0.081 (0.050)	0.011 (0.043)	0.206*** (0.053)	–0.065 (0.051)	–0.039 (0.051)	0.258* (0.156)
Land tenure	0.153 (0.623)	2.857 (81.530)	0.972** (0.416)	1.133*** (0.407)	0.576 (0.376)	–0.818** (0.340)	–0.150 (0.322)	–0.404 (0.881)
Landholding size	0.007 (0.026)	–0.033 (0.067)	0.031 (0.028)	0.006 (0.011)	0.024 (0.033)	0.018 (0.021)	–0.028 (0.019)	0.094 (0.068)

(Continued)

Table 12.4 (Continued)

	Odisha				*West Bengal*			
Variable	*Change of Plantation Date*	*More Irrigation*	*More Pesticides*	*Change in Crop Variety*	*Change of Plantation Date*	*More Irrigation*	*More Pesticides*	*Change in Crop Variety*
Irrigation	2.255*** (0.299)	−0.0374 (0.371)	1.097*** (0.159)	1.075*** (0.143)	0.249 (0.259)	0.336 (0.228)	−0.135 (0.230)	2.119*** (0.685)
Location of land (head reach)	0.312 (0.291)	0.352 (0.392)	-0.557*** (0.204)	-0.605*** (0.177)	-0.306 (0.392)	0.264 (0.192)	-0.011 (0.210)	0.450 (0.538)
Ownership of television	−0.120 (0.280)	−0.275 (0.374)	0.694*** (0.173)	0.449*** (0.155)	0.349* (0.210)	0.290 (0.195)	0.205 (0.208)	0.850* (0.515)
House type	0.925* (0.554)	0.371 (0.386)	0.035 (0.231)	0.153 (0.202)	−0.160 (0.213)	−0.175 (0.167)	−0.183 (0.170)	−0.294 (0.390)
Weather information	0.163 (0.391)	0.868** (0.344)	0.594** (0.261)	0.349* (0.211)				
Institutional credit					−0.381 (0.2990)	−0.463* (0.2490)	−0.205 (0.2580)	−1.763*** (0.5650)
Constant	3.481 (3.936)	2.099 (81.620)	−2.567 (1.862)	−1.744 (1.695)	0.117 (0.931)	1.526** (0.733)	1.128 (0.737)	2.331 (1.822)
Number of observation	603				499			
Wald $Chi^2(60)$	234				186			
Log likelihood	−511				−539			
Prob > Chi^2	0.000				0.000			

Note: Standard errors are given in the parentheses.
$^*p < 0.10$, $^{**}p < 0.05$, $^{***}p < 0.01$.

Table 12.5 Determinants of Household Adaptation Strategies in Odisha and West Bengal (pooled regression)

Variable	*Change of Plantation Date*	*More Irrigation*	*More Pesticides*	*Change in Crop Variety*
Age of household head	0.0238***	-0.0202***	-0.00310	0.0164**
	(0.00887)	(0.00643)	(0.00629)	(0.00732)
Farming experience	-0.0134*	-0.00803	0.00217	-0.00253
	(0.00788)	(0.00528)	(0.00536)	(0.00628)
Gender ratio	-0.187***	0.252***	0.0368	-0.0182
	(0.0562)	(0.0439)	(0.0505)	(0.0586)
Family size	-0.137***	0.0458	-0.0440	-0.249***
	(0.0426)	(0.0332)	(0.0350)	(0.0406)
Work force	-2.108***	5.227***	0.537*	0.438
	(0.406)	(0.312)	(0.325)	(0.427)
Education of household head	0.0141	-0.0368***	0.00486	0.0151
	(0.0182)	(0.0136)	(0.0138)	(0.0151)
Log of farm income	-0.180***	-0.0850***	-0.0734**	0.0294
	(0.0431)	(0.0294)	(0.0309)	(0.0369)
Log of non-farm income	0.0951**	0.0254	-0.0218	0.0947***
	(0.0419)	(0.0313)	(0.0331)	(0.0356)
Land tenure	0.386	-1.143***	0.330	0.318
	(0.270)	(0.218)	(0.229)	(0.265)
Land holding size	0.0371	0.0171***	-0.000539	0.00982
	(0.0245)	(0.00643)	(0.00824)	(0.0103)
Irrigation	0.955***	0.198	0.677***	1.286***
	(0.150)	(0.128)	(0.120)	(0.129)
Location of land (head reach)	-0.142	0.122	-0.334**	-0.260*
	(0.156)	(0.127)	(0.136)	(0.149)
Ownership of television	0.105	0.0626	0.515***	0.648***
	(0.146)	(0.120)	(0.122)	(0.129)
House type	0.214	-0.253**	-0.199*	0.198
	(0.169)	(0.109)	(0.120)	(0.146)
Institutional credit	-0.165	1.387***	-0.0427	-2.556***
	(0.195)	(0.164)	(0.177)	(0.376)
Constant	1.903***	-2.154***	0.477	-1.016*
	(0.610)	(0.434)	(0.476)	(0.549)
Number of observations	1102			
Wald Chi² (60)	860.24			
Log likelihood	-1403.1576			
Prob > Chi²	0.000			

Note: Standard errors are given in the parentheses.
*$p < 0.10$, **$p < 0.05$, ***$p < 0.01$.

information can take timely action like changing planting dates and crop switching and make necessary provisions of irrigation, pesticides and insecticides and soil conservation.

The study finds that the aged household heads negative impact on migration and changing the number of cattle. Aged people are more likely to migrate less perhaps because they are more attached to their land and because of a lack of physical strength. Similarly, aged people are likely to make fewer changes to the number of cattle. This may be also due to their more attachment to cattle and less capability to adaptation. On contrary, having more household laborers was found to help in changing planting dates.

Farmers having their own land are found to irrigate more perhaps because provision of irrigation requires considerable investment and farmers who do not have their own land do not find it rewarding in investing in others' land for irrigation. Similarly, a lack of institutional credit has been found to cause greater migration may be due to the reason that the farmers are then left with hardly any options to adapt. Selling of cattle and shifts from farming activities to other activities also largely occur due to lack of institutional credit.

12.5 Conclusion and policy implications

Traditional water-harvesting systems such as tanks and ponds can contribute immensely to counter the adverse effects of climate change. It is assumed that farming households first perceive the changes in climate and subsequently adopt various strategies to reduce the adverse effects. Given this backdrop, the present study attempts to understand whether farming households in the command area of the tanks perceive climate change and, if so, whether they adopt any measures to counter the adverse effects. The empirical results show that educated heads with more wealth have a larger probability to perceive climate change. Similarly, the households with higher farm income, irrigation facilities, timely weather information and tail-end land have also a higher probability of perceiving the same. At household level, response to climate change is determined directly by non-farm income, landholding size and family size. These factors enhance the adaptation of various strategies for countering the adverse effects of climate change. Households with larger wealth, tail-end land or irrigation facilities are more likely to respond to climate change. Besides, access to weather information influences households' decisions positively.

Education, wealth, location of the land (role of the tank) and timely provision of weather information are found to influence most of the adaptation choices. In general, restoration of tanks is viewed as an important adaptation strategy by the farmers in rain-fed agricultural regions, which are highly vulnerable to frequent climatic shocks. In other words, a traditional water-harvesting system such as tanks has the inherent potential to enhance the ability of the ecosystem and households' resilience to cope with various climatic shocks such as droughts and stabilize social and economic conditions and the ecology.

Notes

1 Rain-fed agriculture contributes more than 75% of total cropped area in the world, whereas one-third of population of the developing countries lives in rain-fed regions (World Bank, 2008).

2 These conditions comprise ability to self-organize, buffer disturbance and learn and adapt (Tompkins and Neil Adger, 2005).

References

Acosta-Michlik, Lilibeth, and Victoria Espaldon. "Assessing Vulnerability of Selected Farming Communities in the Philippines Based on a Behavioural Model of Agent's Adaptation to Global Environmental Change." *Global Environmental Change* 18, no. 4 (2008): 554–563.

Acosta-Michlik, Lilibeth, Ulka Kelkar, and Upasna Sharma. "A Critical Overview: Local Evidence on Vulnerabilities and Adaptations to Global Environmental Change in Developing Countries." *Global Environmental Change* 18, no. 4 (2008): 539–810.

Below, Till, Astrid Artner, Rosemarie Siebert, and Stefan Sieber. "Micro-Level Practices to Adapt to Climate Change for African Small-Scale Farmers." *A Review of Selected Literature* 953 (2010). International Food Policy Research Institute (IFPRI), Washington DC, USA, pp. 1–20.

Bradshaw, Ben, Holly Dolan, and Barry Smit. "Farm-Level Adaptation to Climatic Variability and Change: Crop Diversification in the Canadian Prairies." *Climatic Change* 67, no. 1 (2004): 119–141.

Census of India. 2011. Registrar General & Census Commissioner of India; New Delhi.

Croppenstedt, Andre, Mulat Demeke, and Meloria M. Meschi. "Technology Adoption in the Presence of Constraints: The Case of Fertilizer Demand in Ethiopia." *Review of Development Economics* 7, no. 1 (2003): 58–70.

Deressa, Temesgen, Rashid M. Hassan, and Claudia Ringler. "Measuring Ethiopian Farmers' Vulnerability to Climate Change across Regional States." International Food Policy Research Institute (2008).

DFID, U. K. (1999). *Sustainable livelihoods guidance sheets*. London: DFID, 445.

Downing, Thomas E. "Vulnerability to Hunger in Africa: A Climate Change Perspective." *Global Environmental Change* 1, no. 5 (1991): 365–380.

Gbetibouo, Glwadys Aymone. "Understanding Farmers' Perceptions and Adaptations to Climate Change and Variability: The Case of the Limpopo Basin, South Africa." Vol. 849. *Intl Food Policy Res Inst* (2009).

Gleick, Peter H. "Water and Conflict: Fresh Water Resources and International Security." *International Security* 18, no. 1 (1993): 79-112.

Gleick, Peter H., and Charles W. Howe. "Water in Crisis: A Guide to the World's Fresh Water Resources." *Climatic Change* 31, no. 1 (1993): 119–122.

Hassan, Rashid M., and Charles Nhemachena. "Determinants of African Farmers' Strategies for Adapting to Climate Change: Multinomial Choice Analysis." *African Journal of Agricultural and Resource Economics* 2, no. 311-2016-5521 (2008): 83–104.

Hausman, Jerry A., and David A. Wise. "A Conditional Probit Model for Quantitative Choice: Discrete Decisions Recognizing Interdependence and Heterogeneous Preferences." *Econometrica* 46, no. 2 (1978): 403–426.

Iglesias, Ana, and M. I. Minguez. "Modelling Crop-Climate Interactions in Spain: Vulnerability and Adaptation of Different Agricultural Systems to Climate Change." *Mitigation and Adaptation Strategies for Global Change* 1, no. 3 (1997): 273–288.

IPCC. "Climate change 2001: The scientific basis. Contribution of Working Group I to the Third Assessment Report of the Intergovernmental Panel on Climate Change." In *Climate Change 2001: The Scientific Basis*, J. T. Houghton, Y. Ding, D. J. Griggs, M. Noguer, P. J. Van der Linden, X. Dai, K. Maskell, and C. A. Johnson (eds.). Cambridge University Press: New York, pp. 1–881. (2001).

Kandlikar, Milind, and James Risbey. "Agricultural Impacts of Climate Change: If Adaptation Is the Answer, What Is the Question?" *Climatic Change* 45, no. 3 (2000): 529–539.

Kapetanaki, G., and Cynthia Rosenzweig. "Impact of Climate Change on Maize Yield in Central and Northern Greece: A Simulation Study with CERES-Maize." *Mitigation and Adaptation Strategies for Global Change* 1, no. 3 (1997): 251–271.

Kelkar, Ulka, Kapil Kumar Narula, Ved Prakash Sharma, and Usha Chandna. "Vulnerability and Adaptation to Climate Variability and Water Stress in Uttarakhand State, India." *Global Environmental Change* 18, no. 4 (2008): 564–574.

Kumar, K. S., and Jyoti Parikh. "Climate Change Impacts on Indian Agriculture: The Ricardian Approach." In Dinar et al. (eds.), *Measuring the Impact of Climate Change on Indian Agriculture*, Technical Paper, 0253-7494 ; No. 402, World Bank, Washington, DC, USA, pp. 141–184. (1998).

Kurukulasuriya, Pradeep, and Robert Mendelsohn. *"How Will Climate Change Shift Agro-Ecological Zones and Impact African Agriculture?" The World Bank* (2008).

Lin, C.-T.J., K.L. Jensen, and S.T. Yen. "Awareness of foodborne pathogens among US consumers." *Food Quality and Preference* 16 (2005): 401–412

Lin, Justin Yifu. "Education and Innovation Adoption in Agriculture: Evidence from Hybrid Rice in China." *American Journal of Agricultural Economics* 73, no. 3 (1991): 713–723.

Liu, Chunling, Dominic Golding, and Gang Gong. "Farmers' Coping Response to the Low Flows in the Lower Yellow River: A Case Study of Temporal Dimensions of Vulnerability." *Global Environmental Change* 18, no. 4 (2008): 543–553.

Lutz, E., S. Pagiola, and Y. C. Reiche. "The costs and benefits of soil conservation: The farmer's viewpoint." *The World Bank Research Observer* 9 (1994): 273–295.

Mendelsohn, Robert. "The Impact of Climate Change on Agriculture in Developing Countries." *Journal of Natural Resources Policy Research* 1, no. 1 (2008): 5–19.

Mendelsohn, Robert O., and Ariel Dinar. *Climate Change and Agriculture: An Economic Analysis of Global Impacts, Adaptation and Distributional Effects.* Cheltenham, UK, Northampton, MA, USA: Edward Elgar Publishing (2009).

Mutekwa, V. T. "Climate Change Impacts and Adaptation in the Agricultural Sector: The Case of Smallholder Farmers in Zimbabwe." *Journal of Sustainable Development in Africa* 11, no. 2 (2009): 237–256.

Ndambiri, H., C. Ritho, S. G. Mbogoh, P. Nyangweso, S. Ng'ang'a, E. Muiruri, M. J. Kipsat, P. Kubowon, F. Cherotwo, and P. Omboto. Analysis of farmers' perceptions of the effects of climate change in Kenya: the case of Kyuso district (2012).

Nhemachena, Charles, and Rashid Hassan. "Micro-Level Analysis of Farmers Adaption to Climate Change in Southern Africa." International Food Policy Research Institute (2007).

Norris, Patricia E., and Sandra S. Batie. "Virginia Farmers' Soil Conservation Decisions: An Application of Tobit Analysis." *Journal of Agricultural and Applied Economics* 19, no. 1 (1987): 79–90.

Pattanayak, Subhrendu K., D. Evan Mercer, Erin Sills, and Jui-Chen Yang. "Taking Stock of Agroforestry Adoption Studies." *Agroforestry Systems* 57, no. 3 (2003): 173–186.

Shiferaw, Bekele, and Stein T. Holden. "Resource Degradation and Adoption of Land Conservation Technologies in the Ethiopian Highlands: A Case Study in Andit Tid, North Shewa." *Agricultural Economics* 18, no. 3 (1998): 233–247.

Shultz, S., J. Faustino, and D. Melgar. "Agroforestry and soil conservation: Adoption and profitability in El Salvador." *Agroforestry Today* 9 (1997): 16–17.

Tarleton, Margaret, and Doug Ramsey. "Farm-Level Adaptation to Multiple Risks: Climate Change and Other Concerns." *Journal of Rural and Community Development* 3, no. 2 (2008): 47–62.

Tenge, A. J., J. De Graaff, and J. P. Hella. "Social and Economic Factors Affecting the Adoption of Soil and Water Conservation in West Usambara Highlands, Tanzania." *Land Degradation & Development* 15, no. 2 (2004): 99–114.

Tietenberg, T. Environmental and Natural Resource Economics. Pearson/Addison Wesley (2006).

Tompkins, Emma L., and W. Neil Adger. "Defining Response Capacity to Enhance Climate Change Policy." *Environmental Science & Policy* 8, no. 6 (2005): 562 571.

UKCIP. Climate adaptation: Risk, uncertainty and decision-making. UK Climate Impact Programme (UKCIP) Technical Report May (2003).

UNEP. Vital Water Grphics: An Overview of the State of the World's Fresh and Marine Water. United Nation Environment Programme. Nairobi, Kenya (2002).

UNESCO. Securing the Food Supply. World Water Assessment Programme. United Nations. United Nations Education Scientific and Cultural Organization (Paris) (2001).

Waha, Katharina, Christoph Müller, Alberte Bondeau, Jan Philipp Dietrich, Pradeep Kurukulasuriya, Jens Heinke, and Hermann Lotze-Campen. "Adaptation to Climate Change through the Choice of Cropping System and Sowing Date in Sub-Saharan Africa." *Global Environmental Change* 23, no. 1 (2013): 130–143.

Wehbe, Monica, Hallie Eakin, Roberto Seiler, Marta Vinocur, Cristian Ávila, and Cecilia Marutto. "*Local Perspectives on Adaptation to Climate Change: Lessons from Mexico and Argentina.*" AIACC Working Paper No. 39, The AIACC Project Office, International START Secretariat, Washington, DC. (2006).

World Bank. *World Development Report 2008*. Washington, DC: World Bank (2008).

Wu, JunJie, and Bruce A. Babcock. "The Choice of Tillage, Rotation, and Soil Testing Practices: Economic and Environmental Implications." *American Journal of Agricultural Economics* 80, no. 3 (1998): 494–511.

Part IV

Health

13 Determinants of child survival at the household level

An insight of the method of factor analysis

Arun Kumar Sharma and Rohini Dutta (Ghosh)

13.1 Introduction

The UN Sustainable Goals target to reduce child mortality to not more than 25/1000 live births in every country by 2030 (United Nations Children's Fund [UNICEF], World Health Organization [WHO], World Bank, and United Nations [UN], 2015). However, in India one-fourth of the world's neonatal deaths occur at home (UNICEF, 2008), mostly in rural areas (IIPS, 2017). The factors associated with child mortality start operating from the time of conception, continue during pregnancy and cover the neonatal period (Dubey et al., 2015). Many of the deaths take place even before birth. It is estimated that globally 2.64 million stillbirths occur every year; mostly in South Asia and sub-Saharan Africa (WHO, 2004; Cousens et al., 2011), and one in five pregnancies results in induced abortion. The under-five mortality in India was recently recorded as 50/1000 (IIPS, 2017), and India couldn't achieve the Millennium Development Goals to reduce under-five mortality by two-thirds between 1990 and 2015.

Factors of child mortality in India are complex. First, there are various socioeconomic, maternal, demographic and environmental matters associated with child mortality (National Institute of Medical Statistics, Indian Council of Medical Research, and UNICEF, 2012). Second, despite widespread improvement in health care facilities and the introduction of various maternal and child care programs, there is a marked diversity and inequality in access to health care, resulting in the stagnation of the decline of child mortality in some regions (Ghosh, 2012). This inequality is the result of various factors affecting child survival both at the macro-level (related to political, community and economic factors) and at the micro-level or household level (household environment, income, educational, maternal, entitlements or norms regarding the distribution of resources within the family, among others). These determinants of child survival are again affected by huge socioeconomic variations, directed and influenced by cultural beliefs (Ghosh, 2012).

In this context, this chapter aims to explore whether some maternal, demographic, health care and socioeconomic factors influence child mortality separately at three successive stages: (a) prenatal deaths defined as all deaths before the birth, (b) neonatal deaths defined as deaths between birth and 28 days and (c) post-neonatal deaths defined as deaths between one month and completion of five years. The factors (latent or hidden) associated with child mortality at the household level are explored using the technique of principal component analysis (PCA). This technique is used to identify the latent factors and to retain the originality of the data, reducing the multicollinearity between these variables. The chapter is divided into two parts. The first part discusses the factor analysis (FA) technique, and the second part applies FA to a data set on experience of fetal and child deaths at the household level in a peri-urban setting of Uttar Pradesh state, having the highest child mortality as recorded by the latest report NHHS-4, 2017.

13.2 Factor analysis

Factor analysis (FA) is part of multivariate research methods. Originally developed by Spearman and Thurstone in psychology (Bernard, 2012), it is used widely in social sciences for several purposes, such as data reduction, scale construction, and hypotheses testing in simultaneous equation models, also called structural equation models (Greene, 1976). Social scientists and statisticians have developed various techniques for both exploratory analysis and confirmatory analysis. This chapter aims at explaining the methods and procedures used in applying FA and illustrating them with the application of FA to data on child mortality. For beginners, the various questions regarding FA are answered in a simple manner.

13.2.1 *What are factors?*

Conceptually, they are the theoretical constructs that cannot be measured directly but that influence a number of variables/measurements used in an empirical study. They are identified through modeling relationships between factors and variables. For example, the Human Development Index (HDI) may be viewed as a factor. As it is well known, the United Nations Development Programme constructs the HDI on the basis of measures of life expectancy, mean and expected years of schooling and a decent standard of living (gross national income per capita, measured in terms of purchasing power parity [PPP]). HDI is a linear function of certain transformations of them. Yet, it has a special heuristic significance in the theoretical discourse on development. We can call it a construct or a factor. We cannot collect data on this directly. We collect data on mortality, schooling, income and prices. The theory of development needs construction from the HDI. None of the other measures can serve the same theoretical purpose that a single factor like the HDI does. Similarly, the Gender Development Index is a factor that is constructed on the basis of measures of mortality separately among males and females producing estimates of life expectancy for two sexes, years of schooling among males and females, and gross national income per capita (PPP$). In surveys researchers collect data on a large number of variables/measures pertaining to the theme of their research. Using exploratory FA, they may discover several such factors that may or may not be linked with each other. However, while in case of the HDI we define it on the basis of socioeconomic indicators, factors are unearthed by analysing linear relationships between measures in the form of covariance or correlation matrices.

Algebraically, in much of the literature on the subject, factors are linear combinations of measurements. They are constructed through weighted sum of all or a few of the original measurements. Weights may be equal or unequal. Variables may be correlated or uncorrelated with each other. The values of the variables are the observed values or normalized values. The number of factors is chosen by the researchers and may be the same as the number of variables or less. In the former case, FA is called PCA, and in the latter case, it is called the FA. While PCA may be considered as a deterministic model the FA contains random error. It may be noted that if the number of factors is less than the number of variables the factors taken together would not explain full variation in the data but only a part of it.

It must be noted that results of FA are interpreted by the researcher. The analysis may work in some situations, and it may not work in others. At the end of FA, one may find that the data were not suited to FA and that there are no interpretable factors in the data. Then the analysis is wasted.

Therefore, first of all, one has to see if the data permit or require FA. FA requires large data, that is, data on a number of variables from a reasonably large sample. The variables are assumed to be the effects of a few underlying factors, and for running the analysis, the sample size should be adequate. It is recommended that the sample size should be 10–15 times the number of variables/measurements that are to be factor analyzed. Different rules of thumb are used by different researchers regarding sample-to-variables ratio, varying from 3:1 to 20:1 (Williams et al., 2010). A 10:1 ratio may generally be acceptable; that is, if the number of variables is 20, the minimum sample size has to be around 200. Survey data normally meet this requirement, but small sample data from laboratory experiments or randomized control trials about effectiveness of a drug may not permit the application of FA.

Second, the variables should be neither perfectly correlated nor completely independent. There has to be a pattern of relationships. Imagine that all the variables are perfectly correlated. Does the application of FA make any sense? The answer is no. All the variables tell the same story. Each of them is the perfect reflection of the same factor. If there is no correlation between any pair of variables the variables may be taken to be the reflection of as many factors as the number of variables (Tacq, 1997). In other words, if all the variables are uncorrelated (i.e., have zero inter-correlations), then there is no common factor; each of them may be considered to be a separate factor. To test whether the correlations are diffused (neither perfect nor zeros), two tests are used: Kaiser–Meyer–Olkin (KMO) and Bartlett's test of sphericity. KMO compares partial correlations (correlations between pairs of variables when the effects of all other variables have been removed) with observed correlations. Its value lies between 0 and 1; 0 implies that the sum of partial correlations of a variable with others is large relative to the sum of zero-order correlations, while 1 implies that the pattern of correlations is robust. In the former case, FA would make no sense, indicating a diffused pattern of correlation; in the second case, one can defend that factors are distinct and reliable. The results may be shown in the form of an anti-image matrix in which diagonals are the KMO's measures of sampling adequacy of individual variables and other terms are negatives of partial correlations. Through calculation of the KMO one can decide whether to go for FA or not (Field, 2006).

Ideally, the KMO of an individual variable should be above .50. If it is less than that it may be dropped from the analysis. KMO of any jth variable (x_j) is calculated as sum of squares of all intercorrelations, divided by sum of squares of intercorrelations, plus sum of squares of all partial correlations of x_j with all other variables removing the effect of the rest:

$$\mathrm{KMO}(x_j) = \frac{\sum r_{ij}^2}{\sum r_{ij}^2 + \sum u_{ij}^2} \quad i \neq j.$$

In the preceding equation r_{ij} refers to correlations and u_{ij} to partial correlations. Overall KMO is calculated for all variables using the preceding formula ($i \neq j$). Again, a rule of thumb would suggest that this KMO should be higher than .6. Yet there is a caveat here: the advanced statistical applications are not routine, mechanical exercises. They are judgemental and subjective.

Bartlett's test of sphericity is used to test whether the correlations do not differ significantly from zero; that is, the hypothesis of correlation matrix showing an identity matrix cannot be rejected.

Thus, factors are functions of the original variables and are influenced by several variables. Sometimes (but not always) it is possible to treat them as theoretical constructs that unearth the hidden structure in the data and provide a parsimonious representation of original data. Sometimes it is not possible and the analysis is just wasted. Algebraically, they may be seen as the linear combinations of all the measured variables (there are nonlinear FA techniques also, but they are mostly of theoretical interest and are not practically used). For example, for independent factoring the values of the variables may be expressed as (Kim and Mueller, 1991):

$$\text{Linear combinations}: X_1 = b_{11} F_1 + b_{12} F_2 + \ldots + d_1 U_1$$

$$X_2 = b_{21} F_1 + b_{22} F_2 + \ldots + d_2 U_2$$

$$X_n = b_{n1} F_1 + b_{n2} F_2 + \ldots + d_n U_n$$

By definition F_1, F_2, and so on are factors, and U_1, U_2, and so on are unique factors (i.e., factors unique to any variable). bs are called factor loadings and may be interpreted as correlations between factors and variables (i.e., the extent to which a factor influences a variable or a variable influences the factor). In PCA, with which it is customary to start the analysis, there is no unique factor; the number of factors is same as the number of variables, and thus, all the variation in the data are explained. The analysis is based on the following assumptions:

$$\text{cov}(F_1, F_2) = \text{cov}(F_i, U_j) = \text{cov}(U_j, U_k) = 0$$

The decomposition of variance of any variable X_i:

$$\text{var}(X_i) = b_{i1}^2 + b_{i2}^2 + \ldots + d_i^2$$

This variance shared by any variable X_i with all the factors (i.e., $b_{i1}^2 + b_{i2}^2 + \ldots$) is called the communality of the variable. A variable that shares less variance with the factors has more of the unique variance. On the basis of the preceding, each factor may also be expressed as a linear combination of the original variables.

Variables are grouped into factors on the basis of their factor loadings. Kline (1994) suggests that all interpretations of factors, based on factor loadings, should, however, be validated using external criteria. In other words, the factors must have good predictive validity. Sometimes the differences between loadings of a factor on two or more variables are not so pronounced. To produce a clearer image of which variables affect a factor, varimax rotation is used. It maximizes the variance of factor loadings and facilitates distinction between high and low effects. Although varimax rotation is the commonly used method for deciding factoring, there are some other methods also which are used in special cases. For example, a method that maximizes variances in the rows is called quartimax. Equimax is a compromise between the two: varimax and quartimax. If varimax rotation is used then the number of variables correlating with a factor is minimized. This helps in interpreting the results and forming the factors. This is the most commonly used technique of rotation.

After running the FA, one may compute the scores (values) of all the factors for each subject, because if factors are the theoretical constructs one may like to get the quantitative measures of factors for all the subjects of the study (respondents). These values are called factor scores. Factors are theoretical constructs that may have supposedly given rise to the observed data. For more advanced analysis, variables may be replaced by factors. Thus, after completing the FA, one may treat factors as new variables for further analysis. If one wants to work with original variables only but wants to identify only a limited number of variables (as the number of factors), one may identify variables with which factors have the highest loading. They are called surrogate or substitute variables. In scale construction based on a number of items, summated scores may also be obtained by averaging values of all the surrogate variables, that is, variables that have the highest correlations with different factors. Scores of the first factor may be used as the value of the index that best summarizes the information contained in the variables (Hair et al., 2006).

Yet, the first step in FA is to review the literature and explore if the subject of the study has ever been analyzed using FA and whether one is dealing with a large number of correlated variables that have an underlying factor structure. The next step is to measure the variables for all the study subjects/respondents. Using SPSS or any other statistical package, one may produce the correlation matrix and study the pattern of correlations. As mentioned earlier, correlation coefficients or covariances are the inputs for FA. The application of principal component method can help in deciding the minimum number of underlying factors and some other issues such as how much variation in the data is explained by these factors.

One of the outcomes of the FA is the matrix of factor loadings. These loadings are the correlations between factors and variables. Logically speaking, a factor with which a variable loads highly is the factor that influences this variable. Fixing a minimum value of loadings (irrespective of sign) one can know what set of variables are most influenced by a common factor. This knowledge is used in "naming" or interpreting factors. One may also involve five or six experts in understanding the pattern of loadings and naming the factors.

The theory is FA is simple but its application requires a good understanding of matrix algebra. Basically, factors are the best linear combinations of variables that are based on certain statistical principles such as maximum likelihood estimation. They can be found by equating mod of correlation matrix minus lambda times the identity matrix to zero (Anderson, 1974):

$$|R - \Lambda I| = 0.$$

R is the correlation (or covariance) matrix. Lambda is an unknown parameter of the model and the equation produces multiple solutions—same as the number of variables. Then the values of lambdas are arranged in descending order, and they are interpreted as the eigenvalues or the variances of the factors. They refer to variance explained by different linear combinations of variables. The sum of all lambdas is the same as the number of variables (i.e., the total variability of the data). The sum of first k lambdas shows what part of the total variability is explained by these k factors (Cooper and Weekes, 1991).

The next step is to produce eigenvectors corresponding to these lambdas (factors). For each lambda, putting matrix products of correlation matrix minus lambda times the identity matrix, and the column vector of coefficients (eigenvectors) equal to zero gives the factor loadings or eigenvectors:

$$(R - \Lambda I)x = 0.$$

If there are p variables the dimension of the correlation matrix is p×p and the product of this matrix with a column vector consisting of p coefficients produces a p×1 column vector.

The relative importance of different factors derives from how much variation in the data they can explain. When normalized, each variable has a variance of 1. Thus, the total sum of squares of the data may be taken to be the same as the number of variables. A small number of factors explain a large part of the total variability in the data. They are the factors to be chosen for further analysis. One may use a number of criteria for deciding the number of factors: (a) factors with variance of more than unity and (b) a small set of factors which explain a major part of variation in the data (say, 70%). The principle of parsimony suggests that when the additional factor does not add much to explanation of variability in the data one should stop adding to factors. After performing a PCA, one may decide how many factors one would like to retain. A scree plot that plots variance of lambdas arranged in descending order is also used for visual inspection and decision.

The condition of only one factor underlying all or most of the items is called the condition of unidimensionality. In the construction of scales in psychology and health, it is important to ask whether there is *strict unidimensionality* of the scale items (i.e., whether there is only one factor which loads highly on most of the items), or there is an *essential unidimensionality* (where there is one more minor factor with high loadings on one or more items). It has been observed that evidence for unidimensionality depends on sample size, communality, distribution of responses, and the decision-making methods, such as chi-square test or maximum likelihood method (Slocum-Gori and Zumbo, 2010).

13.2.1.1 Methods of factoring

There are several technical methods of factoring, such as PCA, unweighted least squares, generalized least squares, maximum likelihood, principal axis factoring, alpha factoring and image factoring (Kim and Mueller, 1991). All the methods produce the same degree of fitness between the model and the data but they differ in the ease with which factors can be identified. In most cases, they will produce similar results, but for advanced students of FA, it will be important to distinguish between these methods. Unless the scholar is well versed with the algebra of different methods and there are specific needs for doing so, PCA with or without rotation may be used.

Like the original variables, the factors may also be made to be correlated or uncorrelated. The choice depends on the researcher and the complexity of relationships. There are separate techniques to obtain dependent (oblique) or independent (orthogonal) factors. Depending on one's understanding of the factoring of data, a researcher may choose one or the other type of factor methods. Normally, for statistical purposes, orthogonal factors are of greater importance, but if theory says that the underlying factors may be overlapping, then one may go for oblique factoring. In case of multivariate analysis, researchers regularly face the problem of multicollinearity. In such cases, orthogonal factoring comes as a big help in data analysis.

There are several ways of judging the success of FA. As mentioned earlier, the extent to which the factoring reproduces the observed correlation matrix is the true indicator of success of FA. Thus, the statistical way of measuring the goodness of applicability of FA is the table of "reproduced correlation matrix". The diagonals of the table show the communalities of the variables. The lower left triangle of the matrix shows the reproduced correlations and the upper right triangle shows the residuals, that is, differences between empirical correlations and reproduced correlations. If most of the residuals are small one can say that the FA is applicable (Malhotra and Dash, 2011). What it means is that there are distinct common factors (constructs) that underlie the observed correlations.

In practice, a successful FA must produce meaningful and discernible factors. Thus, the most important point is to take the expert advice on whether factors can be named and have a certain theoretical meaning. If factors are not found to represent some aspect of reality which seems plausible in the light of the theory, FA is wasted. One may also use the FA on split samples separately to explore the consistency of results and to test whether a model combining measures and factors is plausible. One may also use confirmatory factor analysis (CFA) to test the validity of a specified model or to select one from the several alternative models. CFA, however, should not be done on the basis of the same sample. It may be done on a part of the sample on which an exploratory FA (EFA) has not been done. Some scholars suggest that the dependent variable(s) should not be mixed with independent variables. In social science and management practices, however, many variations to these rules are observed.

The two major aims of FA are (a) to unearth the underlying factors and (b) to develop indices based on FA. There is no alternative to FA for the first aim. However, in the case of the second, one may use common sense, substantive theory and a simple weighted average of all the relevant variables. The FA discussed earlier is called R factor. An alternative to this is Q FA that analyzes a correlation matrix between individuals or subjects rather than variables. This results in the grouping of individuals on the basis of multiple measurements on them.

In the case of CFA, that is, FA for testing a model containing factors and variables, a chi-square test of goodness of fit (GOF) may be used to identify which model fits the data best. It may be noted that CFA can only tell whether a model is plausible and not that it is true. Chi-square is calculated as F_{ML} multiplied by N (Brown, 2015). Here, ML stands for maximum likelihood estimates and F for F-ratio. It compares the original covariance matrix with the estimated covariance matrix. For large values of chi-square (i.e., for $p < .05$ or $.01$) the null hypothesis (H_0) of imperfect fit is rejected in favour of the alternative hypothesis (H_1) of complete fit. Root mean square error of approximation (RMSEA) and comparative fit index (CFI) are other measures of fitness of the model (Suhr, internet; Prudone, internet). Long

(1988) discusses the various methods used in CFA, such as unweighted least squares, scale dependency, generalized least squares and maximum likelihood methods. Structural equation modeling is often considered to be equivalent to path analysis but is more general than that. Unlike EFA, CFA is based on assumptions and theories about the nature of relationships between variables and factors.

Correspondence analysis is a technique similar to PCA which can be used for categorical variables. It displays data into two-dimensional graphic forms. It is often used for dividing the chi-squared statistic for the table into two orthogonal factors. It can provide discrimination measures for each variable and a graphical display of all cases according to the two factors or dimensions. In social sciences, it is, however, not so popular as FA.

Kerlinger and Lee (2000) suggest that sometimes one may need to go for second-order FA. Giving the example of g test of intelligence they suggest that it may be important to examine factoring of factors. But this is rarely done in substantive works in social sciences. This concerns only mathematical theoreticians.

13.3 Analysis

The earlier-discussed technique of FA is used to explore socioeconomic, demographic and health care factors associated with perinatal, neonatal and post-neonatal. It was thought that FA can help if one begins with all variables/measures which can be posited to influence child mortality. Then the variables are factor analyzed and factors are formed. The variables which are influenced by the same factor as a component of child mortality are considered to be most closely associated with child mortality (either as consequences or determinants). It is notable that child mortality is affected by various factors initiating and affecting even before the birth of the child. The intricate web of various biological and cultural factors interact with the child survival status. FA has earlier been used to identify groups or clusters of deaths (Field, 2006). For example, a recent study in Yemen found that three socioeconomic status indicators—wealth, education and housing quality—were associated with maternal and child health (Alosaimi et al., 2016). Another study used a PCA in China to construct socioeconomic status indices for prenatal care and examined their association to perinatal care and its outcomes (Nwaru et al., 2012). The major component of child mortality in India is high neonatal mortality, despite improvement in access to healthcare and nutrition (Ghosh, 2012). This study applies PCA to some household variables affecting child mortality in a state with one of the highest child mortality in India—Uttar Pradesh.

13.3.1 Study area

The study was done in the peri-urban area of Kanpur city of Uttar Pradesh which has a very high under-five child mortality. The state is characterized by illiteracy, cultural restrictions on women's movement and stigma related to reproductive and sexual health. Data was collected from five villages, in close proximity to a community health center (CHC) and with access to urban health facilities. This area is about 25 km from the city of Kanpur, with a population of diverse socioeconomic backgrounds.

13.3.2 Research design

Information was gathered using the structured interview schedule from all women in the selected villages situated at a radius of 2 to 5 km from the CHC. The schedules were developed by referring to questionnaires used in the National Family Health Survey (IIPS, 2007). Questions were prepared in local language (Hindi) and a pilot survey was done to test the validity of the questionnaires. Learning from the experiences of the pilot study, some modifications were made in the order and wordings of the questions and the changes were again tested in the field.

Information on age, pregnancy history, socioeconomic background and health care services were recorded. It may be noted that in the absence of any birth and death records enumerating age was difficult; the area lacks education and literacy among women, particularly among middle and lower castes. Therefore, estimation of age was done from the reported age at menarche, marriage, first pregnancy, birth interval and date of the childbirth. Local historical events and important dates were also referred. Data on socioeconomic characteristics like respondent's caste (social class), education, and work status was collected. Possession of more than or equal to four Bighas of land, which was considered sufficient to feed a family of five (pilot survey information), was taken as one of the economic indicators in the community. Poverty status was recorded from the possession of Below Poverty Line (BPL) Card, and the household Standard of Living Index (SLI) was created by the method used in NFHS-3. It was an additive scale based on poverty level, possession of agricultural land and bank account, housing, environmental condition, presence of electricity, animal possession, presence of kitchen, toilet facility and material possessions like television, refrigerator and so on. Also, using observational technique, the field assistants were asked to observe the housing, environment condition and household material possession, and an observational socioeconomic score (observed SES) was created. This scale was matched with the SLI index and households were categorized as high, medium and low in terms of SES.

Fieldwork was carried out for three months from the beginning of January to end of March 2008. All married women (<50 years) from the five chosen settlements were included in this study. Women who had achieved menopause on the date of the survey were excluded. A total of 983 women were found to be fulfilling the criteria fixed for the selection of respondents in this study. However, some incomplete schedules due to non-response to certain important questions were dropped from the analysis and the final analysis was done on 912 women.

Table 13.1 presents the summary of all the demographic, health care and socioeconomic variables considered for PCA to understand their association with mortality in the three successive stages. The mean age of the women in the study area was 31.86, with a mean number of pregnancies of 3.69. The mean total live births was 3.18 with a family size of 5.08. Most of the women respondents belonged to Scheduled Castes (SCs), Scheduled Tribes (STs) and castes named as other backward classes, that is, Other Backward Castes (OBCs; 70.2%). Literacy was recorded as 64% among women and 77.7% among the men. Most of the women lived at a distance of about 4 km from the community health center (33.7%). Only 46% of women had full antenatal checkups during pregnancy, and most of the deliveries happened at home (77%). The mean birth weight of newborn babies was 2.83 kg, and only 9.4% of the newborn babies were fully immunized after birth. Exclusive breastfeeding was found among 45.5% of children.

Table 13.1 Background Characteristics of the Study Area

Socioeconomic and Demographic Variables	*N (%)/Mean*
Age	31.86
Number of pregnancies	3.69
Number of Live births	3.18
Members in the family	5.08
Caste	
Other caste (SC/ST/OBC)	647 (70.2)
General caste	274 (29.4)
Women's education	
Non-literate	332(36.0)
Literate	589(64.0)
Husband's education	
Non-literate	205(22.3)
Literate	716(77.7)
Distance of health care (Kms)	
2	169(18.3)
3	202(21.9)
4	310(33.7)
5+	240(26.06)
Antenatal care	
Yes	46.2
No	53.8
Place of delivery	
Home	77.0
Others	33.0
Birth weight	2.83
Immunization of new borne	
Fully immunized	9.4
Partially or not immunized	90.6
Breastfeeding	

(Continued)

Table 13.1 (Continued)

Socioeconomic and Demographic Variables	*N (%)/Mean*
Yes	45.5
No	54.5
Possession of land	
Less than 4 bigha	88.3
More than 4 Bigha	11.7
Income (previous month)	2950.71
Expenditure (previous month)	2796.20
Socioeconomic score*	
<50	359(39.0)
50-54	378(41.0)
>=55	184(20)
Observed socioeconomic status**	
Low	377 (40.9)
Medium	342(37.1)
High	202(21.9)
Poverty level	
Below Poverty level	437(47.4)
Above Poverty level	484(52.6)
Women who experienced perinatal mortality (Pregnancy wastage)	31.3
Women who experienced neonatal mortality (Death between birth and 28 days)	13.2
Women who experienced post-neonatal mortality (Death between 1 month and 60 months)	10.2

* The score calculated from the material possession
** Observation of the condition of living by the field assistant

The socioeconomic indicators reveal that 88% of the household possess less than four Bighas of land. The SLI index calculated from the material possession of each household indicated that only 20% of households have a score more than 55. They have all the necessary amenities. 39% have a score less than 50. On the basis of observed SES, 21.9% respondents were placed in the high category and 40.9% of the respondents were categorized into low; 37% were placed in the medium category. Of households, 47% were placed in the below poverty level. Prenatal mortality (miscarriage, abortions and stillbirths) is observed to be as high as 31%. Neonatal mortality and postnatal mortality rates are experienced by 13.2% and 10.2% respondents, respectively.

Table 13.2 shows the values of the KMO and Bartlett's test for the entire list of variables included in the FA. Both statistics show that the application of FA to data is adequate. The KMO is .749, which is much above the threshold of 0.5 and justifies the use of FA. The p-value of the difference of correlation matrix from identity matrix is .001.

The principal component analysis (PCA) method was used to analyze data on prenatal, neonatal and post-neonatal death experience and

Table 13.2 Kaiser-Meyer-Olkin and Bartlett's Test

Kaiser-Meyer-Olkin Measure of Sampling Adequacy.		*.749*
Bartlett's Test of Sphericity	Approx. Chi-Square	29.832
	df	231
	Sig.	.000

demographic, socioeconomic and service delivery variables. In all, 22 variables were subjected to FA. They covered almost all important measures of relevance to this study. Column 1 of Table 13.3 lists all these variables.

In all, there are five components that have eigenvalue above 1.000. Therefore, in the subsequent analysis, five factors were included. Scree plot also suggests that there are only five major factors (Figure 13.1). In all, they explain close to 56.31% of all the variability in the data (Table 13.3).

The KMO of prenatal, neonatal and post-neonatal death experiences are .796, .595 and .525 respectively. The KMO of all these variables are above .5, and this also permits the use of FA. Their communalities are .473, .521 and .529 respectively (data not shown here), which show what proportion of variances they share with the five factors explored in the study are 47.3%, 52.1% and 52.9%, respectively.

Table 13.4 shows the factor loadings, which may be interpreted as correlations between factors and different variables. Variables that have a correlation of .4 or above (absolute value) with any factor are shown in Table 13.5. Perinatal death was a part of Factor 5 and caste (.403),

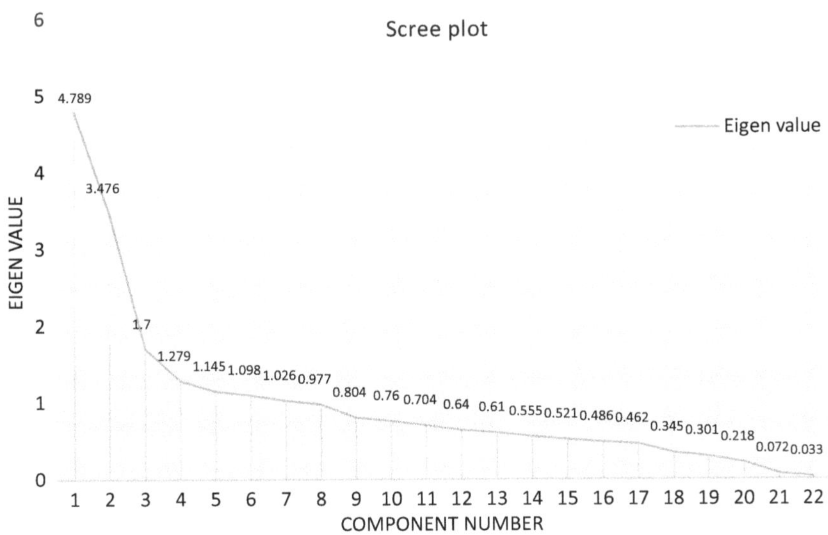

Figure 13.1 Determinants of Child Survival at the Household Level: An Insight of the Method of Factor Analysis

Table 13.3 Total Variance Explained

Total Variance Explained									
Component	*Initial Eigenvalues*			*Extraction Sums of Squared Loadings*			*Rotation Sums of Squared Loadings*		
	Total	*% of Variance*	*Cumulative %*	*Total*	*% of Variance*	*Cumulative %*	*Total*	*% of Variance*	*Cumulative %*
1	4.789	21.767	21.767	4.789	21.767	21.767	4.552	20.690	20.690
2	3.476	15.798	37.565	3.476	15.798	37.565	3.479	15.812	36.502
3	1.700	7.728	45.293	1.700	7.728	45.293	1.690	7.684	44.186
4	1.279	5.815	51.108	1.279	5.815	51.108	1.347	6.124	50.310
5	1.145	5.204	56.311	1.145	5.204	56.311	1.320	6.001	56.311
6	1.098	4.989	61.300						
7	1.026	4.666	65.966						
8	.977	4.440	70.406						
9	.804	3.655	74.060						
10	.760	3.454	77.515						
11	.704	3.201	80.716						
12	.640	2.908	83.624						
13	.610	2.775	86.399						
14	.555	2.521	88.920						
15	.521	2.367	91.287						
16	.486	2.207	93.495						
17	.462	2.099	95.594						
18	.345	1.570	97.164						
19	.301	1.370	98.534						
20	.218	.989	99.523						
21	.072	.327	99.850						
22	.033	.150	100.000						

Extraction Method: Principal Component Analysis.

Table 13.4 Factor Loadings from the Rotated Component Matrix

Rotated Component Matrix[a]					
Factors	*Component*				
	1	*2*	*3*	*4*	*5*
Age	.190	.697	.033	.176	-.156
Number of pregnancies	-.056	.916	.066	.053	-.208
Total live births	-.149	.936	.076	.003	-.027
Members in the family	-.016	.718	.000	.064	.025
Caste	-.602	.067	.009	.150	.403
Education	.629	.315	.011	.282	.201
Husband's education	.636	-.171	-.025	-.209	-.244
Distance of health centre	-.105	-.093	.016	-.101	.508
Antenatal care	-.217	.233	-.022	.648	-.054
Immunization	.042	-.028	-.702	.038	-.010
Breastfeeding	-.034	-.124	.810	.120	.022
Delivery place	.319	-.182	.039	-.596	.173
Birth weight	.012	.083	-.011	-.413	.014
Possession of land	.585	.029	-.008	-.067	-.071
Income of last month	.798	.159	.025	.090	.447
Expenditure of last month	.779	.153	.024	.102	.466
Poverty level of HH	.638	.030	-.037	-.080	-.139
Socioeconomic core	-.662	.076	.055	.060	.134
Observed SES	.820	.031	-.025	-.084	-.029
Perinatal mortality	-.222	-.256	.023	-.135	.485
Neonatal mortality	.039	-.367	-.611	.186	-.033
Postnatal mortality	.092	-.501	.385	.315	-.040

Extraction Method: Principal Component Analysis.
Rotation Method: Varimax with Kaiser Normalization.
[a] Rotation converged in 8 iterations.

distance of health center (.508), income (.447) and expenditure (.466) were positively correlated with this factor. Immunization was negatively and breastfeeding status was positively correlated with deaths of children between birth and 28 days, as noted in Factor 3. Factor 2 indicates the positive association of age, number of pregnancies, total number of live births and number of members in the family with post-neonatal deaths of children between 1 and 5 years of age (Tables 13.4 and 13.5). Factor 4 includes ante natal care, place of delivery and birthweight and does not

Table 13.5 Variables Affecting the Five Factors

Factors	*Classification based on Absolute Value of Factor Loading ≥ 0.4*								
F1	Caste	Education	Husb.edu	land	HH.in.	HH.exp	PL	SES	Obsses
F2	Age	Number of pregnancies	Total live births	Mem. in the family	Postnatal mortality				
F3	Imm.	Breast feeding	Neonatal mortality						
F4	ANC	Delivery place	Birth weight						
F5	Caste	Distance of health centre	Income	HH.exp	Perinatal mortality				

Imm.= Immunization, Mem.=Members, HH.exp= Household expenditure, Obsses= Observed socioeconomic status seem to affect other factors. Interestingly. Factor 1 consisting of socioeconomic characteristics of households such as caste, education, size of landholding, poverty level and observed socioeconomic score also seems to be independent of child mortality.

13.4 Discussion and conclusion

This chapter applied PCA to a set of 22 variables. The result of the analysis indicated the complex interaction of myriads of demographic, health care and socioeconomic factors in three successive stages of child deaths. The latent factors as discussed in the FA technique surfaced in this analysis of the data collected at the household level. In this chapter, dependent and independent variables are all combined and the variables/measures which fall in the same factor as the dependent variables are used to explain them.

The PCA identified five factors (latent variables) that influence the original 22 variables and explain more than 56% of the variability in the data. High rates of miscarriage and abortions were noted mainly for the reason of son preference, prevalent in India. In northern India, high-income families, mostly belonging to the upper castes, prefer male children for carrying out their cultural legacy (Ghosh and Sharma, 2012). Other studies also found that educated families and high income are associated with strong preference for male children, resulting in sex determination and elimination of female fetus (Sen, 1992, 2003; Subramanian and Selvaraj, 2009). A similar trend was also observed in this study area.

Child immunization and exclusive breastfeeding are important agendas of most development programs related to better child survival. Preference of pre lacteals or "first food" is widely prevalent in India (56% as recorded in NFHS-3, 2007), due to strong cultural beliefs and taboos associated with childbirth. A similar phenomenon was also noted in this study—Factor 3 indicating that the status of exclusive breastfeeding and immunization were correlated with child deaths between birth and 28 days after delivery. The introduction of various types of food given after birth and early supplementation often result in multiple illnesses among neonates, resulting in deaths (Mohon et al., 2008). Factor 3 also indicated similar outcome in this study. The FA found Factor 2, comprising of maternal demographic factors, affected deaths of children between 1 to 5 years. There is evidence of this association of demographic characteristics and child mortality (Fall et al., 2015; Sinha et al., 2016; United Nations, 2016) elsewhere around the world.

Thus, FA used in this chapter as one of the techniques of multivariate analysis revealed various latent factors associated with child mortality in the three successive stages, i.e., prenatal, neonatal and post-neonatal stages. The determinants of child mortality form a network, acting and interacting within themselves. They include environment, culture and socioeconomic characteristics of a particular place rather than socioeconomic characteristics of the women/households. This chapter unearthed the latent variables with which child mortality is most closely associated at various stages.

References

Alosaimi, A. N., R. Luoto, A. W. Al Serouri, B. I. Nwaru, and H. Mouniri. "Measures of Maternal Socioeconomic Status in Yemen and Association with Maternal and Child Health Outcomes." *Maternal and Child Health Journal* 20, no. 2 (2016): 386–397.

Anderson, T. W. *An Introduction to Multivariate Statistical Analysis* New Delhi: Wiley Eastern Private Limited (1974).

Bernard, H. R. "The Science in Social Science." *Proceedings of the National Academy of Sciences of the United States of America* 109, no. 51 (December 18, 2012): 20796–20799. http://www.jstor.org/stable/pdf/41830619.pdf?acceptTC=true.

Brown, T. A. *Confirmatory Factor Analysis for Applied Research*. New York: Guilford Press (2015).

Cooper, R. A., and A. J. Weekes. *Data, Models and Statistical Analysis*. New Delhi: Heritage Publishers (1991).

Cousens, S., H. Blencowe, C. Stanton, D. Chou, S. Ahmed, L. Steinhardt, A. A. Creanga, O. Tunçalp, Z. P. Balsara, S. Gupta, L. Say, and J. E. Lawn. "National, Regional, and Worldwide Estimates of Stillbirth Rates in 2009 with Trends since 1995: A Systematic Analysis." *Lancet* 377, no. 9774 (2011): 1319–1330.

Dubey, M., U. Ram, and F. Ram. "Household Levels of Infant and Under-Five Mortality for Crossover between Life Expectancies at Ages Zero, One and Five in India: A Decomposition Analysis." *PLoS One*, Vol. 18, No. 10 (2015).

Fall, C. H., H. S. Sachdev, C. Osmond, M. C. Restrepo, C. Victora, R. Martorell, A. D. Stein, S. Sinha, N. Tandon, L. Adair, I. Bas, S. Norris, and L. M. Richter. Association between Maternal Age at Childbirth and Child and Adult Outcomes in the Offspring: A Prospective Study in Five Low-Income and Middle-Income Countries (COHORTS Collaboration). *Lancet Global Health* 3, no. 7 (2015): 366–377.

Field, A. *Discovering Statistics Using SPSS*. London: Sage Publications (2006).

Ghosh, R. "Child Mortality in India: A Complex Situation." *World Journal of Paediatrics* 8, no. 1 (2012): 11–18.

Ghosh, R., and A. K. Sharma. (2012) "Missing Female Fetus: A Micro Level Investigation of Sex Determination in a Periurban Area of Northern India." *Health Care for Women International* 33, no. 11 (2013): 1020–1034.

Greene, J. G. (1976). "A Factor Analytic Study of Climacteric Symptoms." *Journal of Psychosomatic Research* 20 (1976): 425–430.

Hair, J. F., W. C. Black, B. J. Babin, R. E. Anderson, and R. I. Tatham. *Multivariate Data Analysis*. New Delhi: Pearson Prentice Hall (2006).

International Institute for Population Sciences. (2007). *National Family Health Survey, India: 2005–06*. Vol 1. Mumbai: ICF Macro (2017).

International Institute for Population Sciences. (2017). *National Family Health Survey, India: 2015–16*. Vol 1. Mumbai: ICF Macro (2017).

Kerlinger, F. N., and H. B. Lee. *Fundamentals of Behavioral Research*, Boston: Cengage Learning (2000).

Kim, J.-O., and C. W. Mueller. (1991). *Introduction to Factor Analysis: What It Is and How to Do It*. Newbury Park, CA: Sage Publications (1991).

Kline, P. *An Easy Guide to Factor Analysis*. New York: Routledge (1994).

Long, J. S. *Confirmatory Factor Analysis*. Beverly Hills, CA: Sage Publications (1988).

Malhotra, N. K., and Satyabhushan Dash. *Marketing Research*. New Delhi: Pearson Prentice Hall (2011).

Mohan P., S. D. Iyengar, K. Agarwal, J. C. Martines, and K. Sen. "Care-Seeking Practices in Rural Rajasthan: Barriers and Facilitating Factors." *Journal of Perinatology, Suppl*, 2 (Supplement, 2008): 31–37.

National Institute of Medical Statistics, Indian Council of Medical Research, and United Nations Children's Fund. *Infant and Child Mortality in India: Levels, Trends and Determinants*. National Institute of Medical Statistics, Indian Council of Medical Research and UNICEF India country office, New Delhi, India (2012).

Nwaru, B., R. Klemetti, H. Kun, W. Hong, S. Yuan, X. Wu, and E. Hemminki. Maternal Socio-Economic Indices for Prenatal Care Research in Rural China. *European Journal of Public Health* 22, no. 6 (December 2012): 776–781.

Prudon, P. "Confirmatory Factor Analysis: A Brief Introduction and Critique." (2013). Retrieved from http://home.kpn.nl/p.prudon/GOF.pdf

Sen, A. (1992) "Missing Women." *British Medical Journal* 304, no. 6827 (1992): 587–588.

Sen, A. (2003). "Missing Women—Revisited." *British Medical Journal* 327, no. 7427 (2003): 1297–1298.

Sinha, S., A. R. Aggarwal, C. Osmond, C. H. Fall, S. K. Bhargava, and H. S. Sachdev. "Maternal Age at Childbirth and Perinatal and Under Five Mortality in a Prospective Birth Cohort from Delhi." *Indian Journal of Paediatrics* 53, no. 10 (2016): 871–877.

Slocum-Gori, S. L., and B. D. Zumbo. (2010). "Assessing the Unidimensionality of Psychological Scales: Using Multiple Criteria for Factor Analysis." *Social Indicators Research* 102, no. 3 (2010): 443–461.

Subramanian, S. V., and S. Selvaraj. (2009). "Social Analysis of Sex Imbalance in India: Before and After the Implementation of the Pre-Natal Diagnostic Techniques (PNDT)." *Journal of Epidemiology and Community Health* 63, no. 3 (March 2009): 245–252.

Suhr, D. D. *Exploratory or Confirmatory Factor Analysis*. (2006). Retrieved from http://www2.sas.com/proceedings/sugi31/200-31.pdf

Tacq, J. *Multivariate Analysis Techniques: Social Science Research*. London: Sage Publications (1997).

United Nations Children's Fund, World Health Organization, World Bank, and United Nations. *Level and Trends of Child Mortality. Estimates Developed by the UN Interagency Group for Child Mortality Estimation*. Report (2015). Retrieved from http://www.who.int/maternal_child_adolescent/documents/levels_trends_child_mortality_2015/en/

United Nation Children's Fund. *The State of the World's Children 2008, Child Survival*. UNICEF (2008).

United Nations. (2016). "Goal 3: Ensure Healthy Lives and Promote Well-Being for All at All Ages." Retrieved from http://www.un.org/sustainabledevelopment/health/

Williams, B., T. Brown, and A. Onsman, "Exploratory Factor Analysis: A Five-Step Guide for Novices." *Australasian Journal of Paramedicine* 8, no. 3 (2010): 1–13. Retrieved from http://ro.ecu.edu.au/jephc/vol8/iss3/1

World Health Organization. *International Statistical Classification of Diseases and Related Health Problems, 10th Revision, Instruction Manual*. 2nd ed. Geneva: World Health Organization (2004). Retrieved from http://www.who.int/classifications/icd/ICD-10_2nd_ed_volume2.pdf

14 Access to drinking water and the health outcome

Vikash Gautam

14.1 Introduction

Billions of people worldwide face a lack of access to safe drinking water (World Health Organization, 2013). In India, estimates suggest that of the total urban population about 21.3% and of the total rural population about 50% of the households depend on a tube well and a hand pump as their principal source of drinking water (Pushpangadan, 2003).[1] This means paucity of safe drinking water is an important health issue in both rural and urban areas. Moreover, sufficiency and quality of drinking water are even more questionable as nearly 80% of deaths in India are due to waterborne diseases (De, 2015).

If cheaply or freely available water is unsafe, insufficient and of a substandard quality, self-selection on the desired nature of water is likely to be an important determinant of waterborne diseases. The chapter, in such a scenario, discusses three self-selection criteria for households: selection of safe drinking water; selection of safe and sufficient drinking water and selection of safe, sufficient and good-quality drinking water. Safe drinking water includes water from those sources which are protected such as bottled water, tap water and so on. Drinking water is sufficient if its availability is not a constraint throughout the year. Good-quality drinking water is without any defect in taste, smell or due to any other reasons.

The analysis attempts to address the debate on the effective channels of improvement in health, which is critical for healthy human capital. The three channels to which the debate pertains to are self-motivation, social structure and policy intervention. These channels are often not mutually exclusive. For example, self-motivation is generally reflected in the household's income, size, education level, location choice, efforts and expenditures. The efforts and expenditures, in our case, would be reflected in the monetary amount a household pays for water provision, installation of water source in the premises, purification and storage. Household size, education level and choice of location are also likely to be influenced by the social structure. Besides, groupings based on caste are an important consideration in the Indian social structure. The effect of policy intervention is reflected in household income via employment availability and household location being influenced by better public provision of amenities. A stand-alone significant role of each of three channels is theoretically obvious, but it remains to be seen whether they are jointly important in the current state of the country.

The empirical exercise is performed using the 69th round of National Sample Survey (NSS) data. The survey collected comprehensive information on quality and sources of drinking water, sanitation and hygiene measures, and housing conditions of 53,393 rural and 42,155 urban households. The evidence suggests that the three channels—self-motivation, social structure and policy intervention—are jointly important. Specifically, self-selection in access to drinking water, class structure, education level, household size, location, efforts and expenditures play important roles in determining the occurrence waterborne diseases.

The reminder of this chapter is organized as follows: Section 14.2 outlines the access to drinking water and health outcomes as a self-selection problem. Section 14.3 discusses the empirical model and the sample followed by the estimation results in Section 14.4. Finally, our conclusions are presented in Section 14.5.

14.2 Access to drinking water and health outcomes as a self-selection problem

We outline the model for self-selection of safe drinking water leading to a waterborne health outcome in equations 14.1–14.8. The other two selection problems follow similarly. A household must first decide whether to opt for safe drinking water as its provision is costly. This decision process is unobservable, denoted by y_1^*, and is contingent on a set of characteristics x_1 and private information ε_1 as

$$y_1^* = x_1' \beta_1 + \varepsilon_1, \qquad \varepsilon_1 \sim N\left(0, \sigma_1^2\right). \tag{14.1}$$

We observe a dichotomous y_1 corresponding to y_1^* defined for a threshold τ as

$$y_1 = \begin{cases} \text{safe drinking water provision} & \text{if} \quad y_1^* > \tau \\ \text{safe drinking water non-provision} & \text{if} \quad y_1^* \leq \tau \end{cases}. \tag{14.2}$$

The health outcome (y_2^*) of the individual/household is contingent on a set of characteristics x_2 and private information ε_2 as

$$y_2^* = x_2' \beta_2 + \varepsilon_2, \qquad \varepsilon_1 \sim N\left(0, \sigma_2^2\right). \tag{14.3}$$

We observe a dichotomous y_2 corresponding to y_2^* defined for a threshold μ as

$$y_2 = \begin{cases} \text{healthy} & \text{if } y_2^* > \mu \\ \text{diseased} & \text{if } y_2^* \le \mu \end{cases}. \tag{14.4}$$

For simplicity, let us call those who have the provision of safe drinking water as the treatment group, whereas those who do not have provision of safe drinking water constitute the control group. On average, in examining health outcomes, individuals in the treatment group are likely to differ from those in the control group in both observable and unobservable ways. The differences in observables can be controlled by covariates in x_2. The differences in unobservables, on the other hand, are based on self-selection; that is, this difference occurs because the treatment cannot be randomly assigned to individuals. This implies that in our setup the errors, ε_1 and ε_2, are correlated with the following structure:

$$\begin{bmatrix} \varepsilon_1 \\ \varepsilon_2 \end{bmatrix} \sim N\left[\begin{bmatrix} 0 \\ 0 \end{bmatrix}, \begin{bmatrix} 1 & \sigma_{12} \\ \sigma_{12} & \sigma_2^2 \end{bmatrix}\right], \tag{14.5}$$

where σ_1^2 is normalized to 1.

The effect of self-selection can be seen for a healthy outcome as follows:

$$\begin{aligned} E\left[y_2 | y_1^* > 0\right] &= E\left[x_2' \beta_2 + \varepsilon_2 | x_1' \beta_1 + \varepsilon_1 > \tau\right] \\ &= E\left[x_2' \beta_2 | x_1' \beta_1 + \varepsilon_1 > \tau\right] + E\left[\varepsilon_2 | x_1' \beta_1 + \varepsilon_1 > \tau\right] \\ &= x_2' \beta_2 + E\left[\varepsilon_2 | \varepsilon_1 > -x_1' \beta_1 + \tau\right]. \end{aligned}$$

In such a case, Heckman (1979) compiled that $\varepsilon_2 = \sigma_{12}\varepsilon_1 + \xi$, where the random variable ξ is independent of ε_1. Thus,

$$\begin{aligned} E\left[y_2 | y_1^* > 0\right] &= x_2' \beta_2 + \sigma_{12} E\left[\varepsilon_1 | \varepsilon_1 > -x_1' \beta_1 + \tau\right] \\ &= x_2' \beta_2 + \sigma_{12} E\left[\varepsilon_1 | \varepsilon_1 > -x_1' \beta_1 + \tau\right] \end{aligned}$$

Since ε_1 is distributed as standard normal, denoting normal probability density function as $\phi(.)$ and normal cumulative density function as $\Phi(.)$, we can write

$$E\left[y_2 | y_1^* > 0\right] = x_2' \beta_2 + \sigma_{12} \frac{\phi\left(-x_1' \beta_1 + \tau\right)}{\left[1 - \Phi\left(-x_1' \beta_1 + \tau\right)\right]}.$$

Using symmetry in standard normal and denoting the ratio $\phi(.)/\Phi(.) = \lambda(.)$ as the inverse Mills ratio, we get

$$E\left[y_2|y_1^* > 0\right] = x_2'\beta_2 + \sigma_{12}\frac{\phi(x_1'\beta_1 - \tau)}{\Phi(x_1'\beta_1 - \tau)}$$

$$= x_2'\beta_2 + \sigma_{12}\lambda(x_1'\beta_1 - \tau). \quad (14.6)$$

For $\sigma_{12} = \rho\sigma_2$, where ρ is coefficient of correlation between ε_1 and ε_2, equation 14.6 can be equivalently written as

$$E\left[y_2|y_1^* > 0\right] = x_2'\beta_2 + \rho\sigma_2\lambda(x_1'\beta_1 - \tau). \quad (14.7)$$

Equation 14.7 shows that the internal validity of the health outcome model would be biased unless it considers the presence of self-selection in the provision of safe drinking water.

An important concern in estimating the model relates to the identification of the two equations 14.1 and 14.3. This, in turn, is contingent on the kind of form $\lambda(x_1'\beta_1 - \tau)$ assumes. Figure 11.1 presents the plot of $\lambda(x_1'\beta_1 - \tau)$ and its components, $\phi(x_1'\beta_1 - \tau)$ and $\Phi(x_1'\beta_1 - \tau)$, as a function of $x_1'\beta_1 - \tau$ which is number of standard deviations $x_1'\beta_1$ is above or below the selection point τ. Let us assume that $x_1'\beta_1$ is fixed and then consider the effect of changes in τ. As τ exceeds $x_1'\beta_1$ (leftward movement in the figure), $\phi(x_1'\beta_1 - \tau)$ becomes larger than $\Phi(x_1'\beta_1 - \tau)$, generating values

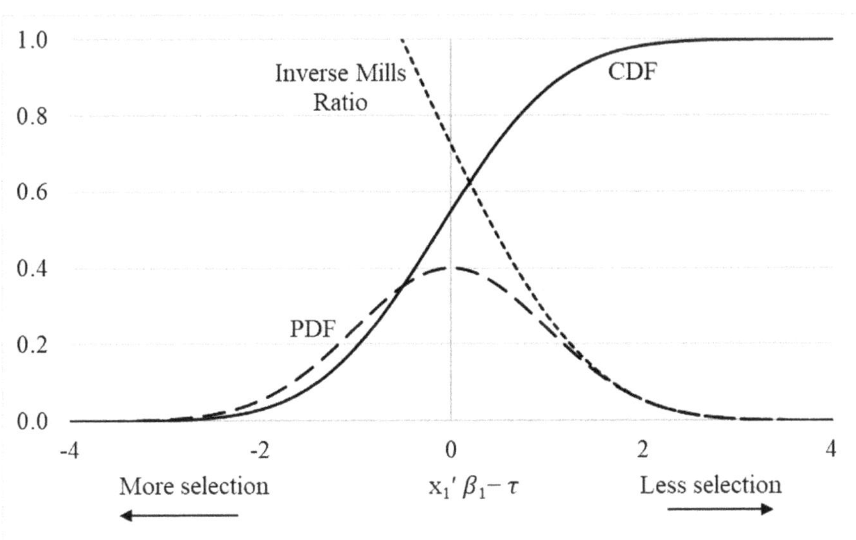

Figure 14.1 Inverse Mills Ratio and its Components

of $\lambda(x_1'\beta_1 - \tau)$ greater than 1. In this case, selection is more extreme. As τ decreases (rightward movement in the figure), $\phi(x_1'\beta_1 - \tau)$ becomes smaller than $\Phi(x_1'\beta_1 - \tau)$, generating values of $\lambda(x_1'\beta_1 - \tau)$ smaller than 1. At extreme $\phi(x_1'\beta_1 - \tau) \to 0$ and $\Phi(x_1'\beta_1 - \tau) \to 1$. So $\lambda(x_1'\beta_1 - \tau) \to 0$. This implies that the selection is negligible, leading to $E[y_2|y_1^* > 0] \to x_2'\beta_2$. With these considerations, Puhani (2000) noted that $\lambda(x_1'\beta_1 - \tau)$ is approximately linear over a wide range of its arguments; that is, $\lambda(x_1'\beta_1 - \tau) \approx a + bx_1'\beta_1$ for some constant *a* and *b*. Equation 14.7 would, thus, evolve as

$$E[y_2|y_1^* > 0] \approx \rho\sigma_2 a + x_2'\beta_2 + \rho\sigma_2 b x_1'\beta_1. \tag{14.8}$$

Equation 14.8 suggests a multicollinearity problem if $x_1 = x_2$. Therefore, estimation of the self-selection model requires that at least one regressor in the equation 14.1 be excluded from the equation 14.3. We keep this issue in hindsight while carrying out the estimation exercise. The estimates for the model can be derived by either following Heckman's two-step method or by the maximum likelihood method. They are discussed in detail in Long (1997) and Cameron and Trivedi (2005).

14.3 Empirical model and the sample

The empirical model corresponding to the selection rule in equation 14.1 can be specified as

$$\begin{aligned} X_i = {} & \alpha_0 + \alpha_1\, \textit{household income} + \alpha_2\, \textit{social groups} \\ & + \alpha_3\, \textit{education level} + \alpha_4\, \textit{household size} \\ & + \alpha_5\, \textit{location of household} \\ & + \alpha_6\, \textit{water charges} \\ & + \alpha_7\, \textit{proximity to water source} + \varepsilon_1, \end{aligned} \tag{14.9a}$$

where X_i= {access to safe drinking water (Model I); access to safe and sufficient drinking water (Model II); access to safe, sufficient and good quality drinking water (Model III)}.

The empirical model corresponding to the outcome rule in equation 14.3 and for the selection rule in Model I can be specified as

$$\begin{aligned} \textit{Health outcome} = {} & \beta_0 + \beta_1\, \textit{household income} \\ & + \beta_2\, \textit{social groups} + \beta_3\, \textit{education level} \\ & + \beta_4\, \textit{household size} + \beta_5\, \textit{location of household} \\ & + \beta_6\, \textit{water purification} + \beta_7\, \textit{water storage} \\ & + \beta_8\, \textit{water sufficiency} + \beta_9\, \textit{water quality} + \varepsilon_2. \end{aligned} \tag{14.9b}$$

For the selection rule in Model II, *water sufficiency* is omitted from the set of regressors and for the selection rule in Model III, *water quality* along with *water sufficiency* is omitted from the set of regressors. These omissions keep the models tractable. In addition, in all three empirical models, *water charges* and *proximity to water source* are taken as exclusion restrictions; that is, they appear in the selection equations but not in the outcome equations.

The data for the estimation are taken from the 69th round of the NSS on "Drinking Water, Sanitation, Hygiene and Housing Condition in India", 2012. This survey collected comprehensive information on quality and sources of drinking water, sanitation and hygiene measures, and housing conditions of the households. It covered the entire Indian Union, including the rural areas of Nagaland and Andaman and Nicobar Islands which were not given due attention in earlier NSS surveys. The sample consists of 53,393 households in rural India and 42,155 households in urban India. The variables are constructed as follows:

Access to safe drinking water: The NSS mentions 13 different sources of drinking water that we classify as a dummy variable where safe drinking water takes the value 1, and unsafe drinking water takes the value 0. The safe drinking water category consists of bottled water, piped water into dwelling, piped water to plot, public tap or standpipe, tube well or borewell, protected well and spring. The unsafe drinking water category consists of unprotected well, rainwater collection, tank/pond water, unprotected spring, other surface water and water from other sources like tanker truck, cart with small tank or drum.

Access to safe and sufficient drinking water: It is defined as a dummy which assigns a value of 1 to households with safe and sufficient drinking water throughout the year and 0 otherwise.

Access to safe, sufficient and good-quality drinking water: It is defined as a dummy variable that takes a value of 1 if drinking water is safe, sufficient and without any defect in taste, smell or due to any other reasons and 0 otherwise.

Health outcome: It is a dummy variable that takes the value of 1 if a household does not suffer from any waterborne disease and 0 otherwise. We identify occurrence waterborne disease if any member of a household is suffering from stomach problem, skin disease and fever other than due to malaria.

Household income: Income determines the economic status of a household. The NSS does not collect information on income directly as it is frequently misreported. Instead, it collects information on monthly per capita consumption expenditure (MPCE). We take its natural logarithm as the proxy for household income.

Social groups: Social groups or caste is an important feature that controls for behavior and perception of similar people in dealing with socioeconomic issues. This variable is represented by three dummies: *scheduled tribe (ST) dummy* taking a value of 1 for the ST category and 0 otherwise, *scheduled caste (SC) dummy* taking a value of 1 for the SC category and 0 otherwise and *other backward class (OBC) dummy* taking a value of 1 for the OBC category and 0 otherwise. Households belonging to others, which predominantly consists of the general category, are considered as the base category.

Education level: Education level of the household reflects the awareness about the importance of sanitation and health. We take the education of the household head as the proxy for it. The NSS provides education of the household head in 10 groups. We constructed three dummies for these 10 groups: *primary dummy* taking a value of 1 for having education till primary and 0 otherwise, *secondary dummy* taking a value of 1 for having education till secondary and 0 otherwise and *above secondary dummy* taking a value of 1 for having more than secondary education and 0 otherwise. Households without formal schooling are taken as base category.

Household size: Household size implies total number of members living in a household. It represents family support network within a household, and therefore, a bigger household size should enable better access to drinking water. However, with bigger households, the chances of a waterborne disease should also increase as the scope of contagion increases.

Location of the household: It is a dummy variable taking a value of one for urban households and zero for rural households. Urban areas generally have better drinking water facility because of awareness, better public provision, availability of more alternative sources and so on.

Water charges: It is the natural logarithm of the amount of money paid for water provision. Households would be willing to pay higher charges for access to safe, sufficient and good-quality drinking water.

Proximity to water source: It is a dichotomous variable that takes a value of 1 if drinking water is available within the household premises and 0 otherwise. Water source within household premise is more likely to be safe, sufficient and of good quality.

Water purification: If drinking water is properly purified, the possibility of bad health reduces. We include a dummy variable that takes a value of 1 if a household purifies drinking water; else, it is 0.

Water storage: Water storage in containers can lead to a sizeable reduction in waterborne diseases. We include a dummy variable that takes a value of 1 if households store drinking water in a metallic or nonmetallic container; else, it is 0.

Table 14.1 presents the summary statistics of these variables. It reveals significant heterogeneity in the sample. Importantly, we can note that as the selection criterion becomes more stringent, the proportion of households remaining in the treatment group increases. This gives an a priori indication that self-selection can be crucial in the observed health outcome. Also, there are significant differences between households with no waterborne diseases and households with waterborne diseases in most of the characteristics.

Table 14.1 Summary Statistics

	No Waterborne Disease	*Waterborne Disease*
Mean (standard deviation) for non-qualitative variables:		
Ln MPCE	8.61* (0.72)	8.57 (0.64)
Household size	4.23* (2.21)	5.13 (2.49)
Ln Water charges	1.36* (2.02)	0.89 (1.73)
Percentage for qualitative variables:		
Access to safe drinking water	53.11	46.89
Access to sufficient drinking water	53.44*	46.56
Access to good quality drinking water	54.45*	45.55
Access to safe and sufficient drinking water	53.69*	46.31
Access to safe, sufficient and good quality drinking water	54.66*	45.34
ST	53.01	46.99
SC	47.58*	52.42
OBC	53.41	46.59
Other Castes	56.59*	43.41
No formal education	47.44*	52.56
Primary education	48.29	51.71
Secondary education	53.90*	46.10
Above secondary education	63.13*	36.87
Urban household	60.02*	39.98
Rural households	48.18*	51.82
Water source within household premises	55.62*	44.38
Water source outside household premises	50.34	49.66
Water purification	59.17*	40.83
Water storage	54.02*	45.98

*Note: Represents difference from 'waterborne disease' group at 5% significance level. The total number of observations is 95,548. MPCE = monthly per capita consumption expenditure; ST = Scheduled Tribe; SC = Scheduled Caste; OBC = Other Backward Caste.

14.4 Estimation results

Table 14.2 presents estimates for the determinants of the health outcome with self-selection, which pertain to equations 14.9a and 14.9b. Panel A reports the estimates for the selection equation whereas Panel B reports the estimates for the outcome equation. Across all the model specifications, the self-selection criterion is significant at 5%. This implies that simple probit-type models would be biased.

In both Panels A and B, MPCE is positively related to access to better drinking water and the health outcome across all the models.[2] It indicates that households with improved economic status have more ability to pay for better drinking water. This reduces the chances of getting a waterborne disease.

The estimates of social group variables suggest that ST and OBC households are less likely to have access to better drinking water, whereas SC

Table 14.2 Determinants of Health Outcome with Self-Selection

Panel A: Selection Equation

	Model I	*Model II*	*Model III*
Ln MPCE	0.240*** (11.43)	0.109*** (5.73)	0.138*** (7.97)
ST	-0.385*** (10.89)	-0.439*** (14.35)	-0.305*** (10.39)
SC	0.028 (1.77)	0.042 (1.55)	0.021 (0.87)
OBC	-0.099*** (3.72)	-0.110*** (5.03)	-0.058** (2.93)
Primary education	-0.079* (2.48)	-0.107*** (4.06)	-0.100*** (3.92)
Secondary education	-0.037 (1.07)	-0.065 (1.27)	-0.037 (1.35)
Above secondary education	0.016 (0.37)	0.003 (0.08)	0.003 (0.09)
Household size	0.044*** (8.37)	0.013** (2.92)	0.012** (2.96)
Location	0.315*** (10.56)	0.226*** (9.39)	0.149*** (7.26)
Ln Water charges	0.063*** (7.08)	0.038*** (4.95)	0.014** (2.94)
Proximity to water source	0.425*** (15.91)	0.423*** (20.94)	0.205*** (10.04)

Panel B: Outcome Equation

	Model I	*Model II*	*Model III*
Ln MPCE	0.071*** (3.95)	0.051* (2.54)	0.072*** (3.49)
ST	0.097** (2.72)	0.122* (2.39)	0.104* (2.23)
SC	-0.116*** (4.74)	-0.112*** (4.26)	-0.120*** (4.34)
OBC	-0.006 (0.30)	-0.043 (1.89)	-0.025 (1.09)
Primary education	0.005 (0.18)	-0.005 (0.88)	-0.008 (0.27)
Secondary education	0.101*** (3.64)	0.083** (2.76)	0.087*** (2.80)
Above secondary education	0.179*** (5.15)	0.179*** (4.81)	0.173*** (4.39)
Household size	-0.120*** (27.00)	-0.116*** (20.43)	-0.115*** (19.89)
Location	0.210*** (9.22)	0.281*** (11.38)	0.216*** (6.96)
Water purification	0.182*** (9.67)	0.203*** (9.99)	0.193*** (8.52)

(Continued)

Table 14.2 (Continued)

Panel A: Selection Equation			
	Model I	*Model II*	*Model III*
Water storage	0.208***	0.169***	0.213***
	(6.00)	(4.38)	(5.79)
Water sufficiency	0.090***		
	(3.64)		
Water quality	0.249***	0.242***	
	(10.09)	(8.63)	
ρ	0.303	0.351	0.368
Prob > chi^2 (H_0: $\rho = 0$)	0.044	0.039	0.043
Prob > chi^2 (model)	0.000	0.000	0.000

Note: Model I corresponds to selection on access to safe drinking water. Model II corresponds to selection on access to safe and sufficient drinking water. Model III corresponds to selection on access to safe, sufficient and good quality drinking water. Estimation is with robust standard errors. Absolute z-statistics are in parentheses. *, ** and *** represent level of significance at 5%, 1% and 0.1%, respectively. MPCE = monthly per capita consumption expenditure; ST = Scheduled Tribe; SC = Scheduled Caste; OBC = Other Backward Caste.

households are more likely to have similar access to drinking water as the general category (the reference group). In terms of the health outcome, the chances of getting a waterborne disease are lower for ST households and higher for SC households. For OBC households, the chances are similar to that of the general group. A structural interpretation for this pattern would be incongruous. It, however, controls for behavior and perception of similar people about water use and waterborne diseases.

The educational background of each household measures the marginal influence of schooling attainment, relative to lower levels of school attainment. For example, anyone who has finished secondary school will necessarily have studied "till primary" and "till secondary"—so the cumulative influence of education is the sum of coefficients at the current and earlier levels of education. The estimates reveal two important points. First, households having lower education are less likely to prefer better drinking water over ordinary drinking water. This is plausible as households with lower education are less aware of the disadvantages of ordinary drinking water use, and for them, it is a cheaper substitute to costlier better drinking water. Second, at a given education level, a water-related health outcome is a more serious concern than access to better drinking water as the potential cost of the water-related health outcome is immediate. It is commonly noted in the perception-based studies that the immediate costs are discounted at a much lower rate than the future costs (Loewenstein and Prelec, 1992).

The coefficient associated with household size suggests that an extensive family support network enables access to better drinking water. However, the coefficient in the outcome equation suggests that with more members in a household, the likelihood of waterborne disease also increases as the carriers of contagion are also large.

The coefficient of location dummy suggests that the urban areas have access to better drinking water and they are less likely to be affected by waterborne diseases than the rural areas. This is because the level of awareness, public provision of drinking water and disease-prevention services, among others, are better in urban areas.

The variables in the selection equation which are used as exclusion restriction—water charges and proximity to water source—turn out to be positive and significant. They suggest that households opting for the private provision by paying a higher cost for drinking water and having a water source in their premise are more likely to have access to better drinking water.

The variables relating to purification, storage, sufficiency and quality of drinking water in the outcome equation turn out to be positive and significant. Although these variables turn out to be as expected, the important point to note is that each of these processes and provisions, at the margin, has a positive effect on the water-related health outcome. For example, even when drinking water is without any defect in taste and smell, it is sufficient and stored; its purification before consumption would have a significant positive effect on the water-related health outcome.

14.5 Conclusion

Against the backdrop that cheaply or freely available water is often unsafe, insufficient and of a sub-standard quality, this chapter discusses three self-selection criteria for incidence of waterborne diseases in households: the selection of safe drinking water, the selection of safe and sufficient drinking water and the selection of safe, sufficient and good-quality drinking water. In an empirical exercise, performed using the 69th round of NSS data, the evidence suggests that self-selection in access to drinking water, class structure, education level, household size and location play important roles in determining the occurrence of waterborne diseases. Furthermore, all the processes and provisions of water consumption, such as choice of source, purification, storage, sufficiency and quality, at the margin, have a positive and significant effect on water-related health outcomes.

The analysis adds to the debate on effective channels of improvement in health that pertain to self-motivation, social structure and policy intervention. It suggests that each of these channels is important even when the other channels are controlled for, and better health outcomes would require active participation at all levels.

Notes

1 The estimates are compiled by Pushpangadan (2012) based on NSS 54th round data, 1999.
2 Here and henceforth, access to better drinking water implies access to safe and/or sufficient and/or good quality drinking water.

References

Cameron, A. Colin, and Pravin K Trivedi. *Microeconometrics: Methods and Applications*. New York: Cambridge University Press (2005).

De, A. "Top 5 Water Related Diseases in India" (2015). Retrieved from http://www.onlymyhealth.com/top-water-related-diseases-in-india-1300271205

Heckman, J. "Sample Selection Bias as a Specification Error." *Econometrica* 47, no. 1 (1979): 153–161.

Loewenstein, George, and Drazen Prelec. "Anomalies in Intertemporal Choice: Evidence and an Interpretation." *The Quarterly Journal of Economics* 107, no. 2 (1992): 573–597.

Long, J. *Regression Models for Categorical and Limited Dependent Variables*. Thousand Oaks: Sage Publications (1997).

Puhani, P. "Heckman Correction for Sample Selection and its Critique." *Journal of Economic Surveys* 14, no. 1 (2000): 53–68.

Pushpangadan, K. "Drinking Water and Well-Being in India: Data Envelopment Analysis." Working Paper 352, Centre for Development Studies (2003). Retrieved from http://www.cds.edu/wp-content/uploads/2012/10/352.pdf

World Health Organization. "Water Cooperation: Facts and Figures." WHO Report (2013). Retrieved from http://www.unwater.org/water-cooperation-2013/water-cooperation/facts-and-figures/en/

15 Role of information in determining the willingness to pay for health insurance

Namrata Gulati and Basudeb Chaudhuri

The study was funded by the European Union NOPOOR project with the grant agreement number being 290752.

15.1 Introduction

High medical expenditure undermines and hits the poor especially hard. According to the 71st round of National Sample Survey (NSS) for India, collected in 2014, medical expenditure incurred in the 15 days preceding the survey for outpatients as a percentage of total expenditure in the same period was as high as 79%. This expenditure is even higher, around 185%, for the poor, belonging to the lowest income quintile and residing in rural India, implying that the poor borrow to pay for their medical expenses. Data from the NSS survey reinforce the view that it is the poor who are the hardest hit because of a medical emergency as they spend most of their savings to meet expenses. Despite recurring high out-of-pocket expenditure, the poor still do not rely on health insurance.

The primary objective of insurance is to transform high out-of-pocket expenditure into health insurance premiums whereby the sum is collected from a much larger insured group of people instead of a small group of people inflicted by illness. If done efficiently, health insurance reduces the financial stress on the poor because of medical emergencies. The main focus of the chapter is to analyze whether there is a scope for health insurance amongst the deprived in a developing country like India. To answer this, this chapter examines the willingness to pay amongst the slum dwellers of Mumbai and adjoining areas and tries to identify the factors that influence decision-making at an individual level. There is a lot of empirical research that illustrates the plight of the poor as a result of high medical expenditure. However, there are very few studies conducted in the slums of the developing world that have attempted to analyse the role of market forces and their potential to help the poor escape the vicious circle of high poverty, high medical expenditure and resultant misery. This work attempts to fill in this gap and proposes policy measures needed to increase awareness of the desirability and benefits of health insurance among the poor.

The significant finding of the chapter is that the poor are often unaware of the need for health insurance scheme or its benefits; even if they are, they do not have the resources to get themselves insured. Hence, measures to spread awareness are as necessary as measures geared towards providing medical services. Small investments in spreading awareness regarding the availability of insurance policies among the poor can have a profound impact.

The remaining part of the chapter is organized as follows: The next section looks the position of health insurance status and health care expenditure in India using data from the 71st NSS round. Section 3 reviews the literature on the demand for health insurance. Section 4 describes our survey design and provides descriptive statistics. Section 5 details the methodology and provides an analysis of the willingness to pay for additional coverage. Section 15.6 gives the results, makes concluding observations and provides policy recommendations.

15.2 Health insurance in India

The health insurance programme in India dates back to the late 1940s, and early 1950s. It was initially introduced for civil servants (Central Government Health Schemes) and workers in the formal sector (Employees' State Insurance Scheme) and, thus, was confined only to workers in the formal sector (PHFI, 2011). Private and voluntary health insurance had insignificant presence until 1986, when the General Insurance Corporation (GIC) of India launched the "Mediclaim" policy. However, there was no other reliable voluntary health insurance provider then, other than GIC.[1] It was only after the liberalisation in 1991 that private-sector players were allowed into the insurance sector in the country. This enabled higher income groups to access quality health care from private tertiary care facilities. However, this was seen to have resulted in increasing the cost of health care facilities for the poor. Community health insurance then emerged as an alternative financing mechanism to protect the poor (Devadasan et al., 2004). Ranson et al. (2003) show that in India these schemes have been successful in engaging only a small proportion of the target population.

To protect the poor from the financial burden resulting from health care needs, the government announced the Universal Health Insurance Programme in 2003. As noted by Ahuja (2008) and Rao (2004), this program did not accomplish much as the poor were not aware of the scheme, and they did not know how to avail of its benefits. This led to the initiation of the Rashtriya Swasthya Bima Yojana in 2007, which was targeted at below-poverty-line households.

In parallel with the central government, many state governments initiated insurance programs to ensure wider coverage. These initiatives specific to states included the Rajiv Aarogyasri (Andhra Pradesh), Kalaignar's Insurance Scheme for Life Saving Treatment (Tamil Nadu), and the Vajpayee Arogyasri and Yeshasvini programmes in Karnataka (PHFI, 2011).

Although the health insurance business is beginning to pick up with the introduction of competition from private players in the insurance sector, health insurance for the poor is still very low. As per the 71st round of National Sample Survey Organisation, about 85% of the Indian population is not insured. Those insured are mostly insured with government-funded insurance schemes like the Rashtriya Swasthya Bima Yojana (RSBY) or similar plans, although there has been a steady rise in the proportion of coverage by some scheme of health expenditure support with a rise in the monthly per capita expenditure (MPCE) level. Table 15.1 gives sector-wise details of the source of insurance in India for different income quintiles.

It is apparent from Table 15.1 that private health insurance companies cater only to the top income quintile of the population staying in urban areas. According to United States Agency for International Development (USAID, 2008), as compared to other Asian countries, spending by the government of India on health care as a percentage of gross domestic product is low, leaving households to bear a large part of health care financing.

The 71st round of the NSS indicates that the medical expenditure for the 15 days prior to the survey for outpatients as a percentage of total expenditure in the same period is as high as 79%. This percentage is much higher for those who are not insured. Table 15.2 gives the details of medical expenditure as a percentage of total expenditure for different income quintiles as well as for those who are insured and those who are not.

It is evident from Table 15.2 that the uninsured poor, belonging to the lowest income quintile, end up spending as much as 185% of their income on health, implying that their expenditure is more than their income. Most of them rely on past savings for meeting this expenditure. One gets similar results if one looks at the expenditure of those who were admitted to the hospital in the last 365 days. This indicates the need for health insurance

Table 15.1 Insurance Choice for Different Quintiles as per MPCE as per NSS 71st Round

Sector	*Not Covered*	*Government Funded*	*Employer Funded*	*Insurance Company*
Rural				
1	89.1	10.1	0.7	0.0
2	88.8	10.7	0.4	0.0
3	87.4	11.9	0.6	0.0
4	83.3	15.9	0.5	0.1
5	81.1	17.0	0.8	0.2
Urban				
1	91.4	7.7	0.6	0.0
2	87.5	10.6	1.3	0.5
3	84.7	12.9	1.3	1.0
4	79.7	13.5	3.3	3.4
5	66.6	15.1	5.6	12.4

Table 15.2 Medical Expenditure for Last 15 Days for Outpatients as a Percentage of Total Expenditure for the Last 15 Days

Income Quintile	*Overall*	*Uninsured*	*Insured*	*Medical Expenditure Reimbursed for Insured as a Percentage of Total Medical Expenditure (percentage)*
1	175	185	109	3.81
2	102	103	100	4.28
3	82	84	75	4.77
4	73	77	59	4.79
5	46	51	35	21.17

Note: This table is based on authors own calculations based on NSS 71st round.

not just to ensure that the uninsured do not end up spending amounts on health care that pushes them into poverty but also preempt the adverse impact this has on their health. Most often, the uninsured do not get necessary health care.

Generally, the uninsured have higher rates of preventable and/or untreated illness. This is reflected in the higher percentage of uninsured who do not take their ailment seriously and do not avail any treatment. This is illustrated in Table 15.3, which shows that a higher percentage of the uninsured population do not take illness for a prolonged period seriously. This may have an adverse impact on their health status in the future.

As illustrated by Figure 15.1, the choice of the medical institution and the amount of money spent for a particular illness for the Mumbai sample is also influenced by health insurance affiliation. Households with health insurance prefer to go to private hospitals for treatment. This raises the question why the health insurance system is not working in the country, especially for the poor. The following section examines literature on the issue to explore the reasons.

15.3 Factors affecting demand for private health insurance

According to Shaw and Ainsworth (1995), the choice of a health insurance plan and the level of participation by households are determined both by

Table 15.3 Whether Ailment Was Considered Chronic (percentages)

Chronic Ailment	*Overall*	*Uninsured*	*Insured*
Yes	51.62	46.74	66.43
No	48.38	53.26	33.57

Note: This table is based on NSS 71st round.

the characteristics of the plan and also by the personal, household and community attributes of the individual making a choice. Amongst the individual attributes, many studies (Scotton 1969; Cameron et al. 1988; Aizuddin et al. 2012; Tirukoti 2015) have found income to be a factor affecting health insurance decisions. Health care expenditure is another critical variable influencing the decision to purchase health insurance, as noted by Kronick and Gilmer (1999). Families with a higher risk of hospitalization are more likely to purchase health insurance. Other factors that significantly affect the decision to purchase health insurance are age, education and gender. Grossman (1972) and Muurinen (1982), who have looked at the role of education in making health decisions, point out that educated people are more likely to be more aware of both the insurance options available as well as the benefits of having health insurance. Age, too, has been found to have a positive impact on the likelihood of having health insurance cover in many studies (Cameron and Trivedi 1991); Burrows et al. 1989; Savage and Wright (2001). Looking at the demand for community-based health insurance schemes in India, Panda et al. (2014) have found that households with higher financial liabilities find insurance more attractive.

As reported by Kimani et al. (2012), studies conducted in many sub-Saharan African countries showed that people employed in the formal sector had better access to health insurance as compared to those employed in the informal sector. Huyen and Van Minh (2014), who examined the factors that affected an individual's willingness to pay for health insurance, found that gender, the education level of the respondent's household, the

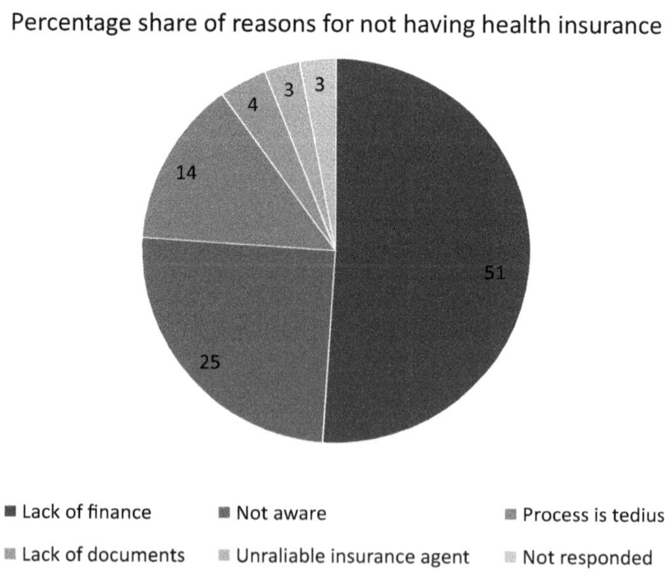

Figure 15.1 Relation Between Money Spent on Major Sickness and Ownership of Health Insurance

respondent's income level and the likelihood of having chronic disease were the major factors determining the willingness among informal sector employees to pay for health insurance.

There are very few studies that have examined the role of awareness. Bawa and Ruchita (2011) reported that most significant factors that impede subscription to health insurance are a lack of funds and a lack of awareness, Kansra and Patharia (2014) found that the five key factors that constrain buying health insurance were procedural formalities, agent-related problems, coverage issues, awareness issues and negative feedback. Another study by Singu and Kaur (2014), found many problems in the health insurance sector, the most important of which was the lack of awareness.

Even though there have been studies that have tried to evaluate the role of various factors influencing a household's decision to buy health insurance services, none of them has looked at factors that affect this decision for the people in the bottom quintile, mainly staying in the slums in India, which this chapter does. This study is also unique in the sense that it gives a simple choice to the households and then tries to estimate their willingness to pay. The details are provided in the methodology section.

15.4 Data and descriptive statistics

To determine the extent of health insurance coverage among the urban and rural poor and to identify the factors that influence a household's decision to invest in health insurance, primary data were collected by conducting a household survey in July 2013 through site visits in different areas of Mumbai and nearby villages. The main target population for the survey was the economically weaker sections of society; hence, there was a focus on slum dwellers.

Households were selected through random sampling. Following this technique, 165 households were surveyed from different slums located in Mumbai and its suburban districts and a few villages of the Raigad and Thane districts of Maharashtra. The sample was representative of people staying in slums in different parts of Mumbai and its outskirts.

Of the surveyed population, around 88% had a monthly income below Rs. 20,000. In 80% of the families, at least one member of the family had completed the secondary level of education. Table 15.4 gives descriptive statistics of the households surveyed. Educational level has been categorised into three types: low education includes members who are illiterate or have studied up to the primary level; medium education consists of members who have studied up to secondary level, that is, they have they studied up to anywhere between classes 6 and 10; and high education comprises members who have studied up to class 12 or have pursued higher education.[2] Occupational status has also been categorized into three groups—primary, secondary and tertiary. This classification includes the members of the working population, and the categories have been defined according to the NSS definition.[3]

Table 15.4 Characteristics of the Surveyed Households (HH)

Variables	*Category*	*% of HH*
Income	<5,000	21.8
	5,000–10,000	41.8
	10,000–20,000	26.7
	20,000–40,000	7.9
	>40,000	1.8
Education	Highly educated	44.2
	Moderately educated	41.2
	Low educated	14.5
Occupational status	Primary	6.7
	Secondary	20.6
	Tertiary	72.7
Insurance Status	Insured	15.2
	Un-insured	84.2

Approximately 84% of the households surveyed were not insured. The reasons for this were looked into. A lack of financial resources is one of the significant constraints preventing households from opting for health insurance. Figure 15.2 illustrates the various reasons given by people for not being insured. Half of the surveyed households pointed out that they could not afford to pay the required premium. Approximately one-fourth of those sampled did not have information on the insurance facilities available to them. A majority of them were averse to insurance because of the tedious procedures involved in getting insured. The fact that 50% of the surveyed households did not opt for health insurance for reasons other than affordability indicates that there is room to improve access to health insurance.

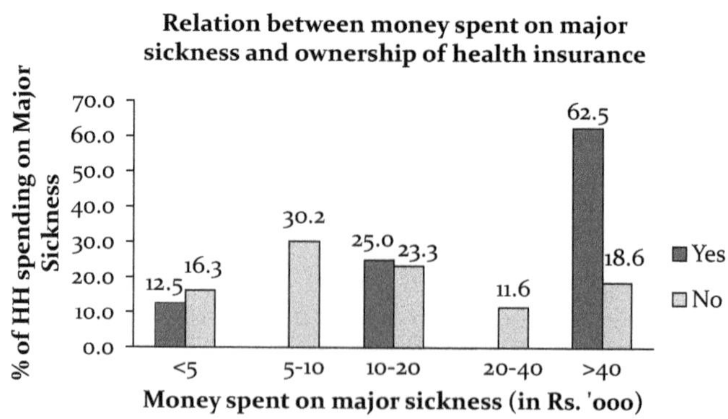

Figure 15.2 Reasons for Not Having Health Insurance

15.5 Methodology

To evaluate the demand and willingness to pay for health insurance, consumers were given a choice between two schemes, which varied with respect to coverage and the premium. Different premiums with related coverage help consumers choose the bundle that gives him the highest utility. To keep things simple, the surveyed population was given only two choices. They had to choose between two schemes: (1) an annual premium of Rs. 1,500 for a coverage of Rs. 50,000 coverage and (2) an annual premium of Rs. 2,500 for a coverage of Rs. 100,000. This coverage was for a family of up to six people with two adults. The coverage was for both inpatients as well as outpatient expenses.

The proposed choices were relatively low premium and low coverage to cater to the needs of the surveyed population, taking into consideration their low-income level. However, these choices also had to be feasible from the seller's perspective. Therefore, quoting a very low premium was not an option. The survey revealed that 17.6% of the households were willing to get insurance for annual coverage of Rs. 50,000 for their family members by paying a premium of Rs. 1,500. The percentage of households interested in an annual coverage of Rs. 100,000 for the premium of Rs. 2,500 was only 4.8%.

As recommended by Ahuja and De (2004), it is essential to understand the demand side by taking into consideration the paying capacity of the poor while discussing existing health insurance schemes available in the country. To understand the demand side, it is imperative to understand how much consumers are prepared to pay for different health insurance schemes. Willingness to pay gives important insight while investigating public preferences for goods for which free-market price is not there (Bacon-Shone and McGhee, 2007). Researchers follow stated preference techniques such as contingent valuation (CV) and choice experiments to elicit the consumer's maximum willingness to pay. In CV, people are directly asked in a survey how much they would they like to pay for certain services. Unlike CV, choice experiments ask people to make choices based on a theoretical situation and then assess the willingness to pay instead of asking people to declare their monetary valuation directly. Willingness to pay is inferred from the theoretical choices or trade-offs that people make.

For health insurance, researchers use CV more often, using the random bidding approach to estimate the demand and the willingness to pay for it. For instance, there are studies by Gustafsson-Wright et al. (2009) in Namibia, by Dror et al. (2007) in India, by Odeyemi (2014) in rural Cameroon and by Huyen and Van Minh (2014) in the urban areas of Viet Nam, among others. The use of the contingent valuation method provides reliable and precise estimates of the willingness to pay (Diamond and Hausman, 1994).

The idea of choice experiments was introduced in a seminal paper by McFadden (1974) and later popularised by Anderson et al. (1992), Haab and McConnell (2003), Train (2009) etc. In the choice model, individuals are given a set of alternatives, and each of the alternative itself is a culmination of many attributes. The participant then chooses his or her most favoured alternative (for a review, see Bennett and Blamey, 2001). Swait and Adamowicz (2001), Bennett (1996), and Hanley et al. (1998) discuss the advantage of the choice model (CM) as compared to the contingent value model (CVM). With welfare maximisation subject to a budget constraint, a consumer will choose the option that gives maximum welfare. As the choice model is based on the "random utility" framework, it makes it possible to identify the trade-offs that each participant makes between attributes. If one such attributes is money, then a person would have to pay to make the change; precisely exploiting this idea, one can measure willingness to pay as the ratio of the parameters associated with the cost of choices. Because of this, we use the choice model technique instead of the contingent valuation model. Besides, we give individuals a simple choice between two alternatives instead of confusing them with a large number of options.

The use of data helps us explore and illustrate the causality from different factors of choice. To this end, we use a mix of the Conditional Logit model and Multinomial Logit Model to estimate the strength of the relationship between different attributes and an individual's choice among various schemes. The Conditional and Multinomial Logit models come under the category of discrete choice models. These models assume that of different options available an individual makes a choice that maximises his or her utility. An individual i chooses among j alternatives that provide different levels of utility. Individuals are assumed to maximise their utility. For instance, take the case of a health insurance model. In this model, the chosen insurance plan would give an individual higher utility compared to other plans. If the utility of individual i choosing insurance plan j is represented as U_{ij}, then insurance plan j will be chosen if and only if $U_{ij} > U_{ik}$ for $j \neq k$. Here, U_{ij} consists of two parts, the deterministic part, which is the observed utility V_{ij} and one random part, ε_{ij}. V_{ij} is the which is the observable part of the utility can be estimated using regression technique. ε_{ij} is the residual part which is not known to the researcher. The equation that is estimated is the following:

$$U_{ij} = V_{ij} + \varepsilon_{ij} \tag{15.1}$$

The probability of individual i choosing alternative j would be given by

$$P_{ij} = \Pr\left(U_{ij} > U_{ik}\right)$$

$$\Pr\left(V_{ij} + \varepsilon_{ij} > V_{ik} + \varepsilon_{ik}\right) \text{ for all } j \neq k.$$

To solve this equation, one needs to impose a probability density function on ε_{ij}. The type of a density function imposed on ε_{ij} would result in a different discrete model. This model assumes that ε_{ij} exhibits the Gumbel distribution (Williams, 1977). Furthermore, this model restricts all the ε_{ij} to be independent and identically distributed. This model helps us answer the question how the individual characteristics and the attributes of alternatives affect an individual's decision when making choices. Thus, the deterministic or the estimable part of the utility of an individual has three components, the first is related to the to the attributes of alternatives, the second relates to the characteristics of the individual who is making the decision and the third relates to the intermingling to attributes of alternatives and the characteristics of the individual making the decision.[4]

In the case of the conditional logit model, the utility of the individual is considered to be a function of only the attributes of choice. In our analysis, the utility is exclusively dependent on the attributes of alternatives health insurance schemes, given by premium and coverage. This portion of the deterministic component can be expressed as linear in parameters, that is $V_{ij} = \beta Z_{ij}$. Here, Z_{ij} is the vector of choice specific attributes, and β is the utility coefficient that has to be estimated. This variable Z_{ij} can vary across choices or across both observations and choices. For example, this part of the utility function would look as follows:

$$V_{ij} = \beta Z_{ij} = \beta_1 Z_{i1} + \beta_2 Z_{i2} + \beta_3 Z_{i3} + \ldots\ldots\ldots \beta_k Z_{ij},$$

where β_k is the coefficient that is estimates and it captures the strength and the direction of the effect of change in attribute k on the utility of an alternative, and Z_{ij} is the value of attribute k for alternative j.

In the specific example we are looking at this portion is the weighted sum of attributes like premium and coverage for the three alternatives, which are No Insurance (NI), Insurance Plan 1 (IP1) and Insurance Plan 2 (IP2):

$$V_{iNI} = \beta Z_{ij} = \beta_1 Premium_{NI} + \beta_2 Coverage_{NI}$$

$$V_{iIP1} = \beta Z_{ij} = \beta_1 Premium_{IP1} + \beta_2 Coverage_{IP1}$$

$$V_{iIP2} = \beta Z_{ij} = \beta_1 Premium_{IP2} + \beta_2 Coverage_{IP2}$$

It is important to note that parameters, β_k, are same for all the three alternatives we are looking at. This is because we assume that the effect of a change in attribute on the probability of choosing a particular alternative is constant across alternatives. Hence, we only estimate k parameters.

The conditional logit estimates a model with an underlying probability function. The probability that the individual chooses insurance plan j can be shown by

$$\Pr\left(Y_{ij} = j\right) = \frac{\exp\left(\beta Z_{ij}\right)}{\sum_{j=1}^{J} \exp\left(\beta Z_{ij}\right)}.$$

In the multinomial logit model, the expected utilities of the households who are the decision makers in our case, is the function of household income, highest education level of the members in the household and so forth. Again, this portion of the deterministic component can be expressed as linear in parameters, that is $\eta_{ij} = \alpha_j X_i$.

So, this portion of the utility function could look like

$$\eta_{ij} = \alpha_j X_i \ = \alpha_1 X_{i1} \ + \alpha_2 X_{i2} \ + \alpha_3 X_{i3} \ + \ldots\ldots\ldots \alpha_m X_{im},$$

where the parameter α_m captures the strength of the covariates on the log-odds of making a given choice.

So, in the multinomial estimation, one of the response categories is taken as a base category while keeping the base category fixed, the log-odds for all choices available is calculated relative to the base category. And then coefficients are calculated assuming that the log-odds is the linear function of the predictors.

Any category can be taken as the base category and then one calculates the odds that an individual falls in category j as opposed to the baseline as

$$\frac{\textit{Probability of being in } j}{\textit{Probability of being in the base category}}.$$

In our example, we have looked at the odds of opting for scheme 1 instead of not being insured and the odds of opting for scheme 2 instead of not being insured.

This form of the logit model is obtained by combining the multinomial and conditional logit formulations. The general model is usually written as

$$V_{ij} + \eta_{ij} = \alpha_j X_i \ + \beta Z_{ij},$$

where X_i captures the individual characteristic that are constant across choices, and Z_{ij} are the set of characteristics that vary across choices (whether they vary by individual or not). The model estimates the parameters with a basic probability function. The probability that an individual chooses insurance plan j is estimated based on the following equation:

$$\Pr\left(Y_{ij} = j\right) = \frac{\exp\left(\alpha_j X_i \ + \beta Z_{ij}\ \right)}{\sum_{j=1}^{J} \exp\left(\alpha_j X_i \ + \beta Z_{ij}\ \right)},$$

where Z_{ij} is a vector of independent variables that vary across both individual i and choice alternative j and is a $k \times 1$ vector of coefficients to be estimated. The utility specific attributes are estimated using the maximum likelihood method.

One of the important advantages of this model, resulting from its being based on the conditional logit model, is that it allows one to evaluate how much an individual is willing to trade-off one attribute of the choice against another. For instance, if the utility of the individual is expressed as the weighted sum of the following *k* attributes,

$$V_{ij} = \beta Z_{ij} = \beta_1 Z_{i1} + \beta_2 Z_{i2} + \beta_3 Z_{i3} + \ldots\ldots\ldots \beta_k Z_{ij},$$

then to evaluate how much an individual is willing to trade-off attribute Z_2 with Z_3 keeping the utility constant, we totally differentiate both sides. As everything else is held constant, so $\frac{\partial Z_2}{\partial Z_3}$ is $\frac{-\beta_3}{\beta_2}$. In our model we have

$$\beta Z_{ij} = \beta_{premium} Premium + \beta_{insurance\ coverage}\ Insurance\ Coverage.$$

Therefore, to assess how much more an individual is willing to pay for more coverage, we calculate the ratio $\frac{-\beta_{insurance\ coverage}}{\beta_{premium}}$.

15.6 Estimation results and discussion

The regression equation used is

$$U_{ij} = V_{ij} + \eta_{ij} = \alpha_j X_i + \beta Z_{ij} + \varepsilon_i.$$

However, U_{ij} is not observable. Instead, we observe an indicator y_i with the following rule: $y_i = 1$ corresponding to *i*th individual if his utility from alternative *j* is highest and 0 otherwise. In our model, as discussed earlier, the individual has three options, which are NI, IP1 and IP2. Therefore, the choice variable, which in our case is insurance choice, takes the value 1 corresponding to the alternative that individual chooses, and 0 otherwise. X_is are the various individual specific characteristics such as income, education, occupation, the percentage of old age persons and family size. These are explained in detail in the following:

Family income is the total monthly family income (in thousand INRs). This is the most important variable as this determines the buying capacity of the household and directly affects willingness to pay.

Years of education is the average number of years of education of the earning member in the family. We have taken the earning member because the decision to buy insurance usually depends on the discretion of the head or the earning member, and since the head of the family, in most cases, is illiterate, we rely on the earning member's decision.

Family size is the number of individuals in the family.

Non-awareness is the dummy that takes the value 1 if the family was not aware of the benefit of insurance before the policy was explained to them in detail.

Inpatient is a dummy that takes the value 1 if the family had experience of major sickness of one of the members and he or she was hospitalised, and zero otherwise.

Z_{ij} are the characteristics specific to the alternatives, which may also vary across choices although in this chapter are assumed to be constant across choices. In our case, these are as follows:

> *Price of insurance,* which is the premium to be paid for insurance. As discussed earlier, the value it takes is Rs. 0 for no insurance, Rs. 1,500 for choice 1 and Rs. 2,500 for choice 2.
>
> *Insurance Coverage* takes value Rs. 0 for no insurance, Rs. 50,000 for choice 1 and Rs. 100,000 for choice 2. In our data, these characteristics do not vary across individuals and vary only across choices.

Then, we estimate the coefficients of this model, which combines both the conditional as well as the multinomial model. The results from the estimation are given in Table 15.5.

It is apparent from Table 15.5 that income is one of the factors influencing the choice of health security schemes. It is evident from specification 2 that insurance coverage is also a significant factor influencing the choice of the household. However, when both are taken together as in specification 3, then the significance of premium reduces, and insurance coverage becomes

Table 15.5 Results from Estimation

Model	*1*	*2*	*3*	*4*
No Insurance (Base Outcome)				
Premium		−0.004*	−0.001***	−0.001***
		(1.89)	(7.34)	(4.32)
Insurance coverage		0.000		
		(1.42)		
Choice 1				
Family income	−0.095***	0.032		
	(5.20)	(1.23)		
Lack of awareness		0.835*	0.910*	
		(1.81)	(1.86)	
Inpatient			0.331	0.419
			(0.76)	(0.87)
Family size				−0.014
				(0.16)
Choice 2				
Family income	−0.270***	−0.097		
	(5.66)	(1.14)		
Not aware		1.345*	1.183	
		(1.80)	(1.49)	
Inpatient			−0960	−1.121
			(0.87)	(0.98)
Family size				0.045
				(0.32)

Note: Absolute t statistics in parentheses.
*p < .10, **p < .05, ***p < .01.

insignificant. This is because of the high correlation between the two variables, which introduces significant multicollinearity and reduces the significance of both the variables. The most critical household level variable is the income of the household. Again, this factor tends to become insignificant in the presence of the premium, given the high correlation between the two factors.

The other important factor is the lack of information. This is evident from the second and third specification. log-odds of choosing policy 1 relative to being uninsured are significant. This suggests that once individuals are informed about the policy and its benefits, they are inclined to buying the policy. This indicates that the demand for health insurance schemes can be developed. However, the log odds of choosing policy 2 relative to being uninsured is insignificant, suggesting that policy 2 is too expensive for them. Even though consumers have a low preference for a high level of premium as indicated by the results earlier, as discussed in the methodology section, one can assess how much more an individual is willing to pay for more coverage. For this, we calculate the following ratio $\frac{-\beta_{insurance\ coverage}}{\beta_{premium}}$, which as per our estimates from specification 3, is equivalent to $\frac{.0000837}{0.004}$, which is very close to 0, substantiating the earlier finding that policy 2 is too expensive for the poor.

15.7 Conclusion

From the preceding analysis, it is apparent that the poor do not opt for health insurance even when they are aware of the benefits of doing so because they lack the resources to do so. This is mainly because there is a lack of health insurance schemes designed to suit the needs of the economically weaker sections. Therefore, this leaves ground for the government to intervene not just in collection and allocation of resources but also in the design of health insurance schemes so that they are more equitable and efficient. Besides, many people have reported not being fully aware of health insurance schemes. A lack of awareness is a major reason for low subscription to health insurance in the country. This impedes the market for health insurance schemes in India. In this situation, the role of agents in bringing about awareness and encouraging individuals to subscribe to health insurance plans needs to be emphasised. Campaigns and other measures to improve awareness of health insurance schemes and their benefits also need to be conducted.

Notes

1 For more on this.
2 This type of classification only takes into consideration household members who are 15 years of age or older. To evaluate a household's education level, weights have been assigned to individual family members. The weight of 0.2 is assigned to individuals with low education level, 0.3 to those with medium education level and 0.5 to high education level.

3 To assign an occupational status to the household from individual status, the following procedure has been followed. If in a household, there are three working members and one member is working in the primary and two in the tertiary sector, and the household income is more from tertiary sector, then the occupational status of that household is defined as tertiary.
4 For more on this look at McFadden (1974), Long (1997) and Train (1998, 2003).

References

Ahuja, Rajeev. "Universal Health Insurance Scheme." *Rural Infrastructure: Sanitation, Housing, Health Care*(2008): 229–234.
Ahuja, Rajeev, and Indranil De. "Health Insurance for the Poor: Need to Strengthen Healthcare Provision." *Economic and Political Weekly* (2004): 4501–4503.
Anderson, Simon P., Andre De Palma, and Jacques-Francois Thisse. *Discrete Choice Theory of Product Differentiation*. Cambridge, MA: MIT Press (1992).
Aizuddin, Azimatun Noor, Saperi Sulong, and Syed Mohamed Aljunid. "Factors Influencing Willingness to Pay for Healthcare." In *BMC Public Health* 12, no. S2 (2012): A37.
Bacon-Shone, J. H., and Sarah M. McGhee. "An Exploratory Assessment of Willingness to Pay for Health Care in Hong Kong." *Hong Kong Medical Journal* 13, no. 5 Suppl 5 (2007): 26–29.
Bawa, Sumninder Kaur, and Miss Ruchita. "Awareness and Willingness to Pay for Health Insurance: An Empirical Study with Reference to Punjab India." *International Journal of Humanities and Social Science* 1, no. 7 (2011): 9.
Bennett, Jeff. "The Contingent Valuation Method: A Post-Kakadu Assessment." *Agenda: A Journal of Policy Analysis and Reform* (1996): 185–194.
Bennett, Jeff, and Russell Blamey, eds. *The Choice Modelling Approach to Environmental Valuation*. Edward Elgar Publishing (2001).
Burrows, C., K. Brown, and M. Ngui. "Who Buys Health-Insurance—A Multivariate-Analysis of ABS Surveys." *Community Health Studies* 13, no. 3 (1989): 352.
Cameron, A. Colin, and Pravin K. Trivedi. "The Role of Income and Health Risk in the Choice of Health Insurance: Evidence from Australia." *Journal of Public Economics* 45, no. 1 (1991): 1–28.
Cameron, A. Colin, Pravin K. Trivedi, Frank Milne, and John Piggott. "A microeconometric model of the demand for health care and health insurance in Australia." *The Review of Economic Studies* 55, no. 1 (1988): 85–106.
Devadasan N., Ranson Kent, Damme Van Wim and Criel Bart. "Community Health Insurance in India: An Overview." *Economic and Political Weekly* 39, no. 28 (2004): 3179–3183.
Diamond, P., and Hausman, J., (1994): "Contingent valuation: is some number better than no number?" *Journal of Economic Perspectives* 8 (1994):45–64.
Dror, David Mark, Ralf Radermacher, and Ruth Koren. "Willingness to Pay for Health Insurance among Rural and Poor Persons: Field Evidence from Seven Micro Health Insurance Units in India." *Health Policy* 82, no. 1 (2007): 12–27.
Grossman, Michael. *The Demand for Health: A Theoretical and Empirical Investigation*. Cambridge, MA: NBER Books (1972).
Gustafsson-Wright, Emily, Abay Asfaw, and Jacques van der Gaag. "Willingness to Pay for Health Insurance: An Analysis of the Potential Market for New Low-Cost Health Insurance Products in Namibia." *Social Science & Medicine* 69, no. 9 (2009): 1351–1359.

Haab, Timothy, and K. McConnell. "*The Econometrics of Non-Market Valuation*." Northampton, Massachusetts: Edward Elgar (2003).
Hanley, N., Wright, R. E. and Adamowicz, W. "Using choice experiments to value the environment." *Environmental and Resource Economics* 11 (1998): 413–428.
Huyen, Doan Thi Thu, and Hoang Van Minh. "Willingness to Pay for Health Insurance among Informal Sector Workers: A Case Study from Hanoi Capital Vietnam." *Vietnam Journal of Public Health* 2, no. 2 (2014): 24–31.
Kansra, Pooja, and G. Patharia. "A Study of Factor Affecting the Demand for Health Insurance in Punjab." *Journal of Management and Science* 2, no. 4 (2014): 1–10.
Kimani, James K., et al. "Determinants for Participation in a Public Health Insurance Program Among Residents of Urban Slums in Nairobi, Kenya: Results from a Cross-Sectional Survey." *BMC Health Services Research* 12, no. 1 (2012): 1–11.
Kronick, Richard, and Todd Gilmer. "Explaining the Decline in Health Insurance Coverage, 1979–1995." *Health Affairs* 18, no. 2 (1999): 30–47.
Long, J. *Regression Models for Categorical and Limited Dependent Variables.* Thousand Oaks: Sage Publications. (1997)
McFadden, D. "Conditional Logit Analysis of Qualitative Choice Behavior." In P. Zarembka (ed.) *Frontiers in Econometrics*. New York: Academic Press (1974), 105–142.
Muurinen, Jaana-Marja. "Demand for Health: A Generalised Grossman Model." *Journal of Health Economics* 1, no. 1 (1982): 5–28.
Odeyemi, Isaac A. O. "Community-Based Health Insurance Programmes and the National Health Insurance Scheme of Nigeria: Challenges to Uptake and Integration." *International Journal for Equity in Health* 13, no. 1 (2014): 1–13.
Panda, Pradeep, Arpita Chakraborty, David M. Dror, and Arjun S. Bedi."Enrolment in Community-Based Health Insurance Schemes in Rural Bihar and Uttar Pradesh, India." *Health Policy and Planning* (2014): czt077.
PHFI. "Public Health Foundation of India; A Critical Assessment of the Existing Health Insurance Models in India" (2011). Retrieved from http://planningcommission.nic.in/reports/sereport/ser/ser_heal1305.pdf
Rao, Sujatha. "Health Insurance: Concepts, Issues and Challenges." *Economic and Political Weekly*, 39, no. 34 (2004): 3835–3844.
Ranson, M. K., Devadasan, N., Acharya, A. and Fanstone, R. *How to Design a Community-Based Health Insurance Scheme: Lessons Learned from a Review of Indian Schemes*. Washington, DC: The World Bank (2003).
Savage, Elizabeth, and Don Wright. *Health Insurance and Health Care Utilization: Theory and Evidence from Australia 1989–90*. (2001). Mimeograph B2 – Mimeograph, University of Sydney, Sydney.
Scotton, Richard Bailey. "Membership of Voluntary Health Insurance." *Economic Record* 45, no. 1 (1969): 69–83.
Shaw R. P., and M. Ainsworth. *Financing Health Services through User Fees and Insurance*. Washington DC: The World Bank (1995).
Shijith V.P., and T. V. Sekher. "Who Gets Health Insurance Coverage in India? New Findingsfrom Nation-Wide Surveys." In *XXVII IUSSP International Population Conference, Busan, Republic of Korea*, vol. 26. 2013.
Singu, Hari Babu, and Gurpreet Kaur. "Factor Barriers to Subscribe Health Insurance Policies among the People of Punjab: A Study." *International Journal of Economic Research* 11, no. 3 (2014): 505–525.

Swait, J., and Adamowicz, W. "The Influence of Task Complexity on Consumer Choice: A Latent Class Model of Decision Strategy Switching." *Journal of Consumer Research* 28 (2001): 135–148.

Tirukoti, Subba Lakshmi. "Determinants of Private Health Insurance in Andhra Pradesh: A Case Study." *International Journal of Behavioural and Healthcare Research* 5, no. 1–2 (2015): 52–72.

Train, K. "Recreation demand models with taste variation." *Land Economics* 74 (1998): 230–239.

Train, Kenneth E. *Discrete Choice Methods with Simulation*. Cambridge: Cambridge University Press (2009).

United States Agency for International Development. "Private Health Insurance in India: Promise & Reality." (2008).

Williams, H. C. W. L. (1977) "On the Formation of Travel Demand Models and Economic Evaluation Measures of User Benefit." *Environment and Planning*, 9A (1977): 285–344.

16 Private and public dimensions to infectious disease risks

A case of Kolkata

Shreejata Samajpati

16.1 Introduction

This chapter focuses on the non-market valuation of health-risks from infectious diseases like malaria. Despite global policy initiatives on disease prevention, diagnosis and treatment, infectious diseases continue to impose a serious public health burden across the tropical disease-prone areas of the globe. Here, we consider the "externality" dimension to infectious diseases, recognize the private and public aspects of the associated health risks and address the research question: How do individuals, living in areas with recurrent disease threats, value risk-reductions from competing prevention measures, viz. a publicly-administered community-level disease control measure as against private preventive choices? Such a comparative empirical assessment of health-risk valuation across public and private treatments, in turn, allows for indirectly testing for the externality dimension to disease prevention, thus, informing policy on the urgency of public action.

Incorporating a health production function approach, the theoretical model explores the public-private interplay of health risks. The welfare analysis illustrates the theoretical measures of value for two kinds of health interventions—(1) a community-level malaria control program that reduces health risks for both the decision-maker and her community and (2) a private-level health intervention product which solely benefits the decision-maker. A field survey[1] involving risk-perception elicitation methods and stated preference techniques of health valuation is designed and implemented in and around the city of Kolkata, West Bengal. The survey elicits information on the prevention strategies that individuals already engage in, records malaria-related experiences and treatment costs that individuals faced in a certain recall period and so on. The thrust of the fieldwork, however, lies in incorporating a Contingent Valuation Method (CVM) component, whereby individuals are offered randomized hypothetical scenarios of malaria control and respondents' willingness to financially contribute for the same are elicited. A between-subject design is implemented with the two treatments, public and private. To ensure the reliability of the risk reductions that the randomized private/public scenarios offer, the survey first elicits respondents' risk perceptions using a novel visual risk scale (colored cards). Then the CVM question is posed and associated risk reductions are pictorially conveyed to the respondents.

The empirical analysis involves a comparison of the measures of value for public and private risk reductions. The results contribute to policy with regard to the following: (1) Demand-side assessments of both kinds of malaria control tools, public and private, offered simultaneously in the field, are obtained for the first time in the literature;[2] (2) the public good/externality dimension to malaria control is indirectly tested, informing policy on the extent of public action urgency;[3] (3) the valuation question and the issue of price sensitivity are explored across various split samples[4] that generate policy insights into various socioeconomic factors and other disease-related attributes that shape private prevention behavior. The viability of a scaled-up public program is discussed by comparing the public treatment willingness to pay estimates with the annual estimated costs of the Kolkata Municipal Corporation incurred on account of malaria vector control.

16.2 Literature

In contemporary policy dialogues, the issue of infectious disease control is treated with great urgency. For instance, The Roll Back Malaria Program of the World Health Organization (WHO 2005, 2008, 2010) comprises a massive control effort. Malaria is preventable and yet the road ahead seems complex. Despite international investments promoting the availability of preventive tools like insecticide-treated bed nets (ITNs) around the world, private prevention demand is often reported to be sub-optimal. The literature, thus, speaks of different pricing incentives and commitment mechanisms to trigger better use of bed nets (Cohen and Dupas 2010; Dupas 2009; Tarozzi et al. 2009). Moreover, given the infectiousness of the disease, private prevention generates positive social spillovers (Gersovitz and Hammer 2004, 2005). This externality dimension indicates the role of the government (in terms of community-level indoor residual spraying programs etc.). Apprehensions also remain on the possibility of behavioral feedbacks across private and public risk-control measures (Klein et al. 2007). But the private–public interplay of health risks and a comparative valuation analysis have seldom been empirically tested. This inspires the key motivation for the theoretical and empirical exercises developed in this chapter.

16.3 Theoretical framework

16.3.1 Model

The theoretical framework is fundamentally akin to Harrington and Portney (1987). But, given the particular interest in health risks from malaria, both at the individual and community levels, some key aspects are additionally incorporated. First, household production of health risks, rather than "sick time", is modeled as in Dickie and Gerking (2007). Second, the model attempts to test for the presence of social preferences that the decision-maker may have with regard to the community she lives in. In the present framework, community-level social preferences pertain to the satisfaction that the decision-maker derives from reduced malaria risks facing the other individuals living in her community. Third, the role of public action, that is, the government, is brought in, alongside private prevention efforts, to reduce health risks from malaria.

The decision-maker is rational and asserted to be one who (1) is aware of the morbidity risks that malaria poses to her and the community, (2) is in the know of the disease being preventable and (3) takes private risk-reducing measures as a necessary safeguard, alongside being informed of the community-level control measures. The individual's perception of malaria risk that she herself faces is denoted as R^S. Perceived malaria risks can be reduced through the purchase of a marketed preventive good *a* and the consumption of a malaria-specific public good, namely, community-level malaria control measures, *g*, that the public health and civic authorities implement.

Thus,

$$R^S = r^S(a; g) \tag{16.1}$$

In the context of health risks from malaria, examples of *a* primarily include the purchase of bed nets, mosquito repellants or window-netting. The parameter, *g*, considered in the individual's optimization problem, may involve an array of community-level malaria-control measures such as vector-control programs or indoor residual spraying (IRS), swamp and canal cleaning drives, provision of health facilities for effective prevention, diagnosis and treatment or knowledge and awareness dissemination.[5] In equation 16.1, the technological or objective relationship between the consumption of the private good *a* and that of *g* is one of substitutes, as in Hori (1975). Each of *a* and *g* results in reduced perceived risks that the individual faces. Thus, the marginal products of *a* and *g* are given as $\frac{\partial r^S}{\partial a} < 0$ and $\frac{\partial r^S}{\partial g} < 0$, respectively. It is also assumed that $\frac{\partial^2 r^S}{\partial a^2} > 0$, implying that for a given level of *g*, the marginal effectiveness of *a* declines as additional units of *a* are successively purchased.[6] Following the same logic, we additionally assume that $\frac{\partial^2 r^S}{\partial a \partial g} > 0$; that is, successive increases in the level of public action, *g*, make the marginal product of a fall. Herein, note that the purchase of *a* does not affect utility directly. Also, if the decision-maker cares for the health (or malaria risks) of the other community members living in her proximity, *g* assumes an additional importance, apart from its role in impacting the individual's own malaria risks, R^S. The individual perceives that the community-level malaria control measures contribute towards reducing the malaria risks facing the other community members, R^C, as well. The social effect of g with regard to influencing others' malaria risks is given as

$$R^C = r^C(g), \tag{16.2}$$

where $\frac{\partial r^C}{\partial g} < 0$.

The individual maximizes the utility function

$$\tilde{U}(x,a) \stackrel{def}{=} U\left[x, r^S(a; g), r^C(g)\right] \tag{16.3}$$

subject to the risk production function in equation 16.1, the social effect function in equation 16.2 and the budget constraint

$$Y = x + p_a a. \tag{16.4}$$

The partial derivatives underlying the utility function 16.3 are assumed as $\frac{\partial \tilde{U}}{\partial x} = \frac{\partial U}{\partial x} > 0$ and $\frac{\partial^2 \tilde{U}}{\partial x^2} = \frac{\partial^2 U}{\partial x^2} < 0$, with respect to the numeraire x. Besides, perceived malaria risks, R^S, impact the well-being/utility of the individual negatively; that is, $\frac{\partial U}{\partial R^S} < 0$, and $\frac{\partial^2 U}{\partial R^{S2}} < 0$, implying that each additional unit of R^S results in successive increases in the reductions in utility U.[7] Moreover, the assumptions specified so far, in conjunction with the properties of the utility function, imply that $\frac{\partial \tilde{U}}{\partial a} = \frac{\partial U}{\partial R^S}\frac{\partial r^S}{\partial a} > 0$, and $\frac{\partial^2 \tilde{U}}{\partial a^2} = \frac{\partial U}{\partial R^S}\frac{\partial^2 r^S}{\partial a^2} + \frac{\partial r^S}{\partial a}\frac{\partial^2 U}{\partial R^{S2}}\frac{\partial r^S}{\partial a} < 0$. Alongside, $\frac{\partial^2 U}{\partial x \partial R^S} > 0$; that is, the marginal disutility from private malaria risks decreases with increased consumption of the numeraire good, which, in turn, implies that $\frac{\partial^2 \tilde{U}}{\partial x \partial a} = \frac{\partial^2 U}{\partial x \partial R^S}\frac{\partial r^S}{\partial a} < 0$. Thus, it follows that the utility function is increasing and concave with respect to the private preventive good a and the numeraire good x.

Besides the previously mentioned assumptions, if the decision-maker has social preferences pertaining to the community she resides in, then $\frac{\partial U}{\partial R^C} < 0$. In the utility function (3), R^C constitutes the component representing social preferences with regard to the overall malaria exposure in the community. The illustration of caring externalities in the context of economic evaluation of health policies in Jacobsson et al. (2005) inspires the model specification in this regard. Note, however, that here the decision-maker—even if having regard for others' well-being—lacks control in influencing the same. Thus, governmental risk-control measures are perceived to be of sole significance in bringing about an improvement in the community's overall health conditions. An alternative way of interpreting the social preference component in the utility function 16.3 is that the decision-maker has altruistic preferences towards the other community members facing malaria risks, where altruism is pure and outcome-oriented as in Francois and Vlassopoulos (2008).[8] Herein, two assumptions need mention to theoretically specify the interaction between community-level malaria risks and the numeraire and that between private and community risks: (1) $\frac{\partial}{\partial R^C}\left(\frac{\partial U}{\partial x}\right) < 0$, implying that the marginal utility from the numeraire declines with increasing levels of malaria risks in the community, and (2) $\frac{\partial}{\partial R^C}\left(\frac{\partial U}{\partial R^S}\right) < 0$; that is, the marginal disutility from private malaria risks increases as malaria threats in the community rise.

In the budget constraint that the decision-maker faces (i.e., equation 16.4), Y stands for the exogenous income and p_a represents the price of the marketed risk-reducing good privately consumed by the individual to avoid malaria risks. x is treated as the numeraire.

The Lagrangian of the optimization problem can be written as follows:

$$L(x, a, g, p_a, Y) = U\left[x, r^S(a; g); r^C(g)\right] + \lambda(Y - x - p_a a), \qquad (16.5)$$

where λ is the Lagrange multiplier associated with the budget constraint. The First-Order Necessary Conditions (FONCs) are given by

$$L_x = \frac{\partial U\left[x, r^S(a; g); r^C(g)\right]}{\partial x} - \lambda = 0, \qquad (16.6)$$

$$L_a = \frac{\partial U\left[x, r^S(a; g); r^C(g)\right]}{\partial R^S}\frac{\partial r^S(a; g)}{\partial a} + \lambda[-p_a] = 0, \qquad (16.7)$$

$$L_\lambda = Y - x - p_a a = 0. \qquad (16.8)$$

Manipulations with equations 16.6 and 16.7 and substitution of λ yield

$$\frac{\partial U}{\partial R^S}\frac{\partial r^S(a; g)}{\partial a} = \frac{\partial U\left[x, r^S(a; g); r^C(g)\right]}{\partial x} p_a$$

$$\text{or } \frac{\frac{\partial U}{\partial R^S}\frac{\partial r^S}{\partial a}}{\frac{\partial U}{\partial x}} = p_a. \qquad (16.9)$$

Equation 16.9 suggests, that at the optimum, the consumption of a is such chosen that the monetized marginal benefit of private prevention is equal to the price p_a. Assuming the Second Order Sufficient Conditions hold at the optimum, the Implicit Function Theorem (IFT) is invoked and equations 16.6, 16.7 and 16.8 are used to solve for the optimal values of x, a and λ, in principle, thus giving

$$x = x^*\left(p_a, Y, g\right); a = a^*\left(p_a, Y, g\right), \text{and}\, \lambda = \lambda^*\left(p_a, Y, g\right).$$

16.3.2 *Comparative statics*

Although the preceding model assumes that a and g are technological substitutes in the individual's risk production function, $r^S(a;g)$, it is of interest to explore how, at the optimum, external shocks to the system through changes in the parameter g impact the optimal economic choice of a, that is, a^*. Such a comparative statics exercise will also allow for studying if, theoretically, the presence (or absence) of the social preference component embodied in the utility function 16.3 has implications for the sign and/or magnitude of the comparative statics. This, in turn, may throw light on the nature and extent of the substitutability between private and public efforts of malaria control under different conditions. For the purpose, $x^*(p_a,Y,g)$, $a^*(p_a,Y,g)$ and $\lambda^*(p_a,Y,g)$ are substituted in the FONCs to get the following comparative statics:

$$\frac{\partial a^*}{\partial g} = \frac{-\left[\frac{\partial^2 U}{\partial R^S \partial x}\cdot\frac{\partial r^S}{\partial g}\cdot p_a - \frac{\partial U}{\partial R^S}\cdot\frac{\partial^2 r^S}{\partial g \partial a^*} - \frac{\partial r^S}{\partial a^*}\frac{\partial^2 U}{\partial {R^S}^2}\frac{\partial r^S}{\partial g}\right]}{|H|} + \frac{-p_a\cdot\frac{\partial^2 U}{\partial R^C \partial x}\cdot\frac{\partial r^C}{\partial g} + \frac{\partial r^S}{\partial a^*}\frac{\partial^2 U}{\partial R^C \partial R^S}\frac{\partial r^C}{\partial g}}{|H|}, \tag{16.10}$$

where $|H|$ is the Hessian matrix.

It follows from the assumptions that $\frac{\partial a^*}{\partial g} < 0$, implying that the decision-maker chooses an increased amount of private preventive measures, $a*$, when there is a decrease in the level of community-level malaria control efforts that the public health authorities implement. The comparative statics corroborates that private and public preventive measures are substitutes irrespective of whether social preferences are accounted for. Additionally, it emerges that the presence of concern for fellow community member's health makes the substitution result even stronger. For solely self-interested individuals without concern for the overall health conditions of the community, $\frac{\partial a^*}{\partial g} = \frac{-\left[\frac{\partial^2 U}{\partial R^S \partial x}\cdot\frac{\partial r^S}{\partial g}\cdot p_a - \frac{\partial U}{\partial R^S}\cdot\frac{\partial^2 r^S}{\partial g \partial a} - \frac{\partial r^S}{\partial a^*}\frac{\partial^2 U}{\partial {R^S}^2}\frac{\partial r^S}{\partial g}\right]}{|H|} < 0$. Also, $\left|\frac{\partial a^*}{\partial g}\right|$

| without social preferences < $\left|\frac{\partial a^*}{\partial g}\right|$ |with social preferences, implying the substitution result to be even stronger under social preferences towards the community. Recall that, in the present framework, the spillovers of private preventive actions *a* on the community are not modeled. Rather, the decision-maker enjoys satisfaction from the increased well-being of her fellow community residents, brought about only by government-administered community-wide malaria control measures.

16.3.3 *Welfare analysis*

The economic substitutability between *a* and *g* shown above brings in some associated questions relevant to the welfare analysis of malaria-control policies. For instance, would the individual value two types of risk-reducing measures, viz. private and public health interventions, differently? In clearer terms, these two types may be conceived as (1) new protective goods that generate only private benefits (hitherto non-marketed goods but otherwise similar to the nature of *a*) and (2) community-wide malaria control measures that have both private and public/community-wide dimensions of benefits (like *g*). Second, if economic valuations do differ across private and public goods, what implications, if at all, may emerge with regard to the social preferences of the decision-maker towards the community she lives in? These questions assume policy significance under a budget-constrained public spending scenario which is quite often the reality in malaria-endemic developing countries.

16.3.4 *Valuation of marginal changes in community-level malaria control measures*

Let the indirect utility function be denoted as $V(Y, p_a, g)$ which is defined as

$$V(Y, p_a, g) \stackrel{def}{=} U\left[x^*(p_a, Y, g), r^S\left(a^*(p_a, Y, g); g\right); r^C(g)\right]. \tag{16.11}$$

A compensating surplus approach to valuing changes in the quantities of public goods is adopted (Freeman 1993). For the purpose, willingness to pay (WTP) for changes in the community-wide risk-control measures, when its quantity marginally increases from $\bar{g}$ to g , is defined as the following:

$$V(Y - WTP, p_a, g) = V(Y, p_a, \bar{g}); \text{ where } g > \bar{g}. \tag{16.12}$$

Differentiating both sides of equation 16.12 with respect to *g* as in (Courant and Porter 1981; Harrington and Portney 1987), we get

$$V_Y(.)\left[-\frac{\partial WTP}{\partial g}\right] + V_g(.) = 0 \tag{16.13}$$

Or, $\frac{\partial WTP}{\partial g} = \frac{V_g(.)}{V_Y(.)} = \frac{V_g(.)}{\lambda^*}$, where λ^* is the marginal utility of income at the optimum.

Also, $V_g = \left[\frac{\partial U(.)}{\partial R^S}\frac{\partial r^S}{\partial g}\right] + \left[\frac{\partial U(.)}{\partial R^C}\frac{\partial r^C}{\partial g}\right]$ (by the envelope theorem).

Substituting for the expression of V_g in equation 16.13, we get

$$\frac{\partial WTP}{\partial g} = \left[\frac{\frac{\partial U(.)}{\partial R^S}\frac{\partial r^S}{\partial g}}{\lambda^*}\right] + \left[\frac{\frac{\partial U(.)}{\partial R^C}\frac{\partial r^C}{\partial g}}{\lambda^*}\right]. \tag{16.14}$$

The left-hand side (L.H.S.) of equation (16.14) represents the change in the willingness to pay for an extra unit of the public risk-control measures, g, and is made up of two components. The first bracketed expression on the right-hand side (R.H.S.) reflects the monetized *private* benefits (reduced private health risks) that accrue to the decision-maker in consequence of a change in g.[9,10] The second bracketed expression in equation 16.14 denotes the valuation of the *social* benefits ensuing from the increased provision of the public health intervention, if the decision-maker cares for others' health in the community.[11]

16.3.5 Valuation of private-level risk-control measures

Now consider policy measures concerning the introduction of a new advanced risk-control product, the benefits of which are largely private in nature, that is, are enjoyed solely by the decision-maker who chooses to consume it. Thus, such a good, is essentially akin to a, albeit with the characteristic that it is hitherto non-marketed and possesses advanced effectiveness features. The decision-maker's valuation of such a good will assume significance in order to estimate the benefits of a new policy that may target to achieve increased private-level consumption of a preventive good for an effective combat of malaria threats in the society. As a related matter, note that in the present framework, since a good is deemed private when it only impacts the private health risks of the decision-maker, R^S,[12] the valuation exercise pertaining to the new good essentially translates into the valuation of reduced private risks that the good brings forth.

Suppose L is a new private risk-control product (say, a mosquito-repelling lotion) that may be potentially introduced under a new malaria control plan. Let L be a new parameter in the private risk production function, $r^S(.)$, and is at an initial level, $\bar{L}$. Then the decision-maker's optimal value of utility can be written as

$$V(.) \stackrel{def}{=} U\left[x^*\left(p_a, Y, \bar{L}, g\right), r^S\left(a^*\left(p_a, Y, \bar{L}, g\right); g\right); r^C(g)\right]. \tag{16.15}$$

Note that the valuation procedure followed earlier for the public risk-control program in equation 16.12 is now repeated for the new private good, L. The WTP for an increase in the level of the new good from $\bar{L}$ to L is denoted as Z, and is defined as the following:

$$V(Y - Z, p_a, g, L) = V, \text{ where } L > \bar{L}. \tag{16.16}$$

Differentiating equation 16.16 with respect to L, we get

$$V_Y(.)\left[-\frac{\partial Z}{\partial L}\right] + V_L(.) = 0. \tag{16.17}$$

Or, $\frac{\partial Z}{\partial L} = \frac{V_L(.)}{V_Y(.)} = \frac{V_L(.)}{\lambda^*}$, where λ^* is the marginal utility of income at the optimum.

Also, $V_L = \frac{\partial U(.)}{\partial R^S}\frac{\partial r^S}{\partial L}$ (by the Envelope Theorem).

Substituting for the expression of V_L in equation 16.17, we get

$$\frac{\partial Z}{\partial L} = \frac{\frac{\partial U(.)}{\partial R^S}\frac{\partial r^S}{\partial L}}{\lambda^*} > 0 \tag{16.18}$$

The LHS of equation 16.18 represents the valuation, at the margin, of the private health intervention, i.e., provision of the new preventive good L, offered to the decision-maker. The RHS can be interpreted as the valuation of the change in the private health risks that the policy/good brings forth.

16.4 Field survey

To empirically analyze the public–private comparative valuation question, field data were collected from a household survey in and around the city of Kolkata, India, during September and December 2011. The recurrence of the disease in the survey site and the selection of the area in past malaria-based biological studies inspired the field choice.[13] Using interviewer-administered questionnaires, face-to-face interviews in the setting of the homes of the respondents were implemented. Instruments involved (1) a visual physical device (a pack of 11 bicolored cards used as a 0–10 risk scale) for eliciting malaria-related risk perceptions; (2) a between-subject survey design comprising two health intervention treatments, private and public; (3) the random assignment of participants to one of the treatments with each respondent receiving either 50% or 90% health-risk reductions through the offered intervention; (4) a CVM question; and (5) sections on revealed preference behavior, along with other background questions on family demographics, socioeconomic status, community-specific characteristics and malaria-related awareness and disease history.[14]

Households in the city of Kolkata and its neighboring semi-urban district of the South 24 Parganas, comprised the universe or population. The sample size was 780 (urban: 350; semi-urban: 350; slum: 80).[15] The fieldwork progressed in stages: focus group discussions, pretest sessions and the final survey. A multi-stage simple random sampling procedure was adopted. In the urban site, the Election Commission of India voter rolls (2011) pertaining to a malaria-prone municipal ward (Ward No. 72) comprised the sampling frame from which 350 households (20,887 individuals were coalesced to 3,167 addresses; subsequently, the number of people living in an address were used as weights) were chosen. In the semi-urban area, in the third stage of random sampling, a complete enumeration exercise in the selected village (Kalikapur village) was followed by the random selection of the ultimate 350 units. Also, all 80 households of a slum area within the urban site were studied. A willing adult representative from each of the 780 households selected was interviewed. No screening criterion except for a respondent's age (18 years or older) was implemented. Recovery from malaria does not ensure full immunity to the disease and hence was the plan not to differentiate between prior patients and nonpatients of malaria when determining respondents' eligibility.

A key feature of the survey design was the development and use of a novel visual device for the purpose of risk elicitation. A pack of 11 bicolored cards pasted on cardboard, was developed as a 0–10 risk scale, and to motivate the relevance of the cards to the context of malaria, each card was described as a "Malaria Card" to the respondents. Each of the cards, labeled 0 through 10, was explained to be representing a particular level of belief that malaria would occur (i.e., a certain level of malaria risk). Each card was so constructed that the grid consisted of 10 squares arranged in 5 rows and 2 columns. Some of the squares were colored red while the rest were blue. Blue meant the possibility that malaria would not occur. Red indicated the possibility that malaria would occur. The number of squares colored red on a certain malaria card was used to represent the level of risk. Counting the number of squares colored red and by following the number label that accompanied each card, respondents could comprehend what level of risk each card was representing. For instance, the annual malaria risk of 6/10 (i.e., 6 out of 10 chances) was illustrated by the Malaria Card No. 6 on our risk-elicitation device, on which 6 out of 10 squares were red while the rest of the squares (i.e., 4) were blue. Each respondent was asked to select a card, from among the 11 cards on the cardboard, commensurate with their respective perceived level of malaria risk. Subsequently, the public and private treatments were offered and the associated risk reductions (either 50% or 90%) offered through the health interventions were pictorially conveyed with the help of another card tucked inside the card that the respondent had originally selected. Once the scenarios were explained, the CVM question was posed and the respondents' willingness-to-pay decisions were recorded.

Thus, in effect, the survey involved a 2 × 2 × 4 randomized design: 2 treatments, public and private; 2 kinds of risk reductions, 90% and 50 %; and 4 price levels, INR 55, INR 75, INR 125 and INR 225.

16.5 Econometric specification

The willingness-to-pay functions in the public and private contexts are specified econometrically to develop an empirical framework consistent with the theoretical expressions illustrated in equations 16.14 and 16.18. In the public context, an individual is informed about a new (but hypothetical) community-wide governmental program of vector control. In the private, a hypothetical mosquito-repelling lotion, treated as a new private preventive good, is offered. The empirical exercise attempts to place a value on the associated reductions in the health risks that these two hypothetical contexts bring forth.

16.5.1 Empirical methods: The "Community-Level Health Intervention" treatment

Recall the theoretical specification of equation 16.14:

$$\frac{\partial WTP}{\partial g} = \left[\frac{\frac{\partial U(.)}{\partial R^S}\frac{\partial r^S}{\partial g}}{\lambda^*}\right] + \left[\frac{\frac{\partial U(.)}{\partial R^C}\frac{\partial r^C}{\partial g}}{\lambda^*}\right],$$ where $\frac{\partial WTP}{\partial g}$ is interpreted as the change in the WTP for an additional unit of provision of the governmental malaria control efforts g. The RHS is composed of two parts: the monetized marginal utility of the private dimensions to health-risk reductions that a unit of g brings in and that associated with the community dimension to health-risk reductions which g further achieves. In this regard, note that $\frac{\partial r^S}{\partial g}$ and $\frac{\partial r^C}{\partial g}$ denote the marginal effectiveness of public efforts, in reducing private risks, R^S, and community risks, R^C, respectively.

Instead of making assumptions about the relationship between r^S (or r^C) and g (and hence about their marginal products), the empirical specification of the preceding equation is made operational (keeping in mind the survey protocol) via a direct reduction in the amount of risk. Thus, the marginal products can be substituted by the changes in risk themselves:

$$\frac{\partial WTP}{\partial g} = \left[\frac{\frac{\partial U(.)}{\partial R^S}dR^S}{\lambda^*}\right] + \left[\frac{\frac{\partial U(.)}{\partial R^C}dR^C}{\lambda^*}\right].$$ Furthermore, the risk changes are offered in percentage terms[16] so that, in essence, an individual pays for the percentage reduction in health risks, which is "achieved" or "brought about" via g at a certain stated cost. To facilitate treatment of risk changes

in terms of percentages, in our empirical framework, we do the following algebraic manipulations:

$$\frac{\partial WTP}{\partial g} = \left[\frac{\frac{\partial U(.)}{\partial R^S}.R^S}{\lambda^*}\right]\frac{dR^S}{R^S} + \left[\frac{\frac{\partial U(.)}{\partial R^C}.R^C}{\lambda^*}\right]\frac{dR^C}{R^C}. \tag{16.19}$$

Equation (16.19) can be interpreted as the value of the public efforts, g, comprising $\left[\frac{\frac{\partial U(.)}{\partial R^S}.R^S}{\lambda^*}\right]$; that is, the valuation of proportionate changes (reductions) in private risks, $\frac{dR^S}{R^S}$, and $\left[\frac{\frac{\partial U(.)}{\partial R^C}.R^C}{\lambda^*}\right]$, the valuation of proportionate community risk changes, $\frac{dR^C}{R^C}$. At this juncture, it needs to be mentioned that the field survey is so designed that percentage risk reductions are communicated to be the *same* for the decision-maker and her community. More simply put, in the field, respondents were let known about the hypothetical possibility under which a new vector-control program, g, reduces malaria risk by a certain percentage for the entire community where the respondent lives (thus, benefiting the individual in question and fellow community-members as well). This is to say that $\frac{dR^S}{R^S} = \frac{dR^C}{R^C} = \frac{dR}{R}$[17]. Equation (16.19) can now be expressed as follows:

$$\frac{\partial WTP}{\partial g} = \left[\frac{\frac{\partial U(.)}{\partial R^S}.R^S}{\lambda^*} + \frac{\frac{\partial U(.)}{\partial R^C}.R^C}{\lambda^*}\right]\frac{dR}{R}. \tag{16.19.1}$$

The econometric specification of equation 16.19.1 is given as

$$MWTP^g = \gamma_0 + \gamma_1(risk\_r) + \gamma_2(controls) + \epsilon^g, \tag{16.20}$$

where $MWTP^g$ is the marginal willingness to pay for the public malaria control efforts, g, and the variable *risk_r* denotes the percentage reduction in risks brought about by g; *controls* capture the individual and community-specific characteristics that affect the $MWTP^g$, such as income, education, age, location of the household, household size and so on; and ϵ^g is the random disturbance term representing the unobserved characteristics of the individual. The coefficient γ_1 captures the WTP for a percentage reduction in malaria risk, where $\gamma_1 = \frac{\frac{\partial U(.)}{\partial R^S}.R^S}{\lambda^*} + \frac{\frac{\partial U(.)}{\partial R^C}.R^C}{\lambda^*}$. However, at the moment, we do not obtain separate estimates for the valuation of reductions of health risks that accrue to the decision-maker (i.e., $\frac{\frac{\partial U(.)}{\partial R^S}.R^S}{\lambda^*}$) and to her community ($\frac{\frac{\partial U(.)}{\partial R^C}.R^C}{\lambda^*}$), from the community-level malaria control intervention generating both private and social benefits. To examine if such private and public valuations differ, there is, thus, the need to implement a second treatment, namely, the "private-level health intervention treatment" on a set of individuals. In the latter, an individual is offered risk reductions (in percentage terms) for private risks, R^S, only through the offer of purchasing a new malaria-preventing good, namely a lotion, illustrated in Section 16.5.2.

16.5.2 Empirical methods: The "Private-Level Health Intervention" treatment

In this section, we briefly lay out the empirical specification of the willingness-to-pay function for a new malaria-preventing product, L, pertaining to individuals who are randomly assigned to the "private-level health intervention" treatment in our field survey.

Recall the theoretical specification given in equation 16.18, that is, $\frac{\partial Z}{\partial L} = \left[\frac{\frac{\partial U(.)}{\partial R^S}\frac{\partial r^S}{\partial L}}{\lambda^*}\right]$. Similar to the methodology in Section 16.5.1, risk changes are offered in percentage terms so that, in essence, an individual pays for the percentage reduction in health-risks, which is "achieved" or "brought about" via L at a certain stated price. To bring in percentage forms of risk reductions in our specification, we do the following:

$$\frac{\partial Z}{\partial L} = \left[\frac{\frac{\partial U(.)}{\partial R^S}.R^S}{\lambda^*}\right]\frac{dR^S}{R^S} \tag{16.21}$$

Equation (16.21) can be interpreted as the value of the new private preventive good, L. The RHS, that is, $\left[\frac{\frac{\partial U(.)}{\partial R^S}.R^S}{\lambda^*}\right]$, stands for the valuation of the proportionate changes (reductions) in private risks, $\frac{dR^S}{R^S}$, that L brings forth. The econometric specification of equation (16.21) is given as

$$MWTP^L = \gamma'_0 + \gamma'_1(risk\_r) + \gamma'_2(controls) + \epsilon^L, \tag{16.22}$$

where $MWTP^L$ is the marginal willingness to pay for L and the variable *risk_r* denotes the percentage reductions in malaria risks brought about by

L; *controls* capture the individual and community-specific characteristics that affect the $MWTP^L$, such as income, education, age, location of the household, household size and so on; and ϵ^L is the random disturbance term representing the unobserved characteristics of the individual. The coefficient γ'_1 captures the WTP for a percentage reduction in private malaria risks (i.e., $\frac{\frac{\partial U(.)}{\partial R^S}.R^S}{\lambda^*}$).

Note that the elements in the coefficient γ'_1 are the same as the first additive component in equation 16.19.1, in the "public health intervention" treatment. This facilitates a comparison of the valuation results across the two treatments. Next is a discussion presented on such a comparative exercise and its associated implications.

16.5.3 A cross-treatment valuation exercise

Results from the above two treatments are used to assess the comparative valuation of health risk reductions associated with different preventive strategies. More precisely, the estimates of the structural parameters (gammas) across equations (16.20) and (16.22) are compared. The inclusion of *controls* ensures that differences (if any) in the estimates of the gammas are attributed to the difference in the nature and modes of delivery of the prevention technologies only, which are offered through the two treatments. Furthermore, a random assignment of risk reductions across respondents renders the variable *risk_r* orthogonal to both the observed and unobserved characteristics of the respondents.

16.5.3.1 Probit estimation

In the "community-level health intervention" treatment of our survey, each individual's decision to make a stated financial contribution for the provision of *g* is recorded. For the purpose, a probit equation is specified as follows:

$$Prob(Yes) = Prob\left(MWTP^g > t\right), \tag{16.23}$$

where t = the stated tax amount randomly assigned to the respondent. Using equations 16.20 and 16.23 we get

$$Prob(Yes) = Prob\left[\gamma_0 + \gamma_1(risk\_r) + \gamma_2(controls) + \epsilon^g > \mathrm{t}\right] \tag{16.24}$$

$$\text{Or, } Prob(Yes) = Prob\left[\epsilon^g > t - \gamma_0 - \gamma_1(risk\_r) - \gamma_2(controls)\right] \tag{16.25}$$

Assuming ϵ^g follows a normal distribution with mean 0 and variance σ_g^2, the symmetry of the distribution allows us to write equation 16.26 as follows:

$$Prob(Yes) = Prob\left[\frac{\epsilon^g}{\sigma^g} < \frac{\gamma_0}{\sigma^g} + \frac{\gamma_1}{\sigma^g}(risk\_r) + \frac{\gamma_2}{\sigma^g}(controls) - \frac{t}{\sigma^g}\right] \quad (16.26)$$

Using the maximum likelihood estimation methodology, the coefficients in equation 16.26
can be estimated consistently and efficiently. Note that in probit estimation, the coefficient of t is $\frac{-1}{\sigma^g}$, where σ^g is the variance of the error term. Thus, using Cameron's (1988) method, we can recover the estimate of σ^g from the estimated coefficient of t and hence get unstandardized estimates of γ_0, γ_1 and γ_2. The same estimation methodology is used to obtain parameter estimates for the "private-level health intervention" treatment. Thus,

$$Prob(Yes) = Prob\left[\frac{\epsilon^L}{\sigma^L} < \frac{\gamma_0'}{\sigma^L} + \frac{\gamma_1'}{\sigma^L}(risk\_r) + \frac{\gamma_2'}{\sigma^L}(controls) - \frac{t'}{\sigma^L}\right], \quad (16.27)$$

where t' stands for the stated price of the preventive product that is offered to respondents, randomly assigned to the "private" treatment.

Once probit estimates are obtained, a likelihood ratio test facilitates the comparison of the coefficients, γ_1 and γ_1', whereby we are able to restrict only the structural coefficients while allowing for the unobserved heterogeneity across the two equations to vary.

16.5.3.2 A Likelihood Ratio (LR) test

A comparison between the coefficients corresponding to the variable *risk_r* across the two Probit equations (16.26) and (16.27) can be accomplished by testing the null hypothesis $H_0 : \gamma_1' = \gamma_1$.

Note that the restriction is placed only on the structural parameters corresponding to the variable *risk_r* while the variances of the error terms are allowed to vary. Merely comparing the corresponding estimated probit coefficients would imply making restrictive assumptions regarding the variances of the error terms. The calculation of the restricted and unrestricted likelihoods, as laid out next, shows how the hypothesis test can be carried out.

For the LR test, we assume that $\delta_i = 1$, if the respondent is offered the private preventive product; 0 otherwise. In addition, for the purpose of the LR test, Φ_{pub}[18] and Φ_{pvt}[19] are given as follows:

$$\Phi_{pvt} = \Phi\left\{\frac{1}{\sigma^L}(\gamma_0' + \gamma_1' .risk\_r + \gamma_2' .controls - t'\right\} \text{and}$$

$$\Phi_{pub} = \Phi\left\{\frac{1}{\sigma^g}(\gamma_0 + \gamma_1 risk\_r + \gamma_2.controls - t\right\}.$$

The restricted log-likelihood is given as

$$\log LL_R = \sum_{i=1}^{n}\left[\delta_i \log(\Phi_{pvt}) + (1-\delta_i)\log(\Phi_{pub})\right].$$

The unrestricted log-likelihood is written in a similar manner but without with the restriction,

$$\gamma_1' = \gamma_1 : \log LL_U = \delta_i \sum_{i=1}^{n}\left[\log(\Phi_{pvt})\right] + (1-\delta_i)\sum_{i=1}^{n}\left[\log(\Phi_{pub})\right].$$

Following the LR test, the difference between the two likelihoods follows a chi-square distribution with 1 degree of freedom:

$$\chi^2 = 2\left[\log LL_R - \log LL_u\right] \sim \chi_1^2.$$

A nonrejection of the test would imply that the null hypothesis is valid. This, in turn, would imply that the willingness to pay to reduce malaria risks by a given percentage is the same whether it is achieved through a risk-control intervention of a private nature or through a community-wide vector control program. Also, given the nature of the community-level risk-control program, which benefits all individuals in the community including the decision-maker (i.e., the respondent) herself, a non-rejection of the earlier hypothesis can also be interpreted as the individual not having other-regarding preferences in terms of the malaria risks facing her community. This would imply that the individual cares only about the private aspect of the malaria risks and values either health interventions in a similar manner.

16.6 Results and Policy

16.6.1 Percentage risk reductions

16.6.1.1 Full sample analysis

Recall equations 16.20 and 16.22, corresponding to the public and private treatments, respectively:

$$MWTP^g = \gamma_0 + \gamma_1(risk\_r) + \gamma_2(controls) + \epsilon^g \qquad (16.20)$$

$$MWTP^L = \gamma'_0 + \gamma'_1(risk\_r) + \gamma'_2(controls) + \epsilon^L \qquad (16.22)$$

Considering the basic equation underlying these two specifications, we estimate a binomial probit equation for the full sample comprising 687

Table 16.1 Probit Estimates (percentage risk)—Full Sample

	Model 1	*Model 2*	*Model 3*	
Variable	*Coefficient*	*Coefficient*	*Coefficient*	*Mean*
Constant	0.832**	0.767**	0.744**	
	(0.115)	(0.127)	(0.126)	
=1, if risk reduction is 50%	-0.007	-0.007	-0.006	0.48
	(0.100)	(0.100)	(0.100)	
PRICE	-0.003**	-0.003**	-0.003**	116.87
	(0.001)	(0.001)	(0.001)	
=1, if area is urban		0.123	-	0.51
		(0.100)		
=1, if risk reduction mechanism is public			0.169	0.49
			(0.100)	
Sample Size	687	687	687	

* Significant at 5% level of significance
** Significant at 1% level of significance
Source: Author

respondents (out of 780, 93 accounted for nonresponse). In this basic model, no qualitative distinction is made between g and L (i.e., the two treatments), and hence, the LHS simply stands for *MWTP* for health-risk reductions in the context of malaria. The results of this model estimated for the full sample are presented in Table 16.1. The second and third columns reflect the probit coefficients that result from taking two other variants of the basic specification, one with a dummy variable added for the urban area and the other with a dummy on the public health treatment.

As all three models suggest, the percentage risk changes do not play a significant role in determining the purchase decisions that respondents make. But cost considerations of the public/private health programs significantly influence the likelihood of respondents' saying yes to a particular health intervention. Thus, the coefficient of price is negative and significant (at 1% level of significance) in all three models suggesting that a higher price of a preventive program negatively affects the decision-maker to opt for the same. The coefficients of the urban and public dummies fail to emerge as significant. Given the design of the field survey, the randomly assigned percentage risk reductions and costs of prevention efforts are orthogonal to the observed and unobserved characteristics of respondents; hence, the addition of further controls does not significantly change the risk and cost coefficient estimates.

Now, given our interest in the private–public comparative valuation exercise, the full sample is divided into two subgroups, namely, the set of respondents assigned to the public program as against those offered the private strategy. Equation 16.20 is estimated for the public subgroup and likewise, equation 16.22 for the private. Table 16.2 reports the results from

Table 16.2 Probit Estimates (percentage risk)—Full Sample, Public and Private Treatments

	Public		*Private*	
Variable	*Coefficient*	*Mean*	*Coefficient*	*Mean*
Constant	0.909**		0.75**	
	(0.168)		(0.160)	
=1, if risk reduction is 50%	-0.041	0.48	0.028	0.48
	(0.145)		(0.139)	
PRICE	-0.003**	115.07	-0.003**	118.62
	(0.001)		(0.001)	
Sample Size	337		350	

* Significant at 5% level of significance
** Significant at 1% level of significance
Source: Author.

such a comparative exercise based on 337 "public" sampling units and 350 "private".

As in the full sample analysis, in both the public and private treatments, the coefficient of the percentage risk reductions is not found to be significantly determining respondents' purchase decisions. In each treatment, the coefficient of the cost of health interventions is negative and significantly different from zero at 1%, indicating that price levels offered indeed determine if a respondent would opt for the said intervention or not. Further subsample analyses have been done considering only the urban and the semi-urban subsamples and the reader is referred to the full dissertation at this link.[20]

Notwithstanding the earlier issues, the perceived levels of malaria risks elicited in the survey allow us a possible extension of the empirical plan for exploring the public-private valuation question. The percentage risk changes that respondents received from the public/private health programs, together with the initial malaria risk levels that they estimated, help us compute absolute risk changes for each respondent.

16.6.2 Absolute risk reductions

16.6.2.1 Full sample analysis

A reasonable amount of variation in the perceived risk levels on a scale of 0 to 10 in the survey, when used in conjunction with the two risk changes we offered, that is, 50% and 90%, essentially brings in considerable variation in the absolute risk changes across respondents. The explanatory variable becomes changes in risk *levels* and is no longer in terms of percentages. Table 16.3 contains the results obtained from estimating the basic specification behind equations 16.20 and 16.22 for the full sample

Table 16.3 Probit Estimates (absolute risk)—Full Sample

	Model 1	*Model 2*	*Model 3*	
Variable	*Coefficient*	*Coefficient*	*Coefficient*	*Mean*
Constant	0.539**	0.502**	0.451**	
	(0.130)	(0.137)	(0.140)	
risk reduction	0.091**	0.089**	0.091**	3.49
	(0.025)	(0.025)	(0.025)	
PRICE	-0.003**	-0.003**	-0.003**	116.87
	(0.001)	(0.001)	(0.001)	
=1, if area is urban		0.084	-	0.51
		(0.101)	-	
=1, if risk reduction mechanism is public			0.172	0.49
			(0.101)	
Sample Size	687	687	687	

* Significant at 5% level of significance
** Significant at 1% level of significance
Source: Author.

(687 respondents), without dividing the sample into public and private treatments. Columns 2 and 3 bring in the dummies for the urban area and public treatment, respectively. Notably, the coefficient of the absolute risk changes is positive and significant at 1% in all three specifications. This implies that a larger risk change brought about by a health intervention (private and/or public) significantly increases the likelihood of respondents opting for the said prevention option.

In Table 16.3, the price coefficient is negative and significant at conventional levels in all models, corroborating the law of demand in the present health context. However, in Models 2 and 3, the dummy variables on the urban area and public treatment emerge insignificant in explaining the variation in respondents' purchase decisions.

Next, our interest lies in exploring if dividing the sample into public and private subgroups affects respondents' valuation of risk reductions. As shown in Table 16.4, the risk coefficient pertaining to the public treatment comprising 337 respondents is positive and significant at 1%. The private counterpart is positive and significant at the 5% level, when the model is run on 350 respondents. Thus, in both the public and private treatments a greater amount of risk reductions significantly drives people more towards adopting the health intervention offered. Table 16.4 also depicts that the coefficient of price is strongly significant in both the public and private treatments, implying that people react negatively to the prices associated with any prevention strategy.

Even though the risk coefficient emerges as significant in the full sample, as well as in the constituent private/public subgroups, one may argue that our consideration of absolute risk changes may render the risk estimates biased.

Table 16.4 Probit Estimates (absolute risk)—Public and Private Treatments

	Public		*Private*	
Variable	*Coefficient*	*Mean*	*Coefficient*	*Mean*
Constant	0.55** (0.181)		0.523** (0.187)	
risk reduction	0.123** (0.038)	3.48	0.068* (0.033)	3.50
PRICE	-0.004** (0.001)	115 .25	-0.003** (0.001)	118.44
Sample Size	337		350	

* Significant at 5% level of significance
** Significant at 1% level of significance
Source: Author.

Such a bias may be likely if respondents' perceived risk levels, which we use for computing the absolute risk changes, are correlated to the error term, thus bringing in the issue of endogeneity. However, despite such challenges, the empirical exercises involving absolute risk variations are intended primarily because the ultimate aim lies in comparing the risk coefficient estimates across the private and public treatments. We assert that even in the case when one may believe that both the risk coefficients are biased, the significance of the estimates across both private and public interventions allows for an LR test to facilitate a comparison of the structural parameters corresponding to the absolute risk reductions. Such a test of the difference between structural parameters mitigates the issue of endogeneity to a significant extent.

For the purpose at hand, an LR test is conducted. The structural risk coefficients, when compared across the private and public treatments, produce a high p-value, indicating that the null hypothesis on the equality of the risk parameters cannot be rejected at conventional levels of significance. Thus, individuals differentiated according to the nature of health programs offered, private or public, react in a similar manner to the risk changes that such programs bring forth. This renders an answer to our primary research question by asserting that individuals' valuations of public and private health programs do not differ even when the delivery mechanisms and the scope of benefits of such health interventions diverge.

The econometric exercises drive home two key outcomes: (1) Respondents value health-risk reductions from a private preventive strategy in exactly the same manner as they value the health benefits of a publicly administered malaria-control program in the community, and (2) costs of health interventions (private and public) significantly influence respondents' prevention decisions.

The valuation results generally support the externality dimension to malaria control. Even though the two health interventions, private and public, diverge on account of their delivery mechanisms and scope of benefits, respondents do not significantly view the associated risk reductions differently. The comparative valuation exercise, thus, corroborates the externality dimension to malaria control and indicates towards a possible substitutability between private and community-level efforts, thereby calling for a significantly greater amount of public action to combat malaria.

Although the valuations of risk reductions do not differ statistically across treatments, respondents' willingness to pay elicited separately in the community-level (public) treatment turn out to be substantive in the urban area.[21] On an average, an urban respondent is annually willing to pay INR 58[22] (for absolute risk reductions) and INR 115[23] (for percentage risk reductions) for community-level health benefits of a public risk-control measure. Even when the annual WTP estimate of INR 58 is treated with caution for bias (since it is computed with absolute risk reductions), the figure, nevertheless, provides a lower bound to the WTP. Strikingly, even this lower bound, when extrapolated with the population of Kolkata, 4,496,694 (Census of India, 2011, Govt. of India), produces an aggregated WTP that well surpasses the rough annual cost of the Kolkata Municipal Corporation (KMC) on account of vector control for 2011–2012 (INR 94,260,194). Even though such cost figures are approximate and are not directly comparable to the valuation of specific risk reductions offered in the survey, still the difference between the benefit estimates and costs largely indicates that a scaled-up community-level malaria control effort in the urban malaria-prone area will be economically viable in the city of Kolkata.

The WTP analyses aside, a key finding that emerges from the empirical procedures reflects that private prevention is significantly price-responsive. This, in turn, informs malaria control policies on the need for correct price incentives to induce greater prevention. Issues of health valuation and price sensitivity are further explored across different split samples[24] differentiated on the basis of socioeconomic attributes, disease exposure, actual prevention efforts and perceived malaria risks. Such exercises help analyze the valuation question in greater depth.

Seeking to explore a future route for refining the procedures adopted in the present framework, it is envisaged that, perhaps in future extensions, elicitation of two dimensions to risks from a single respondent, viz. private and community-level risks, could be attempted for. Such a within-subject design may generate a better variation in risks having an important bearing on the comparative valuation question.

Notes

1 See Samajpati (2012).
2 Existing works have mostly considered a single disease control tool as reference, (e.g., ITNs, IRS, and hypothetical vaccines), each time demand has been assessed (Onwujekwe, Hanson, and Fox-Rushby 2003; Cook et al. 2009; John, Walsh, and Moore 1992; Whittington, Pinheiro, and Cropper 2003; Cropper et al. 2004).

3 The importance of our approach gains support from the illustration in Crocker and Shogren (2003) who explain the need for policy makers to consider private choices in order to determine the optimal level of provision of public action such that costs of risk control are minimized.

4 For details on the split-sample analyses, see Samajpati (2012).

5 Jalan, Somanathan, and Chaudhuri (2009) argue that awareness about a public good, say, environmental quality, is itself a public good and is likely to be suboptimally supplied in the market, resulting in a demand for environmental quality that is inefficiently low. This justifies our treatment of government-induced awareness as a public good.

6 Although successive reductions in R^S decline from buying each additional unit of a successively, the algebraic sign of $\frac{\partial^2 r^S}{\partial a^2}$ is positive. The r^S function is decreasing and concave upward.

7 The utility function is downward-sloping and concave downward with respect to R^S.

8 Francois and Vlassopoulos (2008) discuss alternative approaches taken in the literature to describe pure or outcome-oriented altruism. One of the papers cited is Francois (2007), where "agents" having such altruism derive a direct benefit when a good/service that is generally considered socially worthwhile (say, similar to the public malaria control measures g in our framework) is provided. Pure altruism of this nature also means that this direct benefit, which the agent enjoys, is "independent of whether the agent has a hand in producing the good or service or not, something which does not occur with action-oriented motivations".

9 For a better understanding of the valuation of the private benefits ensuing from g, recall that g influences the private risks R^{Self}. Besides, the marginal effectiveness of g in terms of reducing private risks is given as $\frac{\partial r^S}{\partial g}$. Let us now concentrate only on the first bracketed expression on the RHS of equation 16.14 and explore an alternative interpretation of the monetized private benefits from the g change. In this regard, also recall that the FONCs indicate that at the optimum $\lambda^* = \frac{\frac{\partial U^*(.)}{\partial R^S}\frac{\partial r^S}{\partial a}}{p_a}$. Replacing λ^* in $\left[\frac{\frac{\partial U^*(.)}{\partial R^S}\frac{\partial r^S}{\partial g}}{\lambda^*}\right]$ we get $\left[\frac{\frac{\partial U^*(.)}{\partial R^S}\frac{\partial r^S}{\partial g}}{\lambda^*}\right] = \frac{\frac{\partial r^S}{\partial g}}{\frac{\partial r^S}{\partial a}} p_a$.

Thus, solely for the private dimension of the publicly administered malaria control efforts, the decision-maker is willing to pay an amount equivalent to what she would have spent on the private preventive good, a, currently available on the market, to achieve the same amount of private risk reduction that one additional unit of g brings forth.

10 An alternative interpretation of the valuation of the private dimension to the public health intervention: Let $\frac{\partial r^S}{\partial a}$ units of risk are reduced by 1 unit of a. Recalling that the unit price of a is p_a, the cost of a unit reduction in the individual's risk, R^{Self} through the use of the private preventive good, a, is $\frac{p_a}{\frac{\partial r^S}{\partial a}}$.

Thus, the private cost that the individual would have incurred to achieve the same amount of risk reduction as what an additional unit of g brings in (i.e., $\frac{\partial r^S}{\partial g}$) is given as $\frac{\frac{\partial r^S}{\partial g}}{\frac{\partial r^S}{\partial a}} p_a$, which denotes the maximum WTP for private benefits associated with a unit increase in g.

11 As a special case, for solely self-interested individuals, the second component on the RHS reduces to zero and hence, equation 16.14 can be rewritten as $\frac{\partial WTP}{\partial g} = \frac{\frac{\partial U^*(.)}{\partial R^S}\frac{\partial r^S}{\partial g}}{\lambda^*}$.

12 Although from a social planner's perspective provision of such private goods does have a positive social bearing in an aggregate sense and also in the epidemiological sense, recall that for the individual decision-maker in our model, the perception that such private goods benefit others in the community is not felt. In our model, the only way the decision-maker perceives a social effect on the community is when the public good g is provided.

13 See the Urban Health Strategy published by the Government of West Bengal (2008) for details on the policy emphasis on malaria in the city. Sur et al. (2006) report the results of a community-based study in an impoverished urban site in Kolkata to estimate the malaria burden and that of typhoid and explore the risk factors underlying the diseases. Biological research on mosquito breeding and other behavior of the vector galore (see Mandal et al. [2011] for the details).

14 For detailed description of the survey design, descriptive statistics and results, refer to Samajpati (2012). For the sake of brevity, details on the survey design and execution, and summary statistics are omitted from this chapter.

15 Given our between-subject design and randomized treatment plan, a sample size of 686 respondents (i.e., about 340 respondents in each of the private and community-level health intervention treatments) was estimated to be fairly sufficient. Based on response rates from earlier surveys conducted in the area, the survey was planned to be ultimately fielded to a total of 780 respondents, to sufficiently allow for chances of nonresponses (generally about 12% as per instances from prior surveys in Kolkata), thereby generating the targeted number (i.e., 686) of completed surveys.

16 This procedure is akin to the welfare analysis in Dickie and Gerking (2007). Dealing with perceptions on skin cancer risks, in their valuation exercise, the authors use percentage changes in risks to facilitate simplified econometric tests and analysis.

17 Note that although R^S and R^C are modeled separately in our model, the empirical procedure involves reduction of the two risk concepts to a single dimension of risk, namely malaria risk. This is particularly prompted by two reasons. Firstly, the focus in our theoretical and empirical analysis lies clearly on the way benefits accrue as a result of a malaria-prevention strategy and not on the risk-perception levels, per se. In accordance with this, in our survey design, the nature in which benefits accrue (i.e., if the associated health risk reductions have a private or community dimension) are clearly communicated to the respondents through a suitable CVM question across the two treatments. In both the treatments, we elicit perceived private risks R^S on a risk-scale. But the differential framing of the two contexts of health interventions allows for investigating the manner in which people respond to different contexts, namely private and public. Secondly, our methodology of restricting ourselves to elicitation of R^S only (and not R^C separately) is motivated by the existing literature on risk elicitation, across a variety of diseases, where risk questions have been variously framed in both first-person (e.g., Dickie and Gerking [2007] in the context of skin cancer risks.) and third-person formats (e.g., Viscusi [1990] on lung-cancer risks) without any stated differences of these formats on the derived risk-levels. This makes us contemplate that R^S and R^C , even if elicited separately from an individual may be quite close empirically, if not synonymous. For our purpose, we stick to the first-person format and hence, elicit R^S. Even though the risk levels may quite be the same, our model, and hence, the empirical plan expects that, conveying

the different ways benefits of a health intervention ensue (i.e., if the CVM instrument influences R^S/R^C) will have a different bearing on individuals' attitudes towards a particular health intervention.
18 The subscript "*pub*" refers to the community-level treatment.
19 The subscript "*pvt*" refers to the private-level treatment.
20 Samajpati (2012)
21 See Samajpati (2012) for the details.
22 WTP = Risk Coeff. / Price Coeff. Thus, 0.175/.003 = INR 58.
23 .23/.002 = INR 115
24 See Samajpati (2012) for the details.

References

Cameron, Trudy Ann. "A New Paradigm for Valuing Non-Market Goods Using Referendum Data: Maximum Likelihood Estimation by Censored Logistic Regression." *Journal of Environmental Economics and Management* 15, no. 3 (1988): 355–379.

Census of India. https://censusindia.gov.in/2011census/population_enumeration.html Primary Census Abstract Data Tables (India & States/UTs - District Level) (Excel Format) Government of India: New Delhi, India, 2011.

Cohen, Jessica, and Pascaline Dupas. "Free Distribution or Cost-Sharing? Evidence from a Randomized Malaria Prevention Experiment." *The Quarterly Journal of Economics* 125, no. 1 (2010):1–45. doi:10.1162/qjec.2010.125.1.1.

Cook, Joseph, Susmita Chatterjee, Dipika Sur, John Clemens, and Dale Whittington. "Measuring Risk Attitudes among the Urban Poor in Kolkata, India: An Experimental Approach." Evans School Working Paper No. 2009, University of Washington (2009).

Courant, Paul N., and Richard C. Porter. "Averting Expenditure and the Cost of Pollution." *Journal of Environmental Economics and Management* 8, no. 4 (1981): 321–329.

Crocker, T. D., and J. F. Shogren. 2003. "Choosing Environmental Risks." In Henk Folmer and Tom Tietenberg (eds.) *The International Yearbook of Environmental and Resource Economics 2002/2004*. Cheltenham: Edward Elger Publishing, Inc (2003).

Cropper, Maureen L., Mitiku Haile, Julian Lampietti, Christine Poulos, and Dale Whittington. "The Demand for a Malaria Vaccine: Evidence from Ethiopia." *Journal of Development Economics* 75, no. 1 (2004): 303–318.

Dickie, Mark, and Shelby Gerking. "Altruism and Environmental Risks to Health of Parents and Their Children." *Journal of Environmental Economics and Management* 53, no. 3 (2007): 323–341.

Dupas, Pascaline. "What Matters (and What Does Not) in Households' Decision to Invest in Malaria Prevention?" *American Economic Review* 99, no. 2 (2009): 224–230. doi:10.1257/aer.99.2.224.

Francois, Patrick. "Making a Difference." *The RAND Journal of Economics* 38, no. 3 (2007): 714–732. doi:10.1111/j.0741-6261.2007.00108.x.

Francois, Patrick, and Michael Vlassopoulos. "Pro-Social Motivation and the Delivery of Social Services." *CESifo Economic Studies* 54, no. 1 (2008): 22–54. doi:10.1093/cesifo/ifn002.

Freeman, A. Myrick. *The Measurement of Environmental and Resource Values: Theory and Methods*. Washington, DC: Resources for the Future (1993).

Gersovitz, Mark, and Jeffrey S. Hammer. "The Economical Control of Infectious Diseases." *The Economic Journal* 114, no. 492 (2004): 1–27. doi:10.1046/j.0013-0133.2003.0174.x.

Gersovitz, Mark, and Jeffrey S. Hammer. "Tax/Subsidy Policies toward Vector-Borne Infectious Diseases." *Journal of Public Economics* 89, no. 4 (2005) :647–674.

Govt. of West Bengal. The Urban Health Strategy. Edited by Department of Health & Family Welfare and Department of Municipal Affairs (2008).

Harrington, Winston, and Paul R. Portney. "Valuing the Benefits of Health and Safety Regulation." *Journal of Urban Economics* 22, no. 1 (1987): 101–112.

Hori, Hajime. "Revealed Preference for Public Goods." *The American Economic Review* 65, no. 5 (1975): 978–991.

Jacobsson, Fredric, John Carstensen, and Lars Borgquist. "Caring Externalities in Health Economic Evaluation: How Are They Related to Severity of Illness?" *Health Policy (Amsterdam, Netherlands)* 73, no. 2 (2005): 172–182.

Jalan, Jyotsna, E. Somanathan, and Saraswata Chaudhuri. "Awareness and the Demand for Environmental Quality: Survey Evidence on Drinking Water in Urban India." *Environment and Development Economics* 14, no. 6 (2009): 665–692.

John, Kun H., Richard G. Walsh, and Chester G. Moore. "Comparison of Alternative Nonmarket Valuation Methods for an Economic Assessment of a Public Program." *Ecological Economics* 5, no. 2 (1992): 179–196.

Klein, Eili, Ramanan Laxminarayan, David L. Smith, and Christopher A. Gilligan. 2007. "Economic Incentives and Mathematical Models of Disease." *Environment and Development Economics* 12, no. 5 (2007): 707–732. doi:10.1017/S1355770X0700383X.

Mandal, Bithika, Baishakhi Biswas, A. Banerjee, T.K. Mukherjee, J. Nandi, and D. Biswas. "Breeding Propensity of *Anopheles stephensi* in Chlorinated and Rainwater Containers in Kolkata City, India." *Journal of Vector Borne Diseases* 48, no. 1 (2011): 58–60.

Onwujekwe, Obinna, Kara Hanson, and Julia A. Fox-Rushby. "Who Buys Insecticide-Treated Nets? Implications for Increasing Coverage in Nigeria." *Health Policy and Planning* 18, no. 3 (2003): 279–289. doi:10.1093/heapol/czg034.

Samajpati, Shreejata. "Infectious Disease Risks in Developing Countries: A Non-Market Valuation Exercise." PhD dissertation. University of Central Florida. Electronic Theses and Dissertations, 2501 (2012). https://stars.library.ucf.edu/etd/2501

Sur, Dipika, Lorenz von Seidlein, Byomkesh Manna, Shanta Dutta, Alok K. Deb, Banwari Lal Sarkar, Suman Kanungo, Jacqueline L. Deen, Mohammad Ali, Deok Ryun Kim, Vinay Kumar Gupta, R. Leon Ochiai, Ataru Tsuzuki, Camilo J. Acosta, John D. Clemens, and Sujit K. Bhattacharya. "The Malaria and Typhoid Fever Burden in the Slums of Kolkata, India: Data from a Prospective Community-Based Study." *Transactions of the Royal Society of Tropical Medicine and Hygiene* 100, no. 8 (2006): 725–733.

Tarozzi, Alessandro, Aprajit Mahajan, Joanne Yoong, and Brian Blackburn. "Commitment Mechanisms and Compliance with Health-Protecting Behavior: Preliminary Evidence from Orissa, India." *American Economic Review* 99, no. 2 (2009) :231–235. doi: 10.1257/aer.99.2.231.

Viscusi, W. Kip. "Do Smokers Underestimate Risks?" *Journal of Political Economy* 98, no. 6 (1990): 1253–1269.

Whittington, Dale, Armando Castelar Pinheiro, and Maureen Cropper. "The Economic Benefits of Malaria Prevention: A Contingent Valuation Study in

Marracuene, Mozambique." *Journal of Health & Population in Developing Countries* 15, no. 27 (2003): 1–27.

World Health Organization. "Chapter Three: Perceiving Risks." In *The World Health Report 2002—Reducing Risks, Promoting Healthy Life*. Geneva: WHO (2002): 27–45.

World Health Organization. "Malaria Control Today: Current WHO Recommendations." Edited by WHO Roll Back Malaria Department. Geneva: WHO (2005).

World Health Organization. "*World Malaria Report 2008*." Geneva: WHO (2008).

World Health Organization. "*World Malaria Report 2010*." Geneva: WHO (2010).

Part V

Society and policy

17 Analyzing the poverty situation in India

Using a multidimensional approach

Ayush Agrawal, Chitwan Lalji and Debayan Pakrashi

17.1 Introduction

Poverty can be measured, broadly, by two methods, the direct method and the indirect method, also referred to as the income approach. The direct method, which is in concordance with Amartya Sen's (1993) capability approach, defines poverty as "deprivation in the capability to live a good life" and tries to identify the households who don't satisfy a set of specific basic needs, rights, and so on. The indirect method, on the other hand, determines poverty as the fraction of population living below the poverty line or the minimum income level at which some basic needs can be satisfied. In 2015, the international poverty line was revised from US$1.25 a day per person in 2005 purchasing power parity (PPP) to US$1.90 PPP-adjusted dollar a day per person in 2011 PPP terms (World Bank, 2016). Even though the two methods are argued to be complementary to each other (Alkire and Santos, 2014; World Bank, 2016), diversions in results using the two approaches have been found empirically (Sugata and Seth, 2016).

Most countries have so far focused on measuring poverty or identifying the poor using either the income or consumption approach, thus focusing on a single dimension of poverty only. However, it has been suggested that to tackle poverty, "deprivations" that trap people in poverty needs to be addressed (UNDP, 2010; Sugata and Seth, 2016). Individuals often lack adequate access to education, health, nutrition and employment, and therefore poverty measures that focus on a single indicator such as income poverty, fails to capture the multiple disadvantages that households experience. According to the UNDP (2010), "Acceleration in one goal often speeds up progress in others. . . . Given these synergistic and multiplier effects, all the goals need to be given equal attention and achieved simultaneously" p. 3. Therefore, the "need for a multidimensional view of poverty deprivation" was realized and emphasized upon (Sen, 1992, 1993; Foster and Sen, 1997). In 2000, after the Millennium Summit of the United Nations, the member countries committed themselves to achieve, by 2015, eight major Millennium Development Goals (MDGs), namely, eradicate extreme poverty and hunger, achieve universal primary education, promote gender equality and empower women, reduce child mortality, improve maternal health, combat HIV/AIDS, malaria and other diseases, ensure environmental sustainability and develop a global partnership for development, based on the goal of simultaneously eradicating the different disadvantages that contribute to poverty.

In 2013, the World Bank Group announced its twin goals of eradicating extreme poverty by 2030 by reducing the number of extremely poor people to less than 3% of the world population and boosting the "shared prosperity" or per capita income of the poorest 40% of the population in each country. The Report of the Commission on Global Poverty: Monitoring Global Poverty by the World Bank (2016) has in Chapter 2, titled "Complementary Indicators and Multidimensionality", written in depth about the multidimensional nature of poverty. In addition to money, UN agencies are, in the future, expected to take a lead on the non-monetary poverty indicators such as health, education, housing and nutrition (World Bank, 2016). Even though monetary deprivation may indeed impair a person's freedom to enjoy, it has been argued that it fails to capture deprivation in other dimensions of development (Alkire and Seth, 2015).

The most important agenda of any global or national development agency has been to eradicate extreme poverty (Alkire and Seth, 2015). Identifying the "poorest of the poor" is imperative as they are characteristically very different from the moderately poor (Devereux, 2003; Harriss-White, 2005; Alkire and Seth, 2015). The poorest of the poor are often referred to as ultra-poor (Lipton, 1983, 1988), extreme poor, severely poor and destitute (Alkire and Seth, 2015). Monetary approaches have been rampantly used to differentiate the poorest of the poor from moderately poor, a multidimensional framework to measure destitution is, however, scarce (Alkire and Seth, 2015).

In this chapter, we use the Multidimensional Poverty Index (MPI), in the context of extreme poverty or destitution, as discussed in Alkire and Foster (2007), which incorporates three major goals of the MDGs based on 10 different indicators to create an index of in-depth multidimensional poverty by capturing deprivation of households, who are both MPI poor and destitute, using households as the "base" population or the unit of analysis. Moreover, several studies by Oxford Poverty and Human Development Initiative with the help of UNDP (Chakravorty and d'Ambrosio, 2006; Duclos, 2011; Alkire and Foster, 2011; Alkire and Santos, 2014), have made an attempt to collect data and construct the MPI for several countries, thereby making cross-country comparisons based on the MPI possible. We use detailed data from the India Human Development Survey for 2004–2005 and 2011–2012 to identify the households which are multidimensionally and income poor. We undertake a correlation analysis to show that income poverty is indeed not highly correlated with the three indicators of deprivation, namely, health, education and standard of living, used by the MPI to identify households which are multidimensionally poor.

We also use a Linear Probability Model to examine in detail the socio-economic characteristics of the households which are closely associated with being multidimesionally or income poor. The regression analysis finally confirms the results already found using the correlation study. A household may be income poor but not necessarily multidimensionally poor; similarly, a household which is not income poor may be in multidimensional poverty by being deprived in one or more of the three dimensions of poverty not captured by income poverty.

Finally, the MPI is aggregated at the state level for India using data from both waves of the survey to make a detailed comparison at the subnational level over time. The results provide strong evidence that multidimensional poverty has been declining over time. The states with maximum deprivation or destitution are Uttar Pradesh, Bihar, Orissa, Chhattisgarh and Madhya Pradesh. The states with the lowest MPI value are Delhi, Tamil Nadu, Jammu and Kashmir, Punjab and Haryana. Although Delhi is better off than the rest of the Indian states, in terms of MPI for destitution, it still has the lowest decline of just –4.04% from 2004–2005 to 2011–2012. The other states which haven't fared much in terms of decline in MPI between 2004–2005 and 2011–2012 are Haryana, Chhattisgarh and Madhya Pradesh. On the other hand, Tamil Nadu, Andhra Pradesh, Karnataka and West Bengal have seen a drastic decline in the MPI from 2004–2005 to 2011–2012.

The structure of the chapter is as follows: A brief introduction on the need for a multidimensional poverty index is followed by a detailed discussion of the data and methodology. Next we discuss the results at the household and subnational level and finally close the chapter with our conclusions.

17.2 Data and methodology

The India Human Development Survey (IHDS) is a nationally representative panel data survey which has collected detailed household-level information for two years, 2004–2005 and 2011–2012 on a wide range of socioeconomic characteristics such as asset ownership, occupation, wages and salaries, household consumption expenses related to health, education, water, sanitation and hygiene, electricity, fuel and energy use, social customs and marriage practices. Moreover, an eligible woman from each household is identified, and information on gender relationships, marital and fertility history and natal care is documented.

We construct the MPI following the counting approach framework (Atkinson, 2003; Alkire and Foster, 2011) using the detailed IHDS data set. The counting approach has often been preferred over the resource-based approach as the latter method may not be able to capture a household's ability to convert income into well-being (Sen, 2001; Sugata and Seth, 2016). Moreover, the ordinal nature of the MPI helps capture the joint distribution of various dimensions (Sugata and Seth, 2016). The MPI has been computed based on three dimensions of deprivation, namely, health, education and standard of living with each dimension being assigned an equal weightage, that is, one-third weight (Alkire and Foster, 2007). We have used the most extreme indicators in the chapter; that is, we consider the worst form of deprivation of households. These are the households which are both MPI poor and destitute. The construction of the MPI is based on the three dimensions of deprivation, encompassing a total of 10 different indicators, each of which are described briefly.

17.2.1 Education

Even if one individual is educated in the household, its positive externalities can be felt on other household members and their household's decisions. The Education dimension used as an important component of the MPI has been constructed with the help of two major indicators of education, namely, years of schooling and child school attendance.

17.2.1.1 Years of schooling

A household is considered to be deprived in terms of this indicator if not even one household member has completed one or more years of schooling. For this purpose, we created a schooling deprivation dummy, with a value of one if there is no member in the household who has studied for one or more years in school, and zero otherwise. The *"years of schooling deprived"* dummy is given one-sixth weightage in the construction of the MPI.

17.2.1.2 Child school attendance

A school attendance deprivation dummy is similarly constructed, which takes the value of one if the household has a school going child (aged 6–14) but is not enrolled in the school. A zero value is given to the household if even one school-going child is currently enrolled in school. Households with no children in the age group 6–14 (age by which they should finish class 6) have been considered non-deprived and assigned a value of zero (Alkire and Santos, 2010). One-sixth weightage is given to the *"school attendance deprived"* dummy.

Thus, the education dimension receives a weight of one-third in the construction of the MPI, with one-sixth weightage coming from the *"years of schooling deprived"* dummy and one-sixth weightage from *"school attendance deprived"* dummy.

17.2.2 Health

The health dimension in the MPI similarly captures deprivations in two important indicators of health, namely, child mortality and nutrition, with one-sixth weightage coming from each of the two indicators: child mortality and nutrition dummy.

17.2.2.1 *Child mortality*

A household is considered to be deprived in terms of this indicator if two or more children in this household have died. The IHDS dataset provides information on fertility and child history of only the eligible woman of the household. Thus, we have constructed the binary variable taking the value of one if two or more children of the eligible woman in the household have died and zero otherwise. In some households, the eligible woman has not answered questions regarding their fertility history. In such cases, we have considered the values to be missing (Alkire and Santos, 2010). The *"child mortality"* dummy variable is given one-sixth weightage.

17.2.2.2 *Nutrition*

A household is defined to be nutrition deprived if there exists any adult in the household who is severely undernourished; that is, the body mass index (BMI) of the individual is less than or equal to 17 kg/m^2 or any child whose z-score for weight-for-age is below –3 standard deviations from the median of all the households in a state as reference population; that is, they are severely underweight. One-sixth weightage is given to *"nutrition deprived"* dummy.

17.2.3 *Standard of living*

A household deprived of a good standard of living has been classified based on their deprivation in the following 6 indicators; mainly, deprivation in terms of electricity, improved sanitation, safe drinking water, flooring, cooking fuel and assets. Overall, the standard of living dimension is given one-third weight in the construction of MPI, while each of the six different indicators under the standard of living dimension is given one-eighteenth weight.

17.2.3.1 *Electricity*

A household is deprived if there is no electricity in the household. In the IHDS data set, in the fuel and energy use module, the following question is asked to the household: "*Does this house have electricity?*" From here, we computed the *"electricity deprived"* binary outcome variable.

17.2.3.2 *Improved sanitation*

A household is considered to be deprived in this category if it has no improved sanitation facilities like a flush toilet, latrine or ventilated pit and the members defecate in the open. "*Does the household have a toilet of its own? Is there a semi-flush/septic tank toilet? A latrine?*" with options as "*no facility belonging to household (or open fields)*", "*traditional pit latrine*", "*semi-flush (septic tank) latrine*" and "*flush toilet*". From here, we derived the '*improved sanitation deprived*' dummy, which is assigned a value of 1 if there is "*no facility belonging to household (or open fields)*" and zero otherwise.

17.2.3.3 Safe drinking water

A household having no access to safe drinking water, or safe water is not available within a 45-minute round-trip walk is considered to be deprived in terms of safe drinking water. A household is considered non-deprived if it has the following sources of drinking water available: public tap, piped water, protected well, spring or rainwater, borehole or pump and it is within a 45-minute round-trip walking distance. The "*safe drinking water deprived*" dummy was generated with the help of these two questions of the IHDS: "*What is the main source of water for drinking in your house?*" with the following options: piped (public supply); tube well; hand pump; open well; covered well; river, canal, stream; pond; tanker truck; rainwater; bottles; and others and "*How long it would take to walk to the source of water? [One way]*"

17.2.3.4 Flooring

A household having a sand, dirt or dung floor is considered to be a household deprived of appropriate flooring. In the "Interviewer Observations – Housing Quality" module of the IHDS data set, there is a question on the predominant floor type with options: mud; wood, bamboo; brick; stone; cement; tiles, mosaic and others. While creating the "*flooring deprived*" binary variable, households with mud, wood and bamboo flooring were assigned value one, while households with brick, stone, cement, tiles, mosaic and other types of flooring were assigned a value of zero.

17.2.3.5 Cooking fuel

The mode of cooking says a lot about the living conditions of a household. A household using dung or wood as cooking fuel is considered a deprived household, while households using coal/charcoal/lignite are considered non-deprived households. The IHDS covers this aspect of living standard in the "Fuel and Energy" section. Each household is asked whether they use firewood/twigs; dung cake; crop residual/by-product; kerosene; LPG and coal/charcoal. With the help of this question, the binary dummy "*cooking fuel deprived*" variable was created.

17.2.3.6 Assets

A household which owns neither any of these assets (like radio, phone, mobile and so on) nor a car is considered an asset deprived household.

On the other hand, a non-deprived household is one that has at least one of these assets. The detailed information of household ownership of consumer goods, available in the IHDS, helped in the construction of the *"assets deprived"* dummy variable.

The overall deprivation variable is calculated by following the Alkire and Foster (2007) methodology, that is, by summing the three major dimensions, mainly, education, health and standard of living, after assigning them the appropriate weights.[1] A household is finally classified as multidimensionally poor (or "MPI poor") if the household is deprived in at least one-third of the weighted indicators; that is, the overall deprivation value is more than 33.33%. A similar criterion was also used to identify households that are health poor, education poor and poor in terms of standard of living.

The "Great Indian Poverty Debate" remains a question unresolved till date. In India, the Government has given the responsibility to the Planning Commission (Government of India, Planning Commission, 2014) to determine the poverty line. The "Household Consumption" and "Household Consumption Expenditure" module of the IHDS asked 47 questions on lines similar to that of the National Sample Survey household consumption expenditure. 30 questions on the most frequently purchased items (e.g., expenditure on food items, fuel and light, entertainment, telephone, cable and internet, house rent and travel expenses) were asked on a 30-day time frame, while 17 questions (expenses on medical, education, clothing, footware, furniture, crockery, jewelry and ornaments, repairs and maintenance, vacations, social functions like marriages and so on) were asked on an annual time frame. The IHDS constructed a binary variable "POOR", which took the value of one if the household was income poor and zero otherwise. The official Planning Commission's poverty line as of 2005 (Lakdawala Expert Group) and 2012 (Tendulkar Expert Group) was used as the criterion to determine if a household was income poor or not. The official income poverty line for the two years is available, both, state wise and region wise (rural/urban; Government of India, 2007).[2]

To determine which method captures poverty in a holistic manner—the income or the multidimensional poverty approach, we determine the pairwise Spearman's rank correlation between MPI poverty, income poverty and whether a particular household is education poor, health poor or poor with respect to the standard of living at the household level. Spearman's correlation is a nonparametric test that helps find the degree of association (both strength and direction of association) between two variables. The following formula helps to measure the Spearman rank correlation:

$$\rho = 1 - 6\Sigma d_i^2 / n\left(n^2 - 1\right), \tag{17.1}$$

where ρ is the Spearman's rank correlation, d_i is the difference between the ranks of the two indices and n is the number of observations.

The value of Spearman's rank correlation (ρ) lies between –1 and +1, with a higher value of ρ (near –1 and +1) indicating a higher correlation between the two indexes and ρ value close to zero, indicating weak correlation between the two indexes. To estimate the statistical significance of the Spearman's rank correlations, a nonparametric null hypothesis test of the Spearman's rank correlation coefficient was undertaken with null hypothesis "the two indices are independent", as against the alternative hypothesis that the two indices are correlated

To further validate which among the two poverty indicators—multidimensionally poor and income poor—is a better predictor of multidimensional poverty at the household level and to examine the socioeconomic characteristics that are determinants of household-level poverty, we run an ordinary least squares (OLS) regression with the binary variables, that is, a linear probability model with the incidence of household-level poverty as the dependent variable and a number of socioeconomic characteristics as the control variables, as follows:

$$Y_{it} = \alpha + \beta_1 \, \textit{Education deprived}_{\mathrm{it}} + \beta_2 \, \textit{Standard of living deprived}_{\mathrm{it}} + \beta_3 \, \textit{Health deprived}_{\mathrm{it}} + \gamma \, X_{\mathrm{it}} + \lambda S_{\mathrm{it}} + \delta \tau_{\mathrm{it}} + \varepsilon_{\mathrm{it}} \quad (17.2)$$

where Y_{it} is a household i that is either multidimensionally poor or income poor at time t and ε_{it} is the household-specific error term, which is nonsystematic and varies across households. We are particularly interested in β_1, β_2 and β_3 coefficients, which represent the associations of Y_{it} with education, standard of living and health deprived variables, respectively. As a household becomes deprived with respect to certain facilities, for example education, its probability of becoming multidimensionally poor (or income poor) increases by β_1 units. The parameter β_1 is the probability of becoming multidimensionally poor or income poor emanating from the OLS regressions, after netting out the effect of certain household-specific and region-specific characteristics (X_{it}) that may impact the probability of the household being poor. These controls are household head's gender, household size, number of children younger than 6 years of age, number of elders older than 60 years of age, household head's religion (Hindu, Muslim and others), household head's caste (general caste, Other Backward Caste [OBC], Scheduled Caste [SC] and Scheduled Tribe [ST]), primary occupation of the household (agricultural and non-agricultural activity) and area of residence (urban and rural). We also include state (S_{it}) and year (τ_{it}) fixed effects, which capture variations in the incidence of poverty across states and over years. We utilize cluster corrected standard errors across households and present the regression results separately by the year of survey.

Finally, we construct the multidimensional poverty index and further aggregate it at the state-level so that a comparison of multidimensional poverty at the subnational level is possible. The state-wise head-count ratio or incidence of extreme poverty (H) is computed by summing up the proportion of households that are in extreme poverty and destitution in the respective states. The "average proportion of indicators" in which households are deprived is the state-wise intensity of poverty (A) (Alkire and Foster 2007). The state-wise MPI is computed by multiplying the head-count ratio with the intensity of poverty. A higher value of the MPI (or any other index used) indicates a worse situation for the states and a lower value of the MPI (or any other index) indicates a better situation for the states.

The ranking of states in ascending order of the MPI is done, followed by computing the compound annual growth rate (CAGR) of the MPI at two points of time, 2004–2005 and 2011–2012, with the help of the following formulae:

$$\text{CAGR} = \left\{\left[\text{Antilog}\left\{\left(\log Y_i - \log Y_o\right)/n\right\}\right] - 1\right\} \times 100, \tag{17.3}$$

where Y_i is the MPI in 2011–2012, Y_0 is the MPI in 2004–2005 and n is the number of years between (in this case 7 years).

17.3 Results

In this section of the chapter, we first present the results of the correlation analysis and then the regression results, followed by a detailed discussion of multidimensional poverty at the state level and over time.

17.3.1 Correlation analysis

The pairwise Spearman's rank correlation results presented in Table 17.1 shows the degree of association between the incidence of multidimensional poverty, income poverty and whether a household is poor in terms of each of the three dimensions considered in the construction of the MPI, that is, whether it is deprived in terms of education, health or standard of living. The results in Panel A of Table 17.1 for 2004–2005 suggest that there is a high and statistically significant correlation of +0.588 between multidimensional poverty and health deprivation of households. Similarly, +0.419 is the extent of correlation between being multidimensionally poor and standard-of-living poverty, while +0.310 is the correlation between a household being multidimensionally poor and being education deprived. The correlation between multidimensional and income poverty is also only +0.238; which is quite low. Similar results are also observed in Panel B for 2011–2012.

To identify which is a better indicator of a household's deprivation status in terms of the three dimensions of health, education and standard of living, the correlation coefficients of multidimensional and income poverty is compared with each of the three separate poverty indicators. On observing the pairwise correlation of income poverty with health, education and standard of living deprived households, we find that there exists a similar positive and statistically significant correlation. However, the magnitude of

Table 17.1 Spearman Rank Correlation between Multidimensional and Income Poverty and Education-, Health- and Standard of Living–Based Poverty

Variables	*Multidimensional Poverty*	*Income Poverty*	*Education Poverty*	*Health Poverty*	*Standard of Living Poverty*
Panel A: Wave-1 (2004–2005)					
Multidimensional poverty	1.000				
Income poverty	0.238	1.000			
Education poverty	0.310	0.074	1.000		
Health poverty	0.588	0.123	-0.051	1.000	
Standard of living poverty	0.419	0.229	0.136	0.175	1.000
Panel B: Wave-2 (2011–2012)					
Multidimensional poverty	1.000				
Income poverty	0.244	1.000			
Education poverty	0.251	0.033	1.000		
Health poverty	0.489	0.147	-0.017	1.000	
Standard of living poverty	0.504	0.263	0.136	0.200	1.000

Note: Number of observations in 2004–2005 is 30490 households and in 2011–2012 is 26,843 households. All the coefficients are significant at 1% level of significance.

positive correlation between the incidence of multidimensional poverty and the three different dimensions of household-level poverty is significantly higher compared to the income poverty measure. The correlation coefficient for 2011–2012 between education and multidimensional poverty is +0.251 compared to just +0.033 for income poverty. Similarly, the extent of correlation is +0.489 vis-à-vis +0.147 for health poverty and +0.504 vis-à-vis +0.263 for standard-of-living poverty, respectively. The higher correlation coefficients of multidimensional poverty variable with the three separate household-level poverty dimensions suggest that the multidimensional poverty measure is better at capturing the incidence of poverty within a household in terms of health, education and standard of living deprivation than the income measure of poverty traditionally used in the literature. The interpretation is simple: A household that is not income poor may still be education or health poor, so the multidimensional poverty measures fares better in this respect by capturing the household's deprivation in three different dimensions based on ten indicators of development.

17.3.2 Regression analysis

Table 17.2 presents the regression results related to being a multidimensionally and income poor household for 2004–2005 and 2011–2012, both with and without controls. The first two columns relate to the results associated with the incidence of multidimensional poverty for 2004–2005, while the second two columns relate to the same for 2011–2012. The next four columns present the same results for income poverty. From the regression results presented in Table 17.2, it is clear that all the coefficients for the three important poverty dimensions have a significant and positive association with being multidimentionally poor and income poor. On observing the magnitude of the coefficients, it is evident that poverty in terms of health, education or standard-of-living aspect of the household is more closely associated with the likelihood of being multidimensionally poor rather than with being income poor. For example, on observing the health deprivation variable, in 2004–2005, after netting out the effect of the various controls, if a household changes from being a health non-deprived household to a health deprived household, the chances of the household falling in the multidimensionally poor zone are 52.1% higher, while the chances of the household falling in the income poor zone increase by just 4.4%. Similar results are also observed for education and standard-of-living poverty. The results thus confirm that measuring poverty using the multidimensional poverty approach is better than using the more traditional income approach while identifying households that are actually deprived based on different dimensions of development.

The regression results presented in Table 17.2 also suggest that male-headed households are about 2.4% less likely to be income poor but equally likely to be multidimensionally poor. Households with more children after controlling for household size is found to be poorer. Similarly, Muslim

Table 17.2 Regression Coefficients for Incidence of Multidimensional and Income Poverty

Variables of Interest	*Incidence of Multidimensional Poverty*					*Incidence of Income Poverty*		
	Wave I 2004–2005		*Wave II 2011–2012*			*Wave I 2004–2005*		*Wave II 2011–2012*
Health poverty	**0.547***** **(0.005)**	**0.520***** **(0.005)**	**0.343***** **(0.004)**	**0.317***** **(0.004)**	**0.075***** **(0.005)**	**0.038***** **(0.005)**	**0.072***** **(0.004)**	**0.037***** **(0.005)**
Education poverty	**0.604***** **(0.008)**	**0.553***** **(0.008)**	**0.472***** **(0.011)**	**0.445***** **(0.011)**	**0.087***** **(0.011)**	**0.093***** **(0.010)**	**0.004** **(0.014)**	**0.053***** **(0.014)**
Standard of living poverty	**0.295***** **(0.004)**	**0.171***** **(0.004)**	**0.325***** **(0.004)**	**0.259***** **(0.005)**	**0.186***** **(0.004)**	**0.165***** **(0.005)**	**0.178***** **(0.005)**	**0.136***** **(0.005)**
Male head dummy		0.010 (0.006)		0.007 (0.005)		−0.034*** (0.008)		−0.024*** (0.006)
Household size		0.002 (0.003)		0.000 (0.003)		0.089*** (0.004)		0.066*** (0.004)
Household size squared		−0.001*** (0.000)		−0.000* (0.000)		−0.005*** (0.000)		−0.003*** (0.000)
Number of elders aged more than 60 years		−0.017*** (0.003)		−0.014*** (0.003)		−0.011*** (0.004)		−0.000 (0.004)
Number of children aged 0–5 years		0.013*** (0.002)		0.012*** (0.003)		0.046*** (0.003)		0.052*** (0.004)
Religion (Ref: Hindu)								

(Continued)

Table 17.2 (Continued)

Variables of Interest	*Incidence of Multidimensional Poverty*		*Incidence of Income Poverty*	
	Wave I 2004–2005	*Wave II 2011–2012*	*Wave I 2004–2005*	*Wave II 2011–2012*
Muslim	0.030***	-0.008	0.077***	0.042***
	(0.006)	(0.006)	(0.008)	(0.007)
Others	0.004	-0.000	0.006	0.022**
	(0.007)	(0.007)	(0.008)	(0.009)
Category (Ref: General, Brahmins and others)				
OBC	0.027***	0.007	0.032***	0.029***
	(0.004)	(0.005)	(0.005)	(0.005)
SC	0.071***	0.030***	0.100***	0.077***
	(0.005)	(0.006)	(0.006)	(0.006)
ST	0.126***	0.081***	0.213***	0.173***
	(0.008)	(0.009)	(0.010)	(0.011)
House (Ref: Own house)				
Rented	−0.041***	−0.001	−0.082***	−0.010
	(0.006)	(0.006)	(0.006)	(0.007)
Office accommodation and others	0.022*	−0.003	0.006	−0.017
	(0.013)	(0.010)	(0.017)	(0.011)
Primary Household Occupation (Ref: Agricultural activities)				
Non-agricultural activities	−0.051***	−0.007	−0.016***	0.008
	(0.005)	(0.005)	(0.005)	(0.006)

(Continued)

Table 17.2 (Continued)

Variables of Interest	*Incidence of Multidimensional Poverty*		*Incidence of Income Poverty*	
	Wave I 2004–2005	*Wave II 2011–2012*	*Wave I 2004–2005*	*Wave II 2011–2012*
Region (Ref: Rural)				
Urban	–0.129***	–0.051***	0.118***	0.006
	(0.005)	(0.005)	(0.006)	(0.005)
State (Ref: Jammu & Kashmir)				
Himachal Pradesh	0.002	–0.046***	0.031**	0.113***
	(0.015)	(0.015)	(0.015)	(0.013)
Punjab	–0.017	–0.030**	0.034**	0.054***
	(0.015)	(0.015)	(0.015)	(0.012)
Chandigarh	0.106***	0.010	0.062***	0.056***
	(0.029)	(0.033)	(0.024)	(0.015)
Uttaranchal	0.080***	0.036	0.221***	0.168***
	(0.021)	(0.023)	(0.027)	(0.023)
Haryana	–0.032**	–0.020	0.074***	0.115***
	(0.015)	(0.015)	(0.015)	(0.013)
Delhi	0.092***	-0.003	0.130***	0.089***
	(0.015)	(0.015)	(0.017)	(0.013)
Rajasthan	0.068***	0.049***	0.152***	0.079***
	(0.015)	(0.015)	(0.016)	(0.012)

(Continued)

Table 17.2 (Continued)

Variables of Interest	*Incidence of Multidimensional Poverty*		*Incidence of Income Poverty*	
	Wave I 2004–2005	*Wave II 2011–2012*	*Wave I 2004–2005*	*Wave II 2011–2012*
Uttar Pradesh	0.209***	0.186***	0.154***	0.066***
	(0.014)	(0.014)	(0.015)	(0.011)
Bihar	0.186***	0.180***	0.041**	0.172***
	(0.017)	(0.017)	(0.017)	(0.016)
Sikkim	0.014	-0.002	0.023	0.171***
	(0.022)	(0.025)	(0.020)	(0.046)
Arunachal Pradesh	−0.120***	−0.076***	−0.004	−0.052**
	(0.022)	(0.025)	(0.033)	(0.026)
Nagaland	−0.175***	−0.040	−0.093***	0.002
	(0.034)	(0.029)	(0.025)	(0.042)
Manipur	-0.124	0.003	-0.009	0.037***
	(0.105)	(0.023)	(0.035)	(0.014)
Mizoram	−0.010	−0.054	−0.161***	−0.082
	(0.034)	(0.039)	(0.029)	(0.051)
Tripura	−0.014	−0.047*	0.086***	0.063**
	(0.024)	(0.027)	(0.029)	(0.026)
Meghalaya	−0.102***	−0.120***	0.013	0.069
	(0.035)	(0.035)	(0.044)	(0.061)
Assam	0.037**	0.007	0.145***	0.181***
	(0.016)	(0.016)	(0.019)	(0.016)

(Continued)

Table 17.2 (Continued)

Variables of Interest	*Incidence of Multidimensional Poverty*		*Incidence of Income Poverty*	
	Wave I 2004–2005	*Wave II 2011–2012*	*Wave I 2004–2005*	*Wave II 2011–2012*
West Bengal	0.137***	0.041***	0.132***	0.166***
	(0.014)	(0.014)	(0.015)	(0.012)
Jharkhand	0.100***	0.031	0.272***	0.177***
	(0.018)	(0.019)	(0.021)	(0.020)
Orissa	0.171***	0.078***	0.311***	0.213***
	(0.016)	(0.015)	(0.018)	(0.014)
Chhattisgarh	0.063***	0.078***	0.486***	0.254***
	(0.016)	(0.018)	(0.020)	(0.018)
Madhya Pradesh	0.106***	0.107***	0.304***	0.148***
	(0.015)	(0.015)	(0.017)	(0.013)
Gujarat	0.026*	–0.018	0.090***	0.074***
	(0.014)	(0.015)	(0.015)	(0.011)
Daman & Diu	–0.096**	–0.114**	0.016	–0.056**
	(0.038)	(0.046)	(0.035)	(0.024)
Dadra & Nagar Haveli	–0.012	–0.106**	0.313***	–0.058
	(0.042)	(0.044)	(0.052)	(0.038)
Maharashtra	0.035**	–0.016	0.179***	0.171***
	(0.014)	(0.014)	(0.015)	(0.011)
Andhra Pradesh	0.026*	–0.035**	0.047***	0.061***
	(0.014)	(0.014)	(0.014)	(0.010)

(Continued)

Table 17.2 (Continued)

Variables of Interest	*Incidence of Multidimensional Poverty*				*Incidence of Income Poverty*			
	Wave I 2004–2005		*Wave II 2011–2012*		*Wave I 2004–2005*		*Wave II 2011–2012*	
Karnataka		0.015		−0.040***		0.115***		0.110***
		(0.014)		(0.013)		(0.015)		(0.011)
Goa		−0.034		0.021		0.064***		0.104***
		(0.028)		(0.019)		(0.019)		(0.016)
Kerala		−0.005		0.007		0.259***		0.117***
		(0.014)		(0.013)		(0.017)		(0.011)
Tamil Nadu		0.034**		−0.033**		0.189***		0.157***
		(0.015)		(0.014)		(0.017)		(0.013)
Pondicherry		−0.007		−0.052*		0.058**		0.205***
		(0.029)		(0.030)		(0.024)		(0.045)
Constant	−0.112***	−0.034*	−0.117***	−0.083***	0.034***	−0.497***	0.039***	−0.357***
	(0.002)	(0.018)	(0.002)	(0.016)	(0.003)	(0.019)	(0.002)	(0.015)
Observations	30,329	30,327	26,770	26,538	30,329	30,327	26,744	26,527
R-squared	0.536	0.586	0.452	0.485	0.062	0.222	0.079	0.173
Adj. R-squared	0.536	0.585	0.452	0.484	0.062	0.220	0.078	0.171

Note: A household is considered multidimensionally poor if they are deprived in at least one-third of the weighted indicators of education (years of schooling and child school attendance), health (nutrition and child mortality) and standard of living (electricity, improved sanitation, safe drinking water, flooring, cooking fuel and assets). Similarly, health-deprived, education-deprived and standard of living–deprived variables have been constructed. A household is considered to be income poor if its monthly per capita consumption expenditure is less than a certain stipulated amount as determined by the Planning Commission. Cluster corrected standard errors across households have been used. OBC = Other Backward Class; SC = Scheduled Caste; ST = Scheduled Tribe.

households (compared to Hindu households), households from disadvantaged social classes such as SC, ST or OBC (compared to general caste) are more likely to be poor.

17.3.3 State-level results

With the help of the IHDS dataset, the state-wise MPI, taking household as the unit of analysis, for the two years, 2004–2005 and 2011–2012 was calculated and tabulated in Table 17.3.[3] The head-count ratio (H) refer to the proportion of households deprived in terms of 33.33% or more (one-third) indicators used in the construction of multidimensional poverty. Similarly, the intensity of extreme poverty or destitution is the average proportion of indicators in which households are deprived (A%). The MPI therefore not only helps us know the proportion of households deprived of certain health, education or living standard indicators but also tells us the average proportion of indicators in which it is deprived. The income poverty (I), on the other hand, is the proportion of households that are just economically poor. For 2004–2005, the MPI for Delhi is the lowest, followed by Punjab, Jammu and Kashmir, Haryana and Tamil Nadu. Uttar Pradesh has the highest MPI of 0.253, indicating that many households in Uttar Pradesh are deprived in terms of a number of development indicators based on health, education and standard of living. Bihar, Orissa, Madhya Pradesh and Chhattisgarh also have high MPI for 2004–2005. From Table 17.3, we find that the state of Chhattisgarh has the highest proportion of households (58.76%) which are income poor, followed by Madhya Pradesh, Orissa and Jharkhand. Himachal Pradesh has the least proportion of households that are economically poor (with 3.47% only). We find that if we consider only income as a measurement of poverty then Chhattisgarh has the highest proportion of households which are trapped in poverty (I = 58.76%) whereas by using MPI as a measurement of poverty, Uttar Pradesh is found to have the highest proportion of households which are multidimensionally poor (H = 60.15%). From Figure 17.1, we see that with the exception of Delhi, Jharkhand and Chhattisgarh, all states have a higher proportion of households that are multidimensionally poor as compared to the proportion of income-poor households.

While comparing the MPI for 2004–2005 with that for 2011–2012, presented in Table 17.3, we find that both MPI and the likelihood of being MPI poor declined significantly over time for all the Indian states. Delhi's MPI reduced from 0.004 to 0.003, and it still has the lowest state-wise MPI. Tamil Nadu's MPI in 2004–2005 was 0.053. It has reduced drastically to 0.01, improving its rank from being fifth in 2004–2005 to only second in 2011–2012. The other states with low MPI in 2011–2012 are Jammu and Kashmir, Punjab and Himachal Pradesh. States such as Uttar Pradesh, Bihar, Orissa, Chhattisgarh and Madhya Pradesh still continue to have the highest MPI in 2011–2012, although in the case of Orissa, the MPI value has reduced more relative to the other four states, indicating that Orissa is

Table 17.3 State-Wise Multidimensional Poverty Index in India for 2004–2005 and 2011–2012

	Wave I (2004–2005)						*Wave II (2011–2012)*						
State	*H(%)*	*A(%)*	*MPI (H*A)*	*Ranking (on the basis of MPI)*	*Income Poverty (I%)*	*Ranking (on the basis of Income Poverty)*	*H(%)*	*A(%)*	*MPI (H*A)*	*Ranking (on the basis of MPI)*	*Income Poverty (I%)*	*Ranking (on the basis of Income Poverty)*	*CAGR (%)*
Delhi	2 .45	14.69	0.004	1	9.96	6	2.36	14.56	0.003	1	7.12	4	-4.03
Punjab	16.43	21.85	0.036	2	3.54	2	8.92	17.18	0.015	4	6.26	3	-11.76
Jammu & Kashmir	16.63	22.78	0.038	3	5.89	4	8.37	16.22	0.014	3	1.61	1	-13.29
Haryana	18.14	26.45	0.048	4	9.44	5	16.59	20.28	0.034	10	15.11	9	-4.81
Tamil Nadu	19.82	26.74	0.053	5	19.28	12	6.21	16.28	0.010	2	12.91	7	-21.20
Himachal Pradesh	21.09	28.03	0.059	6	3.47	1	10.74	17.05	0.018	5	10.43	6	-15.60
Gujarat	23.36	25.33	0.059	7	10.68	7	16.31	20.34	0.033	9	9.88	5	-7.97
Assam	26.68	28.71	0.077	8	17.32	11	15.32	22.27	0.034	11	22.06	15	-11.02
Andhra Pradesh	28.11	30.72	0.086	9	5.53	3	10.62	17.54	0.019	6	4.18	2	-19.40
Karnataka	28.49	31.03	0.088	10	14.62	8	11.82	19.39	0.023	7	13.16	8	-17.44
Jharkhand	32.35	30.53	0.099	11	35.99	17	26.32	22.50	0.059	14	25.76	17	-7.13
Maharashtra	32.18	31.28	0.101	12	22.80	13	15.62	20.09	0.031	8	19.44	13	-15.53
Uttarakhand	36.89	32.26	0.119	13	28.35	16	23.39	20.98	0.049	13	21.71	14	-11.91
West Bengal	39.11	33.78	0.132	14	16.83	10	19.03	18.37	0.035	12	16.60	11	-17.27

(Continued)

Table 17.3 (Continued)

	Wave I (2004–2005)						Wave II (2011–2012)						
State	*H(%)*	*A(%)*	*MPI (H*A)*	*Ranking (on the basis of MPI)*	*Income Poverty (I%)*	*Ranking (on the basis of Income Poverty)*	*H(%)*	*A(%)*	*MPI (H*A)*	*Ranking (on the basis of MPI)*	*Income Poverty (I%)*	*Ranking (on the basis of Income Poverty)*	*CAGR (%)*
Rajasthan	39.25	34.34	0.135	15	23.72	14	30.97	25.59	0.079	15	15.51	10	-7.37
Chhattisgarh	43.48	35.52	0.154	16	58.76	20	42.13	26.80	0.113	17	35.01	20	-4.33
Madhya Pradesh	48.73	39.03	0.190	17	41.04	19	45.09	28.28	0.128	18	24.42	16	-5.49
Orissa	49.77	39.07	0.194	18	38.51	18	34.27	26.94	0.092	16	28.81	19	-10.11
Bihar	59.24	42.25	0.250	19	15.33	9	50.26	31.62	0.159	19	28.09	18	-6.26
Uttar Pradesh	60.15	42.12	0.253	20	26.30	15	52.00	31.15	0.162	20	18.27	12	-6.17

Note: H (the head-count ratio) is the proportion of households in the state deprived in more than 33.33% of the indicators. A (intensity of extreme poverty) is the average proportion of indicators a state is deprived in. MPI incorporates the proportion of households deprived in 33.33% of the indicators and the average intensity of the deprivation in a state. The compound annual growth rate (CAGR) from the MPI has been calculated with the help of the following formula: CAGR = $\{[\text{Antilog}\{(\log Y_i - \log Y_0)/n\}]-1\}\times 100$. Union Territories (namely, Chandigarh, Goa, Pondicherry, Daman and Diu, Dadra and Nagar Haveli) and some states (mainly, Sikkim, Arunachal Pradesh, Nagaland, Manipur, Mizoram, Tripura, Meghalaya, Goa and Kerala) have also been excluded because number of households surveyed in these areas was too low. Income Poverty (I) is proportion of households which are economically poor.

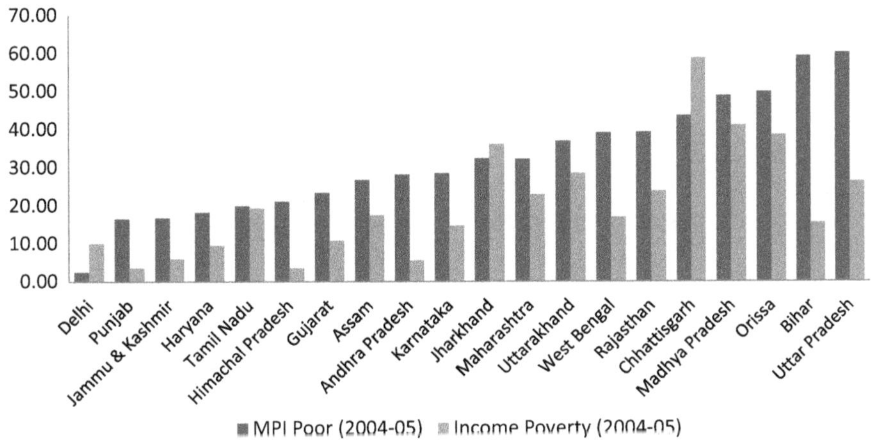

Figure 17.1 State-Wise Comparison of MPI Poor and Income Poor for 2004–2005. Note: MPI poor is the proportion of households in the state deprived in more than 33.33% of the indicators, that is, H% (head-count ratio). Income poor is the proportion of households which are poor on the basis of income, that is, I%.

catching up at a faster pace as against the other four states. This is quite clear from the growth rate (CAGR) which is –10.11%—the highest decline among these five states. Looking at the growth rate (in percentages), it can be seen that in certain states the decline in MPI has been in double digits, indicating that there has been significant improvement in attainment in terms of education, health and standard of living, leading to the reduction in the state-wise MPI. Tamil Nadu, with a decline of 21.20%, leads all the other Indian states with the most improvement in terms of MPI, followed by Andhra Pradesh (–19.40%), Karnataka (–17.44%), West Bengal (–17.27%), Himachal Pradesh (–15.60%) and Maharashtra (–15.53%). These are some of the best performing states in terms of reduction in MPI. Even though Delhi has the lowest MPI as measured during both 2004–2005 and 2011–2012, the rate of reduction of MPI from 2004–2005 to 2011–2012 is the lowest in the case of Delhi with a decline of just 4.03%. Chhattisgarh (–4.33%) has the second-lowest MPI reduction rate. On the other hand, in 2011–2012, Chhattisgarh (I = 35.01%) again has the highest proportion of households which are income poor, although the proportion of poor households has been reduced by more than 20% since 2004–2005. This is followed by Orissa, Bihar and Jharkhand. With only 1.61%, Jammu and Kashmir has the lowest proportion of households that are poor on the basis of income poverty as of 2011–2012. It is evident from Figure 17.2 that using the MPI, apart from Delhi, Tamil Nadu, Assam, Karnataka and Maharashtra, other Indian states have a higher proportion of households that are poor as compared to the proportion of households that are poor only on the basis of the income approach.

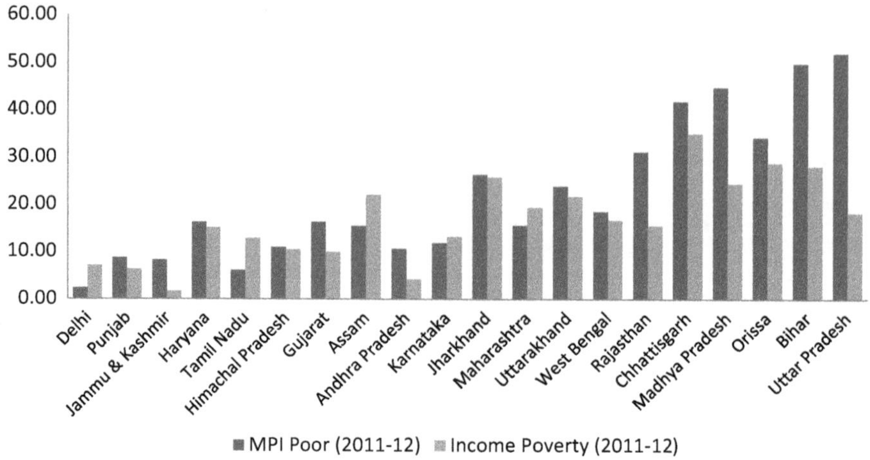

Figure 17.2 State-Wise Comparison of MPI Poor and Income Poor for 2011–2012. Note: MPI poor is the proportion of households in the state deprived in more than 33.33% of the indicators that is, H % (head-count ratio). Income poor is the proportion of households which are poor on the basis of income, that is, I%.

17.4 Conclusion

The indirect or the income-based method of measuring poverty is a resource-based method, and thus, it becomes difficult to identify whether the resource is actually being transformed to well-being. The direct method, on the other hand, used to measure poverty looks into the "deprivation in the capability to live a good life" (Sen, 2001). In this chapter, we have constructed and compared two measures, namely, indicators for multidimensional poverty and income-based measure of poverty, and tried to analyze which of the two is a better measure of poverty. This chapter therefore concentrates on finding the best measure of poverty, at the household level, that is, which of these measures capture the household scenario in a more holistic manner—capturing the deprivation of the household in terms of education, health and standard of living.

With the help of Spearman's rank correlation, we find a statistically significant and positive association between the incidence of multidimensional poverty (when compared with income poverty variable) and education, health and standard of living deprivation at the household level for 2004–2005 and 2011–2012. On running an OLS regression, with and without controls, we find significant coefficients in case of both multidimensionally poor and income poor variables. But the coefficients associated with each of the three dimensions of poverty—health, education and standard of living deprivation confirm that multidimensional poverty is better able to capture deprivation of the household on a wide range of variables compared to income poverty. There is no guarantee that a household which is income rich will also be non-poor based on the three dimensions of poverty, used in the construction of MPI. The results also suggest that Muslim and disadvantaged class categories (i.e., SC, ST, OBC households) are more likely to be multidimensionally poor compared to Hindu households and those from general caste households.

We also calculated state-wise MPI (Alkire and Foster, 2011) using the IHDS for both 2004–2005 and 2011–2012. The MPI for Uttar Pradesh has been the highest, indicating that deprivation of households in terms of education, health and standard of living is highest in the state of Uttar Pradesh, followed by Bihar, Madhya Pradesh, Chhattisgarh and Orissa. Even though there has been significant decline in both the incidence of being multidimensionally poor and in MPI between 2004–2005 and 2011–2012 for all the states considered in our analysis, the decline in MPI has also been low for these four Indian states other than Orissa. The states with the lowest MPI for 2004–2005 were Delhi, Punjab, Jammu and Kashmir, Haryana and Tamil Nadu. In 2011–2012, the states with the lowest MPI were the same, except for Haryana, whose MPI value didn't reduce considerably in 2011–2012. The highest MPI decline on the other hand was observed for states like Tamil Nadu (-21.20%), Andhra Pradesh (–19.40%), Karnataka (-17.44%) and West Bengal (-17.27%). The results indicate that these states have fared particularly well in terms of attainment in the three dimensions considered in the construction of MPI, mainly deprivation in education, health and standard-of-living indicators, finally leading to the reduction in the state-wise MPI.

The overall results presented in this chapter suggest that a more multidimensional approach must be adopted to identify and target households which are actually poor, by using the multidimensional poverty index which focuses on deprivations in terms of three dimensions of development, namely, health, education and standard of living, captured via 10 different indicators. This has already been shown to fare better in terms of identifying the multidimensionally poor compared to the traditionally used income approach. This measure can be easily used to identify the poor in different geographic locations, ethnicity, gender or other social groups, and different developing countries such as Colombia, Mexico and Bhutan have already started using MPI-based poverty reduction and development action plans.

Notes

1 While calculating the MPI, we have excluded the households which have missing information on one or more indicators.
2 Appendix Table 17.A1 shows the state-wise poverty line in the rural and urban area for 2004–2005 and 2011–2012.
3 Union Territories (namely, Chandigarh, Goa, Pondicherry, Daman and Diu, Dadra and Nagar Haveli) and some states (mainly, Sikkim, Arunachal Pradesh, Nagaland, Manipur, Mizoram, Tripura, Meghalaya, Goa and Kerala) have also been excluded because the number of households surveyed in these areas was very low.

References

Alkire, S., and J. Foster. "Counting and Multidimensional Poverty Measurement." *Oxford Poverty and Human Development Initiative*, Working Paper No. 7, Oxford Department of International Development, University of Oxford (2007).

Alkire, S., and J. Foster. "Counting and Multidimensional Poverty Measurement." *Journal of Public Economics*, vol. 95, 2011, pp. 476–487.

Alkire, S., and S. Seth. "Identifying Destitution through Linked Subsets of Multidimensionally Poor: An Ordinal Approach." *WIDER Working Paper* 2015/151 (2015).

Alkire, S., and M. E. Santos. "Acute Multidimensional Poverty: A New Index for Developing Countries." *Oxford Poverty & Human Development Initiative*, Oxford Department of International Development, Queen Elizabeth House, University of Oxford (2010).

Alkire, S., and M. E. Santos. "Measuring Acute Poverty in the Developing World: Robustness and Scope of the Multidimensional Poverty Index." *World Development* 59 (2014): 251–274.

Atkinson, A. B. "Multidimensional Deprivation: Contrasting Social Welfare and Counting Approaches." *The Journal of Economic Inequality* 1 (2003): 51–65.

Chakravorty, S. R., and D'Ambrosio, C. "The Measurement of Social Exclusion." *Review of Income and Wealth* 53, no. 3 (2006): 377–398.

Devereux, S. "Conceptualising Destitution." *IDS Working Paper No. 216*. Institute of Development Studies (2003).

Duclos, J. Y. "UNDP's Multidimensional Poverty Index." *Working Paper I11* (2011). https://ferdi.fr/en/publications/undp-s-multidimensional-poverty-index

Foster, J., and A. K. Sen. *On Economic Inequality with a Substantial Annexe* (2nd ed.). Oxford: Clarendon Press (1997).

Government of India Press Information Bureau. "Poverty Estimates for 2004–05." New Delhi. March (2007).

Government of India, Planning Commission. "Report of the Expert Group to Review the Methodology for Measurement of Poverty." (June 2014).

Harriss-White, B. "Destitution and the Poverty of its Politics—with Special Reference to South Asia." *World Development* 33, no. 6 (2005): 881–891.

Lipton, M. "Poverty, Undernutrition and Hunger." *World Bank Staff Working Papers No. 597*, The World Bank (1983).

Lipton, M. "The Poor and the Poorest. Some Interim Findings." *World Bank Discussion Paper No. 25*, The World Bank (1988).

Sen, A. *Development as Freedom*. New York, NY: Oxford University Press (2001)

Sen, A. K. *Inequality Re-examined*. Oxford: Clarendon Press (1992).

Sen, A. K. "Capability and Well-Being" In: Nussbaum, M., Sen, A.K. (Eds.), *Quality of Life*. Oxford: Clarendon Press (1993), 30–53.

Sugata, B., and S. Seth. "Understanding Standard of Living and Correlates in Slums: An Analysis using Monetary Versus Multidimensional Approaches in Three Indian Cities." Centre for Development Economics, Delhi School of Economics, Delhi, Working Paper No. 263 (2016). http://www.cdedse.org/pdf/work263.pdf

UNDP. "What It Would Take to Reach the Millennium Development Goals" (2010).

World Bank. *Monitoring Global Poverty: Report of the Commission on Global Poverty*. Washington, DC: World Bank (2016). doi:10.1596/978-1-4648-0961-3.

Appendix 17A

Table 17.A1 State-Specific Income Poverty Lines in 2004–2005 and 2011–2012 (in Rs. per capita per month)

		2004–2005 (Lakdawala Methodology)		*2011–2012 (Tendulkar Committee)*	
S.No.	*State/UTs*	*Rural*	*Urban*	*Rural*	*Urban*
1	Andhra Pradesh	292.95	542.89	860	1,009
2	Assam	387.64	378.84	930	1,060
3	Bihar	354.36	435.00	828	1,008
4	Chhattisgarh	322.41	560.00	778	923
5	Delhi	410.38	612.91	738	849
6	Gujarat	353.93	541.16	932	1,152
7	Haryana	414.76	504.49	1015	1,169
8	Himachal Pradesh	394.28	504.49	913	1,064
9	Jammu & Kashmir	391.26	553.77	891	988
10	Jharkhand	366.56	451.24	748	974
11	Karnataka	324.17	599.66	902	1,089
12	Madhya Pradesh	327.78	570.15	771	897
13	Maharashtra	362.25	665.90	967	1,126
14	Orissa	325.79	528.49	695	861
15	Punjab	410.38	466.16	1054	1,155
16	Rajasthan	374.57	559.63	905	1,002
17	Tamil Nadu	351.86	547.42	880	937
18	Uttar Pradesh	365.84	483.26	768	941
19	Uttarakhand	478.02	637.67	880	1,082
20	West Bengal	382.82	449.32	783	981
	All of India	356.30	538.60	816	1,000

Note: UT = Union Territory.
Source: Government of India, Planning Commission, "Report of the Expert Group to Review the Methodology for Measurement of Poverty" (June 2014).

18 Values, perception and the quality of life

An analytical framework and some critical observations from Sundarbans

Diganta Mukherjee and Rajlakshmi Mallik

18.1 Introduction

Climate change and the related issues of vulnerability, environmental justice, mitigation strategies and the measurement and buildup of adaptive capacity—technological, economic and social—is now an issue of global importance requiring immediate attention. The design of mitigation strategies reduces the *sources* and adaptation *moderating the adverse effects* of climate change and at the same time *exploits the beneficial opportunities* created from climate change (IPCC, Intergovernmental Panel on Climate Change), an issue facing every community.

All these concerns regarding sustainable development have been voiced, debated and discussed in several international platforms and forums over the last two and a half decades. Starting right from the Brundtland Commission (1987); The Rio Earth Summit (1992); 2000 UN Millennium Summit; 2002 World Summit on Sustainable Development, which culminated in the decade between 2005 and 2014 being declared the UN Decade of Education for Sustainable Development; and more recently the Copenhagen Summit 2009. The issue has been that of developing strategies that would ensure good and improving quality of life (QOL) across generations, that is, strategies that would result in *sustainable* (environmentally as also socially and economically) improvements in QOL.

Hence, planning for sustainable development, which involves the design of consumption, production and social systems with inherent adaptive capacity, requires as a first step a study of QOL that tries to *define* and *measure* human well-being (individuals and societies) as the ultimate goal of development. The latter is based not only on material well-being but factors in perception and satisfaction along with value systems that define personal priorities. Perception in the sense of satisfaction may be with respect to material condition (both ownership of private goods and access to public goods), but they may also refer to satisfaction in terms of abstract factors, such as a sense of belonging, personal security, social and family bonding, gender equality and so on. Again, perception may refer to how one views, interprets, classifies, associates and identifies things. For example, one may perceive the Gram Panchayat as a representative of the state/government or the dominant political party and individual perceptions may differ. Thus QOL, because of its comprehensive approach to defining human welfare, is compatible with planning for sustainable development that encompasses economic, social and environmental dimensions.

The Sundarbans were declared the World Heritage Site by UNESCO in 1997, and although they are parts of the same forest, the Bangladesh and Indian portions of the forest are listed in the UNESCO World Heritage separately as the Sundarbans and Sundarbans National Park respectively. Sundarbans South, East and West are three protected forests in Bangladesh. This region is densely covered by mangrove forests and is one of the largest reserves for the Bengal tiger. The Sundarbans National Park is a national park, tiger reserve and a biosphere reserve located in the Sundarbans delta in the Indian state of West Bengal.

These are now home to more than 4 million people. The remaining islands are still virgin, covered with dense mangrove forests. (for detailed discussion see Rudra, 2010, Samarakoon, 2004 and Sengupta and Rao, 2003). Apart from difficulties arising out of its natural setting, people living here constantly fight with many odds that have emerged out of wrongly directed management plans being followed since the colonial period (Rudra, 2010). The inefficacy of the plans has been reflected on both fronts—humanitarian as well as environmental—with each gaining importance depending on the political mileage it provided (Danda, 2007, Ch 3, fn 7 and 8). Thus, right from the beginning of human settlement in the region, the Sundarbans has been a case study in dilemmas and contradictions, competing values and interests. The incompatibility of human settlement in an active delta was apparent, but short-term considerations dominated (see also Banerjee, 1998 and Md. Shafi Noor).

18.2 The background of the study

18.2.1 Survey area

The Indian Sunderbans are located in the southern part of the districts of South and North 24 Parganas of West Bengal. Geographically, the area bounded by Hugli river in the west, the Raimangal and Hariabhanga Rivers in the east, the Bay of Bengal in the south and the Dampier and Hodges Line drawn in 1829–1830 in the north is identified as the Indian Sundarban (see the Wikipedia page on Geography of West Bengal for further details).

South 24 Parganas district comprises five subdivisions: Baruipur, Canning, Diamond Harbour, Kakdwip and Alipore Sadar. The Sunderbans area in this district consists of all the eight development blocks (four each) constituting the Canning and Kakdwip subdivisions and a few development blocks, two and three, respectively, from the Diamond Harbour and Baruipur sub-divisions. There no urban municipality areas within the Sunderbans area in either district, and the area under the Sundarban Affairs Department is entirely rural. For our study, we considered four blocks, Canning I,

Canning II, Basanti and Gosaba constituting the Canning Subdivision in the South 24 Parganas region of Sundarbans.

18.2.1.1 Sampling design

The survey covers 320 households, distributed equally over the four blocks (80 households from each block). The sample size was in keeping with the time and budget constraints. The population density and total population of the survey area were also taken into consideration. From each block, at least four Mouzas (the smallest administrative blocks) were selected so as to maximize spatial coverage of the survey area along with geographical variety. From each Mouza a maximum of 20 households were sampled, with closely neighboring Mouzas given less weightage to adjust for similarity. For the Gosaba block, the breakup was more detailed to cover a wider variety of landscape.

18.2.2 Analytical framework: Existing work and the current study

There is a wide range of definitions and interpretations of QOL (Galloway et al., 2005). To secure conceptual clarity attempts have been made to produce typologies of QOL definitions (Farquhar, 1995) and models (Felce and Perry, 1995; Hagerty et al., 2001; Taillefer et al., 2003). Other researchers have tried to investigate the content of QOL in terms of its defining attributes (Meeberg, 1993; Haas, 1999; WHO QOL Group, 1995). Although definitions and positions taken on issues and may vary, there is consensus about the importance of certain core issues. The main debates at the heart of QOL definitions are centered on objective versus subjective approaches; whether QOL is a uni- or multidimensional concept, the role of values, the place of self-evaluation, the cultural context, and QOL as a relative or absolute concept. Depending on the purpose of study, several QOL indicators have been developed over the years—the Human Development Index (HDI), Physical Quality of Life Index (PQLI), Happy Planet Index (2006) based on ecological footprint, EIU's (Economist Intelligence Unit) QOL Index, Mercer's Quality of Living Reports, Popsicle Index.

How QOL is measured clearly relates to how the term is defined and therefore to what is being measured. However, one can make a broad distinction between methods used to measure the *QOL of the general population* and those used to measure the *QOL of individuals*. At the individual level, personal values, as well as life conditions and life satisfaction, interact to determine QOL. The significance of either the objective or subjective assessment of a particular life domain is interpretable only in relation to the importance the individual places on it (Cummins, 1996). Community QOL is a function of the actual conditions in the environment as well as a function of how these conditions are perceived and experienced by the *typical individual* residing within the community (Proshansky and Fabian, 1987).

The current study of QOL takes into consideration the environmental impacts of existing socioeconomic systems and the vulnerability of the systems to global climate change. The study is an exploratory one and is based on a conceptualization of QOL that is defined as a combination of life conditions and satisfaction/perception weighted by scale of importance based on personal views and value systems.

18.3 Perception about personal well-being

The responses to the perception questions were collected on a five-point scale {–2, –1, 0, 1, 2}. We use the pairwise t-test (at 95% level of significance) to identify significant differences between blocks. Below we do this for each dimension on which data was collected. The tables present the sample size (N), average value (Mean), standard deviation (SD), standard error of the mean (S.E.), the upper confidence limit (UCL), and the lower confidence limit (LCL). The pairwise tests are now easily executed looking at consecutive ordered average and checking whether the average of the next block falls within the confidence interval of the previous.

a) Economic Situation (Current and Expected Change): Overall, the feeling in the study area is one of unhappiness with the financial situation, both in terms of current income and performance in saving. But expectations about the future as well as improvement over the last one year in terms of income are slightly more upbeat. Even when we look at the blocks separately, all except Basanti in terms of saving performance (slightly positive), shows on average a negative perception. Overall, we see that Canning 1 and Gosaba are the most pessimistic and Basanti and Canning 2 show slightly more upbeat feelings (Table 18.1).

In the case of future expectations, the positive feeling has been uniform across blocks, with Basanti and Canning 1 being the most positive. The improvement over last year returns mixed feelings, with Basanti and Canning 2 showing net happiness but Canning 1 and Gosaba being slightly downbeat. Comparing blocks, Basanti seems to be most positive about the economic situation and Gosaba seems to be the worst, with the two Canning blocks falling in between in terms of perception about the overall economic situation (past and future).

To check statistical significance, we look at the pairwise t-tests as mentioned above. In most cases, Basanti (B) is significantly more positive than Gosaba (G). The ordering (significant) for these tables are as follows:

Satisfaction with household income: {B > C2 > G, C1}
Expected economic situation five years later: {C1 > G, C2}
Improvement in saving over last one year: {B, C1, C2 > G}

b) Standard of Living: In terms of standard of living the overall perception as well as all the individual blocks' perception is negative, so this area of study can do with a lot of improvement in this area. The perception

Table 18.1 Economic Situation

		Block				
Item	*Descriptives*	*Basanti*	*Canning 1*	*Canning 2*	*Gosaba*	*Total*
Satisfaction with household income	N	79	74	79	80	312
	Mean	–.16	–.91	–.68	–1.08	–.71
	SD	.687	.779	.760	.792	.827
	S.E.	0.077	0.091	0.086	0.089	
	LCL	–0.311	–1.087	–0.848	–1.254	
	UCL	–0.009	–0.733	–0.512	–0.906	
Improvement in economic situation over last one year	N	72	79	77	78	306
	Mean	.14	–.03	.16	–.05	.05
	SD	1.039	.933	1.014	.910	.974
	S.E.	0.122	0.105	0.116	0.103	
	LCL	–0.100	–0.236	–0.066	–0.252	
	UCL	0.380	0.176	0.386	0.152	
Expected economic situation five years later	N	63	49	62	60	234
	Mean	.38	.55	.11	.20	.30
	SD	1.113	.843	1.147	1.038	1.059
	S.E.	0.140	0.120	0.146	0.134	
	LCL	0.105	0.314	–0.176	-0.063	
	UCL	0.655	0.786	0.396	0.463	
Improvement in saving over last one year	N	66	48	62	55	231
	Mean	.02	–.21	–.10	–.64	–.22
	SD	1.074	.824	1.127	.950	1.037
	S.E.	0.132	0.119	0.143	0.128	
	LCL	–0.239	-0.443	–0.381	–0.891	
	UCL	0.279	0.023	0.181	–0.389	

seems to be generally better in Basanti (a very slight negative) and strongly adverse in the other three blocks. We use the same techniques for comparison as in (a) earlier. Now the ordering turns out to be {B > C1, C2 > G} (Table 18.2).

c) Health: Perception of own health status seems to be at about the indifferent level (but slightly below average) in all the four blocks and hence overall. This could be due to a lack of cause or a lack of knowledge regarding relevant issues (Table 18.3).

d) Family and Friends: The study area does not seem to be fortunate in terms of their socializing, both at home and in the neighborhood. It seems that quality time is in short supply for all the blocks, with Canning 2 and Gosaba being the worst hit. Here again, the ordering turns out to be {B > C1, C2, G} (Table 18.4).

e) Psychological Health: The perception of happiness or feeling at peace or without stress does not seem to be very encouraging in our study area. In terms of all three alternative queries that we canvassed on this theme, the response has been, on an average, uniformly negative across blocks.

Table 18.2 Standard of Living

		Block				
Item	*Descriptives*	*Basanti*	*Canning 1*	*Canning 2*	*Gosaba*	*Total*
Satisfaction with standard of living	N	78	73	78	80	309
	Mean	-.03	-.68	-.65	-.90	-.57
	SD	.837	.941	.770	.851	.908
	S.E.	0.095	0.110	0.087	0.095	
	LCL	-0.216	-0.896	-0.821	-1.086	
	UCL	0.156	-0.464	-0.479	-0.714	

Table 18.3 Health

		Block				
Item	*Descriptives*	*Basanti*	*Canning 1*	*Canning 2*	*Gosaba*	*Total*
Perception of own health status	N	77	75	80	80	312
	Mean	-.03	-.01	-.03	-.06	-.03
	SD	.843	.626	.551	.700	.684
	S.E.	0.096	0.072	0.062	0.078	
	LCL	-0.218	-0.152	-0.151	-0.213	
	UCL	0.158	0.132	0.091	0.093	

Table 18.4 Family and Friend

		Block				
Item	*Descriptives*	*Basanti*	*Canning 1*	*Canning 2*	*Gosaba*	*Total*
Satisfaction with time spent with family/ friends	N	78	73	79	80	310
	Mean	.26	-.41	-.61	-.60	-.34
	SD	.904	.895	.838	.976	.968
	S.E.	0.102	0.105	0.094	0.109	
	LCL	0.059	-0.615	-0.795	-0.814	
	UCL	0.461	-0.205	-0.425	-0.386	

Although Basanti has a slightly less negative attitude in terms of these items, the other three blocks are decidedly on the dark side. The overall message is one of general despondency and disquiet. This is quite possibly related to a generally negative attitude regarding the standard of living or the financial situation that we discussed earlier. A lack of social capital, as evidenced by the shortfall of quality time among family and friends, could also have added to this feeling of disquiet. The *work environment*–related issue, on which again the blocks have a uniformly negative perception, adds to the determinants of psychological ill health (Table 18.5).

Here the orderings turn out to be

Happiness over last month: {B > C1, B > G > C2}
Calm and peace of mind over last month: {B > C1 > G, B > C2}
Anxiety over last three months: {B > G > C1, B > C2}

f) Overall: All things put together, the overall sense of satisfaction is in line with the response to the composite question in our survey with all blocks, except Basanti, showing a negative average return while Basanti shows a very small positive one. This is consistent with a general sense of well-being that we gathered from the individual-specific perception items (Table 18.6).

The overall ordering is {B > C1, C2, G}.

18.4 Overall perception about public goods

The perception regarding the availability of public goods in the study area is uniformly bad almost without exception (only Gosaba is positive about

Table 18.5 Psychological Health

		Block				
Item	*Descriptives*	*Basanti*	*Canning 1*	*Canning 2*	*Gosaba*	*Total*
Happiness over last month	N	77	71	79	80	307
	Mean	–.01	–.28	–.42	–.21	–.23
	SD	.639	.759	.653	.791	.724
	S.E.	0.073	0.090	0.073	0.088	
	LCL	–0.153	–0.457	–0.564	–0.383	
	UCL	0.133	–0.103	–0.276	–0.037	
Calm and peace of mind over last month	N	74	74	79	80	307
	Mean	–.11	–.39	–.52	–.59	–.41
	SD	.674	.755	.617	.650	.696
	S.E.	0.078	0.088	0.069	0.073	
	LCL	–0.264	–0.562	–0.656	–0.732	
	UCL	0.044	–0.218	–0.384	–0.448	
Anxiety over last three months	N	75	74	80	80	309
	Mean	–.08	–.89	–.64	–.50	–.53
	SD	1.183	1.093	1.225	1.067	1.175
	S.E.	0.137	0.127	0.137	0.119	
	LCL	–0.348	–1.139	–0.908	–0.734	
	UCL	0.188	–0.641	–0.372	–0.266	

Table 18.6 Overall

		Block				
Item	*Descriptives*	*Basanti*	*Canning 1*	*Canning 2*	*Gosaba*	*Total*
Overall satisfaction – last one year	N	78	74	79	79	310
	Mean	.03	–.45	–.47	–.57	–.36
	SD	.897	.846	.845	1.058	.941
	S.E.	0.102	0.098	0.095	0.119	
	LCL	–0.169	–0.643	–0.656	–0.803	
	UCL	0.229	–0.257	–0.284	–0.337	

availability of educational facilities) with public telephone, transport, electricity, bank and financial institutions, and leisure facilities in the worst state. Basanti is usually less critical about these issues except leisure facilities. These responses are also consistent with findings of the queries related to easy access to the facilities and distance from the facilities (Table 18.7).

Here the significant orderings are

Education: {G > C2 > C1, B > C1}	*Electricity: {B > C2 > C1 > G}*
Health: B, C2 > C1, C2 > G}	Water: {B > C1, G}
Housing support: {B > C1 > G, C2 > G}	Sanitation: {B > C1, G, C2 > C1}
Postal system: {B, C1, G > C2}	Roads: {B, C2 > C1, G}
Public telephone: {B > G, C2 > C1, G}	Judicial system: {B > C1, C2, G}
Public transport: {B > C2 > G, B > C1}	Banks and Financial Institutions: {B > C2 > C1, B > G

Even the other items under discussion fare quite badly (educational facilities seem to be the best among the disappointments). So these numbers paint a bleak picture about the perception regarding the availability of public goods in this area (Tables 18.8 and 18.9).

18.5 Perception about community well-being

a) Business (trading opportunities and demand) and Economy: The perception regarding the business environment of the locality is not very encouraging in the study area. The opinion regarding growth opportunities is uniformly highly negative. Perception regarding the environment is also largely negative, with Basanti and Canning 1 being less critical. So overall there seems to be apathy regarding the economic opportunities here (Table 18.10).

Table 18.7 Overall Satisfaction with Public Goods

		Block				
Overall Satisfaction	*Descriptives*	*Basanti*	*Canning 1*	*Canning 2*	*Gosaba*	*Total*
Education	N	79	80	80	79	318
	Mean	-.06	-.48	-.15	.16	-.13
	SD	1.125	1.018	1.008	1.031	1.066
	S.E.	0.127	0.114	0.113	0.116	
	LCL	-0.308	-0.703	-0.371	-0.067	
	UCL	0.188	-0.257	0.071	0.387	
Health	N	78	80	80	79	317
	Mean	-.37	-.76	-.35	-.59	-.52
	SD	1.163	.958	.982	1.104	1.063
	S.E.	0.132	0.107	0.110	0.124	
	LCL	-0.628	-0.970	-0.565	-0.833	
	UCL	-0.112	-0.550	-0.135	-0.347	
Housing support	N	78	80	80	79	317
	Mean	-.49	-.73	-.64	-1.09	-.74
	SD	.849	.900	.846	.850	.885
	S.E.	0.096	0.101	0.095	0.096	
	LCL	-0.678	-0.927	-0.825	-1.277	
	UCL	-0.302	-0.533	-0.455	-0.903	
Police	N	79	79	79	78	315
	Mean	-.76	-.65	-.65	-.82	-.72
	SD	.977	1.013	.892	1.125	1.003
	S.E.	0.110	0.114	0.100	0.127	
	LCL	-0.975	-0.873	-0.847	-1.070	
	UCL	-0.545	-0.427	-0.453	-0.570	
Postal system	N	79	79	80	78	316
	Mean	-.49	-.72	-.94	-.69	-.71
	SD	1.119	.933	.905	1.097	1.025
	S.E.	0.126	0.105	0.101	0.124	
	LCL	-0.737	-0.926	-1.138	-0.933	
	UCL	-0.243	-0.514	-0.742	-0.447	
Public telephone	N	77	78	80	78	313
	Mean	-1.22	-1.42	-1.26	-1.47	-1.35
	SD	.941	.635	.590	.697	.731
	S.E.	0.107	0.072	0.066	0.079	
	LCL	-1.430	-1.561	-1.389	-1.625	
	UCL	-1.010	-1.279	-1.131	-1.315	
Public transport	N	76	78	80	79	313
	Mean	-.59	-1.33	-1.26	-1.51	-1.18
	SD	1.318	.878	.590	.596	.951
	S.E.	0.151	0.099	0.066	0.067	
	LCL	-0.886	-1.525	-1.389	-1.641	
	UCL	-0.294	-1.135	-1.131	-1.379	

(Continued)

Table 18.7 (Continued)

		Block				
Overall Satisfaction	*Descriptives*	*Basanti*	*Canning 1*	*Canning 2*	*Gosaba*	*Total*
Electricity	N	79	79	79	76	313
	Mean	-.62	-1.20	-.99	-1.50	-1.07
	SD	1.212	.979	.940	.622	1.012
	S.E.	0.136	0.110	0.106	0.071	
	LCL	-0.887	-1.416	-1.197	-1.640	
	UCL	-0.353	-0.984	-0.783	-1.360	
Water	N	78	79	78	77	312
	Mean	-.51	-.99	-.76	-.87	-.78
	SD	1.225	1.080	1.083	1.140	1.141
	S.E.	0.139	0.122	0.123	0.130	
	LCL	-0.782	-1.228	-1.000	-1.125	
	UCL	-0.238	-0.752	-0.520	-0.615	
Sanitation	N	77	79	79	77	312
	Mean	-.62	-1.18	-.80	-.88	-.87
	SD	1.124	.747	.791	1.026	.950
	S.E.	0.128	0.084	0.089	0.117	
	LCL	-0.871	-1.345	-0.974	-1.109	
	UCL	-0.369	-1.015	-0.626	-0.651	
Roads	N	79	79	80	77	315
	Mean	-.59	-.86	-.53	-1.04	-.75
	SD	1.080	.957	.981	.979	1.017
	S.E.	0.122	0.108	0.110	0.112	
	LCL	-0.828	-1.071	-0.745	-1.259	
	UCL	-0.352	-0.649	-0.315	-0.821	
Judicial system	N	75	79	79	79	312
	Mean	-.75	-1.00	-.95	-1.06	-.94
	SD	.902	.862	.815	.965	.891
	S.E.	0.104	0.097	0.092	0.109	
	LCL	-0.954	-1.190	-1.130	-1.273	
	UCL	-0.546	-0.810	-0.770	-0.847	
Banks and financial institutions	N	76	79	80	79	314
	Mean	-.79	-1.47	-1.33	-1.41	-1.25
	SD	1.111	.617	.632	.725	.833
	S.E.	0.127	0.069	0.071	0.082	
	LCL	-1.040	-1.606	-1.468	-1.570	
	UCL	-0.540	-1.334	-1.192	-1.250	
Parks and recreational facilities	N	76	79	80	79	314
	Mean	-1.68	-1.52	-1.55	-1.62	-1.59
	SD	.637	.782	.549	.538	.634
	S.E.	0.073	0.088	0.061	0.061	
	LCL	-1.823	-1.692	-1.670	-1.739	
	UCL	-1.537	-1.348	-1.430	-1.501	

Table 18.8 Facility More than Two Hours Away

Block	*Police*	*Law*	*Pub Health*	*Pvt Health*	*Pub Education*	*Pvt Education*	*Panchayat*	*Bank/ Insurance*	*Postal*	*Bus*	*Electric*	*Telephone etc.*	*Rail*
Basanti	17	71	7	2	3	2	6	9	2	5	13	3	15
Canning 1	41	79	18	6	4	1	1	6	0	3	16	18	27
Canning 2	13	76	9	2	1	0	6	0	0	1	2	2	24
Gosaba	16	79	4	3	1	1	2	2	0	34	13	13	44
Total	87	305	38	13	9	4	15	17	2	43	44	36	110

Table 18.9 Easy Access in Terms of Pricing, Behaviour or Personnel, Procedure

Block	*Police*	*Telephone etc.*	*Law*	*Pub Health*	*Pvt Health*	*Pub Education*	*Pvt Education*	*Panchayat*	*Bank/ Insurance*	*Postal*	*Pub Transport*	*Pvt Transport*	*Drinking Water*	*Electric*
Basanti	20	0	5	9	7	34	8	26	8	35	2	2	33	0
Canning 1	21	7	12	20	15	35	37	25	11	27	8	7	61	8
Canning 2	17	0	13	7	1	14	44	41	0	8	0	1	49	0
Gosaba	14	0	12	2	2	34	27	22	2	11	0	0	52	2
Total	72	7	42	38	25	117	116	114	21	81	10	10	195	10

Table 18.10 Business and Economy

		Block				
Opinion	*Descriptives*	*Basanti*	*Canning 1*	*Canning 2*	*Gosaba*	*Total*
Business environment	N	76	79	80	80	315
	Mean	–.09	.01	–.35	–.33	–.19
	SD	1.009	1.056	1.057	1.167	1.080
	S.E.	0.116	0.119	0.118	0.130	
	LCL	–0.317	–0.223	–0.582	–0.586	
	UCL	0.137	0.243	–0.118	–0.074	
Economic growth	N	75	78	79	77	309
	Mean	–.76	–.86	–.65	–.91	–79
	SD	.836	.833	.769	.891	.835
	S.E.	0.097	0.094	0.087	0.102	
	LCL	–0.949	–1.045	–0.820	–1.109	
	UCL	–0.571	–0.675	–0.480	–0.711	

The significant orderings here are

Business environment: {B, C1 > C2, G}
Economic growth: {C2 > C1}

b) Living Standards and Job Creation: In terms of four indicators regarding the perception about the community living standards, the picture is uniformly bleak across all blocks with a slightly less negative perception regarding the availability of products and services. A very similar picture emerges regarding job-creation opportunities (Table 18.11).

The significant orderings here are

Consumer products and services: {B > C1, C2, G}
Peoples' needs being met: {B > G, C2 > C1 > G}
Quality of life for the majority: {B > C2} (Table 18.12).
The significant orderings here are
Job creation: {C1, C2 > B, G}
Work opportunity: {C1 > G, C2 > B}

c) Economic equity: It's somewhat surprising to see that even though the perception about the economic standard in the community is uniformly low, there is still is a feeling of inequity regarding economic standard in this area. So, according to popular perception, there are haves and have-nots or, rather, have-nots and have-nothings! The feeling of inequity is fairly strong across all blocks (Table 18.13).

The significant orderings in this case turn out to be

Economic equality: {B, C2 > C1 > G}
Equity: {B > G, C2 > C1}

Table 18.11 Satisfaction with Living Standards and Job Creation

		Block				
Satisfaction regarding: (in the village or region)	*Descriptives*	*Basanti*	*Canning 1*	*Canning 2*	*Gosaba*	*Total*
Consumer products and services	N	75	79	80	80	314
	Mean	-.19	-.52	-.60	-.55	-.47
	SD	.926	.904	.908	1.146	.986
	S.E.	0.107	0.102	0.102	0.128	
	LCL	-0.400	-0.719	-0.799	-0.801	
	UCL	0.020	-0.321	-0.401	-0.299	
Economic development (purchasing power and income levels)	N	74	77	80	79	310
	Mean	-.92	-1.16	-.90	-1.39	-1.09
	SD	1.057	.745	.805	.758	.867
	S.E.	0.123	0.085	0.090	0.085	
	LCL	-1.161	-1.326	-1.076	-1.557	
	UCL	-0.679	-0.994	-0.724	-1.223	
Peoples' needs being met	N	74	79	78	78	309
	Mean	-.80	-1.09	-.94	-1.12	-.99
	SD	.844	.754	.795	.967	.849
	S.E.	0.098	0.085	0.090	0.109	
	LCL	-0.992	-1.256	-1.116	-1.335	
	UCL	-0.608	-0.924	-0.764	-0.905	
Quality of life for the majority	N	72	78	79	77	306
	Mean	-.65	-1.10	-.96	-1.18	-.98
	SD	.754	.766	.759	.854	.805
	S.E.	0.089	0.087	0.085	0.097	
	LCL	-0.824	-1.270	-1.127	-1.371	
	UCL	-0.476	-0.930	-0.793	-0.989	

d) Development Policy and Projects: The respondents are all uniformly in support of the development projects in the study area (Table 18.14).

For policies and projects, the significant orderings are

Dam: {G > C1, C2 > B}
Tourism, Water supply and Roads: {C1, C2, G > B}

e) Natural Environment and Conservation Initiative: Respondents across the four blocks are not satisfied with the state of the natural environment and conservation initiatives taken by the local government (Table 18.15).

Here significant ordering is {B > G}.

f) Social Environment (harmony): The respondents from all four blocks have a low opinion regarding the extent of social harmony that is prevalent

Table 18.12 Opinion Regarding Living Standards and Job Creation

		Block				
Opinion Regarding	*Descriptives*	*Basanti*	*Canning 1*	*Canning 2*	*Gosaba*	*Total*
Job creation	N	74	79	79	78	310
	Mean	−1.50	−1.18	−1.04	−1.40	-1.27
	SD	.815	.764	.724	.744	.779
	S.E.	0.095	0.086	0.081	0.084	
	LCL	−1.686	−1.348	−1.200	−1.565	
	UCL	−1.314	−1.012	−0.880	−1.235	
Work opportunity	N	75	79	79	78	311
	Mean	−1.13	−1.05	−.91	−1.27	−1.09
	SD	.859	.714	.720	.733	.765
	S.E.	0.099	0.080	0.081	0.083	
	LCL	−1.324	−1.207	−1.069	−1.433	
	UCL	-0.936	-0.893	-0.751	-1.107	

Table 18.13 Economic Equity

		Block				
Opinion Regarding	*Descriptives*	*Basanti*	*Canning 1*	*Canning 2*	*Gosaba*	*Total*
Economic equality	N	73	78	78	79	308
	Mean	−.82	−1.21	−.97	−1.43	−1.11
	SD	.918	.671	.702	.673	.776
	S.E.	0.107	0.076	0.079	0.076	
	LCL	−1.031	−1.359	−1.126	−1.578	
	UCL	-0.609	-1.061	-0.814	-1.282	
Equity	N	75	79	79	78	311
	Mean	−.96	−1.22	−1.04	−1.18	−1.10
	SD	.877	.827	.669	.879	.819
	S.E.	0.101	0.093	0.075	0.100	
	LCL	−1.158	−1.402	−1.188	−1.375	
	UCL	−0.762	−1.038	−0.892	−0.985	

in this area with Basanti being somewhat less critical. Surprisingly, the verdict on unity is a positive one, with Canning 1 and Gosaba being more emphatic on this score. This may imply that although the area is not that well socially integrated, in the face of some common crisis, people come and act together (Table 18.16).

Here significant orderings are

Social harmony: {B > C1, C2, G}
Unity: {C1, G > B > C2}

Table 18.14 Development Policy and Projects

Opinion Regarding Development Projects	*Descriptives*	*Block*				
		Basanti	*Canning 1*	*Canning 2*	*Gosaba*	*Total*
Dam	N	78	80	79	79	316
	Mean	1.05	1.79	1.68	1.84	1.59
	SD	1.080	.495	.589	.517	.773
	S.E.	0.122	0.055	0.066	0.058	
	LCL	0.810	1.682	1.550	1.726	
	UCL	1.290	1.898	1.810	1.954	
Tourism	N	77	80	80	79	316
	Mean	1.26	1.66	1.54	1.70	1.54
	SD	.801	.762	.871	.837	.833
	S.E.	0.091	0.085	0.097	0.094	
	LCL	1.081	1.493	1.349	1.515	
	UCL	1.439	1.827	1.731	1.885	
Water supply	N	78	80	80	79	317
	Mean	1.32	1.70	1.70	1.76	1.62
	SD	.693	.683	.537	.625	.657
	S.E.	0.078	0.076	0.060	0.070	
	LCL	1.166	1.550	1.582	1.622	
	UCL	1.474	1.850	1.818	1.898	
Roads	N	76	80	80	79	315
	Mean	1.36	1.70	1.68	1.75	1.62
	SD	.687	.644	.591	.707	.673
	S.E.	0.079	0.072	0.066	0.080	
	LCL	1.206	1.559	1.550	1.594	
	UCL	1.514	1.841	1.810	1.906	

Table 18.15 Natural Environment and Conservation Initiative

		Block				
Opinion Regarding	*Descriptives*	*Basanti*	*Canning 1*	*Canning 2*	*Gosaba*	*Total*
Conservation initiative	N	74	77	79	78	308
	Mean	-.62	-.70	-.84	-.88	-.76
	SD	1.094	1.136	.854	1.279	1.100
	S.E.	0.127	0.129	0.096	0.145	
	LCL	-0.869	-0.954	-1.028	-1.164	
	UCL	-0.371	-0.446	-0.652	-0.596	

Table 18.16 Social Environment

		Block				
Opinion Regarding	*Descriptives*	*Basanti*	*Canning 1*	*Canning 2*	*Gosaba*	*Total*
Social harmony	N	73	77	79	78	307
	Mean	-.55	-.97	-.95	-.92	-.85
	SD	.958	.986	.749	1.090	.964
	S.E.	0.112	0.112	0.084	0.123	
	LCL	-0.770	-1.190	-1.115	-1.162	
	UCL	-0.330	-0.750	-0.785	-0.678	
Unity	N	75	79	79	78	311
	Mean	.27	.57	.03	.69	.39
	SD	.890	1.268	1.176	1.199	1.169
	S.E.	0.103	0.143	0.132	0.136	
	LCL	0.069	0.290	-0.229	0.424	
	UCL	0.471	0.850	0.289	0.956	

g) Gender Equality and Child Labor Laws: The opinion of all the four blocks is uniformly low regarding the extent of gender equality that exists in the study area. A similar opinion prevails with regard to laws against child labor and child abuse (Table 18.17).

For equity the significant ordering is {B > C1, G}.

h) Political and Administrative Environment: The political environment (leadership) and governance and administration (including law and order) are also regarded as very poor by the respondents. This low opinion prevails

Table 18.17 Gender Equality and Child Labour Laws

		Block				
Opinion Regarding	*Descriptives*	*Basanti*	*Canning 1*	*Canning 2*	*Gosaba*	*Total*
Gender equality	N	74	78	78	78	308
	Mean	-.68	-.72	-.81	-.71	-.73
	SD	1.206	1.138	.927	1.260	1.134
	S.E.	0.140	0.129	0.105	0.143	
	LCL	-0.955	-0.973	-1.016	-0.990	
	UCL	-0.405	-0.467	-0.604	-0.430	
Equity	N	75	79	79	78	311
	Mean	-.96	-1.22	-1.04	-1.18	-1.10
	SD	.877	.827	.669	.879	.819
	S.E.	0.101	0.093	0.075	0.100	
	LCL	-1.158	-1.402	-1.188	-1.375	
	UCL	-0.762	-1.038	-0.892	-0.985	

across panchayat, land ownership laws and the issue of safety and security. The only exception being the perception about cleanliness that received a positive verdict from three of the four blocks (Canning 2 being the exception and insignificantly below zero) (Tables 18.18 and 18.19).

Table 18.18 Opinion regarding Political Leadership

		Block				
Opinion Regarding	*Descriptives*	*Basanti*	*Canning 1*	*Canning 2*	*Gosaba*	*Total*
Political leadership	N	73	79	78	78	308
	Mean	-.64	-.97	-.90	-1.08	-.90
	SD	.991	.947	.766	1.066	.957
	S.E.	0.116	0.107	0.087	0.121	
	LCL	-0.867	-1.179	-1.070	-1.317	
	UCL	-0.413	-0.761	-0.730	-0.843	

Table 18.19 Opinion Regarding Governance and Administration

Opinion Regarding Governance & Administration		*Block*				
	Descriptives	*Basanti*	*Canning 1*	*Canning 2*	*Gosaba*	*Total*
Panchayat	N	75	79	80	80	314
	Mean	-.37	-.63	-.55	-.73	-.57
	SD	1.050	.865	.940	.927	.951
	S.E.	0.121	0.097	0.105	0.104	
	LCL	-0.608	-0.821	-0.756	-0.933	
	UCL	-0.132	-0.439	-0.344	-0.527	
Safety and security	N	74	77	79	79	309
	Mean	-.35	-.66	-.77	-.90	-.68
	SD	1.164	1.165	.933	1.183	1.128
	S.E.	0.135	0.133	0.105	0.133	
	LCL	-0.615	-0.920	-0.976	-1.161	
	UCL	-0.085	-0.400	-0.564	-0.639	
Trustworthiness of landownership	N	72	79	79	78	308
	Mean	-.49	-1.32	-1.19	-1.38	-1.11
	SD	1.199	1.032	1.241	.996	1.169
	S.E.	0.141	0.116	0.140	0.113	
	LCL	-0.767	-1.548	-1.464	-1.601	
	UCL	-0.213	-1.092	-0.916	-1.159	
Safety	N	75	79	79	78	311
	Mean	.64	.77	.16	.51	.52
	SD	1.226	1.109	1.170	1.365	1.236
	S.E.	0.142	0.125	0.132	0.155	
	LCL	0.363	0.525	-0.098	0.207	
	UCL	0.917	1.015	0.418	0.813	

Here significant ordering is {B > C2 > G, B > C1}.
For governance, the significant orderings are

Panchayat: {B > C1, G}	*Safety: {B, C1, G > C2}*
Safety and security & Trustworthiness of landownership: {B > C1, C2, G}	

All things considered, the *overall rating* of the area has also turned out to be poor in all blocks. with Basanti performing relatively better. By and large, the villagers see no reason to feel proud of their village or neighborhood (Table 18.20).

The overall ordering turns out to be {B > C1, C2, G}.

i) Perception about Nature of Vulnerability to Natural Calamity: In this context, it is worthwhile to note that around 90% of the respondents feel that natural calamities may result in loss of assets and livelihood, and 80% feel that even relocation may be necessary. Thus, it is all the more critical that the respondents feel that government arrangement for protection against such events is woefully inadequate. Such a perception would lead directly to apathy and certain sense of dislocation among the villagers. It would be very difficult to think of long-term plans of investment if this feeling of "temporariness" is persistent (Table 18.21).

To check for any significant difference in the pattern of opinion among blocks, we ran a chi-square test for each of the earlier questions. The following results do not indicate any significant difference.

Natural calamity	*d.f.*	*Chi square*	*p-value*
Destruction of assets	6	10.55254	0.10323
Loss of livelihood	6	11.37915	0.077342
Relocation	6	12.08206	0.060163

Table 18.20 Overall

		Block				
	Descriptives	*Basanti*	*Canning 1*	*Canning 2*	*Gosaba*	*Total*
Pride for the village	N	75	79	77	78	309
	Mean	-.11	-.81	-.94	-.74	-.65
	SD	1.214	1.075	.908	1.323	1.179
	S.E.	0.140	0.121	0.103	0.150	
	LCL	-0.38475	-1.04706	-1.14281	-1.03361	
	UCL	0.164754	-0.57294	-0.73719	-0.44639	

Table 18.21 Result of Natural Calamity

Opinion Regarding Natural Calamities may Result in		*Block*				
		Basanti	*Canning 1*	*Canning 2*	*Gosaba*	*Total*
Destruction of assets	Not possible	0	1	0	0	1
	May be	12	17	7	19	55
	Possible	62	62	71	57	252
Total		74	80	78	76	308
Loss of livelihood	Not possible	0	1	0	0	1
	May be	8	19	12	21	60
	Possible	66	60	65	55	246
Total		74	80	77	76	307
Relocation	Not possible	4	2	1	0	7
	May be	18	20	8	19	65
	Possible	52	58	65	57	232
Total		74	80	74	76	304

In the same vein, the villagers are mostly not satisfied with the *disaster management initiatives* of the local government. More than 90% of the respondents feel that the local government is ill equipped and needs to do more on this front, while only 3% are happy with the arrangements (Table 18.22).

Again, to identify a significant difference of pattern between blocks, we ran a chi-square test. The value of statistics is 31.53002, with d.f. = 9, which is significant (p-value = 0.00024). Basanti seems to have the most positive opinion while Canning1 and 2 and Gosaba are more negative.

j) Perception regarding Modernization: In all four blocks studied, most people think modern arrangements are better than the traditional ones for credit and insurance market arrangements, environmental protection, and conservation, as well as fishing activities. Opinion is mixed regarding activities like health care (where all blocks except Canning 2 are in favor of modern arrangements but Canning 2 prefers the traditional methods) and

Table 18.22 Opinion regarding Disaster Management Initiatives

		Block				
		Basanti	Canning 1	Canning 2	Gosaba	*Total*
Local government prepared for natural calamities	bad	16	25	18	31	90
	need more	45	48	53	46	192
	Ok	13	1	6	0	20
	very well	1	4	2	1	8
Total		75	78	79	78	310

Table 18.23 Traditional Knowledge and Techniques Are Better than Modern

Block	*Health*	*Education*	*Environment*	*Agro Tech*	*Fishing*	*Credit*	*Percentage*
Basanti	31	34	20	33	24	6	0.276
Canning 1	25	50	12	30	23	1	0.263
Canning 2	42	50	4	9	8		0.210
Gosaba	22	18	19	47	29		0.251
Total	120	152	55	119	84	7	

agricultural activities (with Gosaba supporting the traditional methods but the other three blocks preferring the modern ones). Across blocks, it is seen that Basanti is in favor of modern technologies with respect to all the activities under question, while opinions are mixed for the other three blocks (Table 18.23).

18.6 Value systems and priorities

a) Value System: Personal Well-Being: Here we discuss the importance of alternative issues based on priority ranking from recall incidences (Table 18.24).

The priority ranks of attributes that contribute towards personal well-being are more or less on the expected line with *income*, *education*, and *health* claiming the top three positions in that order. The fourth item is *good government*. One can contrast this priority ranking with the respondents' claim of not having enough funds for spending on heads, where *shelter*, *food*, *education* and *health* are the top four recalls (in that order), with *clothing* occupying the seventh position. Out of these five important items, health and education are dependent on public provision and they are duly given

Table 18.24 Personal Priority

Issue/Question	*Value System: Personal*	*Expenditure Shortfall*	*Allocating Windfall*
Income	1		
Education	2	3	
Health	3	4	5
Shelter		1	1
Food		2	2
Clothing		7	4

Table 18.25a Social Priority

Issue	Block				
	Basanti	*Canning 1*	*Canning 2*	*Gosaba*	*Total*
Personal Income	1	2	1	1	1
Education	2	1	2	2	2
Health	3	3	3	3	3
Good Government	4	4	4	4	4

Table 18.25b Rank Correlation between Blocks

	Basanti	*Canning 1*	*Canning 2*	*Gosaba*
Basanti	1			
Canning 1	0.8	1		
Canning 2	1	0.8	1	
Gosaba	1	0.8	1	1

priority. The other three are generically important items: food, shelter and clothing, whose availability is very closely linked to income. Finally, from the 'allocation of a windfall income' question (lottery earning), the revealed priorities are also consistent with the earlier results, with shelter, food, clothing and health featuring in the top five.

b) Value System: Social Well-Being: The top four rankings attributed here are the same as for the personal well-being question. So the respondents' personal and social lenses towards well-being seem to be quite consistent. The ranking is also very consistent across blocks (Tables 18.25a and 18.25b).

The preceding rank correlations among the priority listings show close concordance among the blocks in opinions regarding social priorities.

c) Ranking of Development Projects: The priority allocated by the respondents towards development projects are as follows: (1) Hospital, (2) Bridge, (3) School, (4) Sanitation, (5) Drinking water, and (6) Electricity. Note that 1, 4, and 5 are essentially closely linked with the health issue, 2 and 6 are infrastructural items, and 3 is the determinant of availability of basic education. So here again, the priority over public goods reflects the need for education and health, along with infrastructure, in the social well-being priority ladder as evidenced from the derived rankings (Tables 18.26a and 18.26b).

In terms of opinion regarding development projects, there is good concordance between blocks except for Basanti and Canning 1.

Table 18.26a Priority of Development Projects

Development Project	*Block*				
	Basanti	*Canning 1*	*Canning 2*	*Gosaba*	*Total*
Hospital	1	3	1	2	1
Roads, Bridges	4	1	3	1	2
School	2	4	2	4	3
Sanitation	5	2	5	3	4
Drinking water	3	6	4	6	5
Electricity	8	5	7	8	6

Table 18.26b Rank Correlation between Blocks

	Basanti	*Canning 1*	*Canning 2*	*Gosaba*
Basanti	1			
Canning 1	0.108	1		
Canning 2	0.957	0.297	1	
Gosaba	0.587	0.861	0.710	1

18.7 Concluding remarks

This study captures the role of development priorities based on personal value systems of a population and the influence of perception in developing a measure of QOL. The study is based on a survey in the eastern part of the Sundarbans region in West Bengal, India, and covers the least developed and most vulnerable part of the Indian Sundarbans. Perceptions about the level of well-being and personal rankings of factors such as education, health, transport and communication, roads, credit facilities, income, law and order, gender equality, communal harmony, work environment, community network and participation, social inclusion, business environment and job opportunities, government intervention, and pro-activeness, among others, contributing to a happy individual and society, are factored in for developing a measure of QOL for a geographic region that constitutes one of the most significant biosphere reserves of the world, a population that is highly vulnerable to natural disasters, and a livelihood pattern that is inconsistent with the ecology.

We outline the thought process behind the basic theoretical framework of analysis, the development of research design consisting of the major variables covered by the survey, the scaling techniques used for measuring perception and values, and some observations for policy interventions based on the initial exploratory and descriptive analysis. Some empirical observations regarding the relationship between perceived and objective measures of well-being and value systems and realized outcomes are also presented.

References

Banerjee, A. *Environment, Population and Human Settlements of Sundarban Delta.* New Delhi: Concept Publishing Company (1998).

Cummins, R. A. "The Domains of Life Satisfaction: An Attempt to Order Chaos." *Social Indicators Research* 38 (1996): 303–328.

Danda, Anamitra Anurag (2007), "Surviving in the Sundarbans: Threats and Responses, an Analytical Description of Life in an Indian Riparian Commons." Doctoral dissertation. The University of Twente.

Farquhar, M. "Definitions of Quality of Life: A Taxonomy." *Journal of Advanced Nursing* 22 (1995): 502–508. doi:10.1046/j.1365-2648.1995.22030502. xTaillefer, 2003

Felce D., Perry J. "Quality of Life: Its Definition and Measurement." *Research in Developmental Disabilities* 16, no. 1 (January–February 1995): 51–74. PubMed PMID: 7701092.

Galloway S., D. Bell, C. H. Hamilton, and A. Scullion. *Well-Being and Quality of Life: Measuring the Benefits of Culture and Sport: A Literature Review and Thinkpiece.* Edinburgh: Scottish Executive Social Research (2005).

Geography of West Bengal: Wikipedia

Haas B. A. "Multidisciplinary Concept Analysis of Quality of Life." *Western Journal of Nursing Research* 21, no. 6 (1999): 728–742.

Hagerty M., R. A. Cummins, A. L. Ferriss, K. Land, A. C. Michalos, M. Peterson, A. Sharpe, M. J. Sirgy, and J. Vogel. "Quality of Life Indexes for National Policy: Review and Agenda for Research." *Social Indicators Research* 55, no. 1 (2001): 1–96.

Md. Shafi Noor Islam. "Cultural Landscape Changing due to Anthropogenic Influences on Surface Water and Threats to Mangrove Wetland Ecosystems: A Case Study on the Sundarbans, Bangladesh," Doctoral Thesis. Brandenburg University of Technology at Cottbus.

Meeberg, G. A. "Quality of Life: A Concept Analysis." *Journal of Advanced Nursing* 18 (1993): 32–38. doi:10.1046/j.1365-2648.1993.18010032.x

Proshansky, H. M., and A. K. Fabian. "The Development of Place-Identity in the Child." In C. S. Weinstein and T. G. David (eds.), *Spaces for Children.* New York: Plenum (1987), 21–40.

Rudra, Kalyan (2010) , Dynamics of the Ganga in West Bengal, India (1764–2007): Implications for science–policy interaction, Quaternary International, 2010

Samarakoon, J. (2004), Issues of Livelihood, Sustainable Development, and Governance: Bay of Bengal, *AMBIO: A Journal of the Human Environment* 33, no. 1–2 34–44.

Sengupta S., Rao P. "Adverse Impacts of Climate Change on Sunderban Delta: Perception and Responses of Local Community." *WWF-India, UNFCCC Expert Workshop on Local Coping Strategies for Adaptation*, 12–13 November 2003, New Delhi.

Taillefer M. C., G. Dupuis, M. A. Roberge, and S. LeMay. "Health-Related Quality of Life Modes: Systematic Review of the Literature." *Social Indicators Research* 64 (2003): 293–323. doi:10.1023/A:1024740307643

WHO QOL Group. (1995). "The World Health Organization Quality of Life Assessment (WHOQOL): Position Paper from the World Health Organization." *Social Science and Medicine* 4, no. 10 1403–1409.

19 Sustainability of loan waiver programs in India

Tanika Chakraborty and Aarti Gupta

19.1 Introduction

Intervention in the credit market, through household debt relief, has been a fiscal policy adopted by many governments, both at the central and state levels for many years. Some economists, such as Keynes, were very much in favor of government intervention through fiscal channels during exceptionally harsh economic circumstances. However, not all were in favor of these interventions. On one hand, the economic argument in favor of stimulus programs operating through credit markets rests on the premise that in situations in which households are unable to insure themselves against macroeconomic shocks, such policies will prevent excessive deadweight losses (Bolton and Rosenthal, 2002). They also help reduce high levels of debt which distort consumption and investment decisions of households (Mian et al., 2012). On the other hand, these economic stimulus programs may distort borrower incentives and give rise to moral hazard (Giné and Kanz, 2014). They can create negative externalities and are likely to raise the cost of credit in the long run. Despite this being a controversial argument for many years, loan waivers have been an important tool used by governments. The debate on loan waivers is not whether they should be given but to see whether loan waivers lead to sustainable development.

On July 14 2014, the governments of both states of Telangana and Andhra Pradesh announced a loan waiver scheme of 43,000 crores. The scheme was opposed by several economists, as the government took this decision in spite of being in a severe financial crisis, having a budget deficit of 16,000 crores and no funds for regular schemes. Frequent loan waivers, often announced with a political motivation, have hastened the process of the erosion of the rural credit delivery system (Pandey, 2005).

Since independence, India has been an agricultural economy, with a majority of its citizens being farmers. Loan waiver schemes have been one such political tool used by different governments for decades for vote bank politics. In 1990, then prime minister V. P. Singh announced an agricultural debt relief scheme totaling 10,000 crores for agricultural borrowers. In the same year, a similar scheme was announced by Devi Lal, then chief minister of Haryana, waiving 227.5 crores of farm loans by banks and cooperatives, out of which 162 crores were due to commercial banks. Over the next decade, in spite of the RBI warning that defaults and problems in recovery of dues would affect the credit system, a number of state and central governments came up with debt relief schemes.

In 2008, the Indian government announced one of the largest debt waiver schemes in history. The Agricultural Debt Waiver and Debt Relief Scheme waived Rs. 600 billion spread across 237 districts and reaching 30 million farmers (Kanz, 2012). A complete waiver was given to small and marginal farmers (holding land between 1 and 2.5 hectares). Other farmers with landholding above 2.5 hectare were given 25% waiver. The scheme was introduced to address the increasing suicides among the farmers, by alleviating their miseries. Another goal of the program was to help public and private banks refinance themselves by canceling their nonperforming assets that had accumulated due to direct lending to rural communities over the years (Giné and Kanz, 2014).

Over the next few years, the program received widespread criticism from economists. However, this did not stop governments at the state level from announcing further waiver programs, with Uttar Pradesh being the foremost player. With a history of repeated waivers given out to farmers, it is very important to study debt relief schemes in India. It is necessary to understand the purpose and objectives behind these schemes and to find out whether these objectives were met. In addition, it is important to analyze how households make consumption and investment decisions in anticipation of such waivers.

19.2 Uttar Pradesh Gramin Vikas Rin Maa Yojana (2011)

To understand these issues better, we analyze one of the first and largest loan waiver schemes implemented in the state of Uttar Pradesh, India, called the Uttar Pradesh Sahakari Gram Vikas Bank Rin Maa Yojna. The scheme was announced by the chief minister of Uttar Pradesh in November 2011 and was targeted towards small farmers. Farmers who had taken a loan up to ₹50,000 from the Gramin Vikas bank and had repaid at least 10% of the total amount due, were eligible. The initial budget allocated to the program was ₹1,650 crores. The debt relief was carried over a period of 3 years, and branches of different districts received the relief at different times. By the end of financial year 2014–2015, an amount of ₹1,720.42 crores was given as debt relief for this program covering 7.3 lakh farmers from 70 districts.

From Table 19.1 we see that in the first year, 2012–2013, 73 districts received the waiver.

A total of 833.73 crores of outstanding from these districts was waived off. The district with the highest amount of debt waived off was Sitapur. A loan of ₹151.64 crores was due for the district in 2012–2013, out of which an amount of 108.198 crores was waived off under the debt relief

Table 19.1 Loan Disbursement Details, 2009–2014

	Loan Disbursed	*Outstanding Loan*	*Loan Due*	*Repayment*	*Loan Waived*	*Repayment Including Waiver*	*Repayment Without Waiver*
2009–2010	702.89	3,204.96	1,719.62	746.85		43.43%	43.43%
2010–2011	728.22	3,502.68	1,545.08	774.05		50.10%	50.10%
2011–2012	568.15	3,691.85	2,289.82	395.93		17.29%	17.29%
2012–2013	55.81	3,057.1	3,049.54	1295.27	900	42.47%	12.96%
2013–2014	406.53	2,814.93	2,200.03	1321.44	750	60.06%	25.97%

scheme. Similarly, in 2013–2014, 750 crores of loan were waived off for 30 districts, out of which Shahjahanpur district received the highest amount of 83 crores. In 2014–2015, only four districts received a massive amount of 704 crores, out of which Unnao district had 241.5 crores of their loan waived off.

In Figure 19.1, we show a timeline describing the rollout of the loan waiver program in Uttar Pradesh. As mentioned before the program was announced in the Samajwadi Party manifesto in Nov 2011. We notice that the repayment percentage, that is, the ratio of loans demanded to loans repaid, in 2009–2010 and 2010–2011 was about 50%. In 2011–2012 right before the official manifesto was released, the loan waiver the repayment percentage fell to 17.29%.

Table 19.2 provides a summary of consumption, productivity and other characteristics, by their waiver status. We compare how households allocate and utilize their resources between consumption, investment, bulk expenditures when they receive the waiver, and when they do not. In addition to this, we also compare household behavior when they have knowledge of the waiver as opposed to having no knowledge of the wavier. The results from our primary survey suggest that the average consumption of the full sample is Rs. 38,433 per year. This is lower than the average consumption of households which received the waiver and households with knowledge about the waiver. Productivity, calculated as the rate of total production over land cultivated, is lowest for households that received the waiver.

Another important observation from the table is the behavior of people who have knowledge of loan waiver as opposed to those who do not have knowledge of it. Since loan waivers come in the form of repeated interventions in India and have often been used by political parties as an

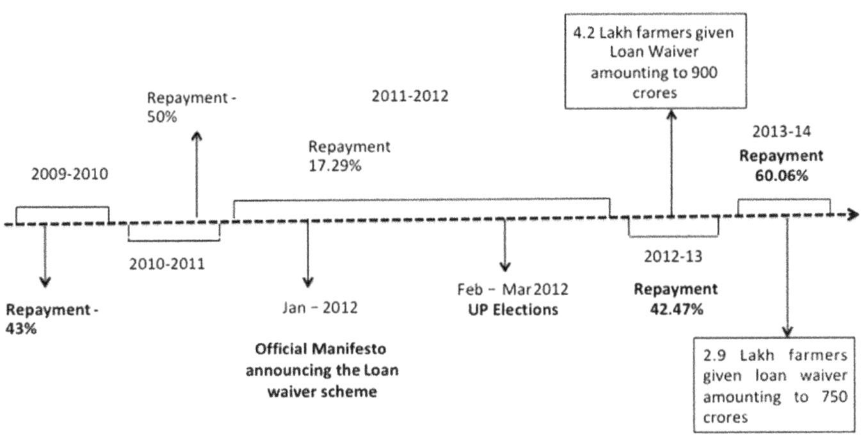

Figure 19.1 Time Line for the Uttar Pradesh Loan Waiver Program

Table 19.2 Outcome Variables by Waiver Status

Variable		*Full Sample*	*Loan Waiver*	*No Loan Waiver*	*Knowledge*	*No Knowledge*
	Obs	Mean	Mean	Mean	Mean	Mean
Consumption	770	38,433	41,479	37,531	42,502	30,725
Productivity	462	32,876	29,397	40,131	29,491	38,691
Total Production	519	47,046	38,769	75,317	39,129	58,511
Income	770	54,860	52,623	63,956	52,864	58,642
HH Loan	770	34,605	24,268	69,621	25,808	51,273
HH Size	770	1.01	1.01	5.32	1.01	1
Wedding	139	0.24	0.22	0.1	0.06	0.27
Bulk Purchases	762	0.13	0.14	0.16	0.09	0.14

Notes: This table gives the mean values of the main variables used for our analysis. The different samples used for the analysis are, full sample, households that received loan waiver, households that did not receive loan waiver, households that are eligible for loan waiver, households that are not eligible for loan waiver, households that have knowledge about the loan waiver program and households that do not have knowledge about the loan waiver program. The variables used for the analysis are yearly consumption in rupee for each household (HH); annual income of the household; rupee value of total produce by the household; value of total production as a fraction of farm size. Household loan refers to the amount of largest loan taken by the household. Household size refers to the number of members in a household. Wedding indicates if there was a wedding in the family in the last one year. Bulk purchases indicate if the household has made a bulk purchase in the last one year.

Source: Primary Data Collection

election-winning strategy, it is likely that people start expecting governments to offer loan waivers during election years. We notice that the yearly consumption of households with knowledge of waiver is almost 28% higher than that of households with no knowledge of the program.

One could argue that this is simply an income effect driven by the possibility that households that have higher consumption are also those households with higher income. People with higher incomes are usually more educated and have access to news which in turn would make them more aware of such programs. However, when we consider the yearly income of both these groups, we find that the mean yearly income of households with no knowledge of the loan waiver is actually higher than that of households with knowledge of loan waiver. In addition, we find that households with knowledge of the loan waiver are 33% more likely to have made a bulk purchase within a year of hearing about the loan waiver.

Table 19.2 indicates that households that either received loan waiver or had knowledge of the program, have lower incomes, higher consumption and lower productivity as compared to their respective counterparts. This indicates possible unproductive utilization of the loan borrowed by the households.

Given the indication of moral hazard in the data, we build a theoretical set-up to understand the possibility of underlying moral hazard when households anticipate loan waiver programs.

19.3 Theoretical model

Rural credit markets in underdeveloped countries like India have primarily three ways of ensuring repayment of loans—screening, monitoring and enforcement. Formal and informal institutions differ in their screening, monitoring and enforcement capabilities. We explore how these characteristics shape the repayment patterns of borrowers. In particular, we explore the role played by strong and weak enforcement by a lending agency, on the repayment patterns of borrowers. Formal institutions have always faced problems with rural credit. Due to the problem of asymmetric information, formal lenders discriminate against small borrowers because of the high cost involved in acquiring information (Wenner, 1995).

On the other hand, informal lenders usually live in the same village as the borrowers and are usually part of the same social network. The better flow of information in these networks makes it possible to have a better screening mechanism that does not involve as high a cost as that of formal institutions. Formal institutions also find it difficult to monitor the households who have taken a loan to ascertain how they use the loan and what mechanisms they adopt to implement the project. On the other hand, informal lenders due to their social proximity and relationship with the borrowers have better access to be able to monitor and influence how the borrower uses the loan (Wenner, 1995).

There are a number of studies that have looked at various mechanisms that could help solve the screening and monitoring problem for formal institutions (Giné and Karlan, 2008; Rajan and Winton, 1995; Stiglitz, 1990). Group credit, for instance, solves the monitoring problem by inducing people who are part of the credit group to monitor their peers (Stiglitz, 1990). The problem of screening can be solved using joint liability, as it induces endogenous peer selection in the formation of groups in a way that is beneficial for increasing repayment rates (Ghatak, 2000).

Finally, we come to the third characteristic of rural credit, which is credible contract enforcement. Many formal institutions find forgiving or refinancing a debt easier than strictly enforcing the contract and foreclosing on a defaulting borrower (Wenner, 1995). The formal institutions in India, especially the state-owned regional banks, are often faced with the problem of a weak legal system. In addition, pervasive views that the bank loans are political patronage make it difficult for the formal institutions to enforce their credit contracts, especially for the rural masses who form the majority of the vote bank in most states. On the other hand, informal moneylenders can rely on social ostracism, interlinked contracts and blatant coercion as effective methods of enforcement (Bakshi and Hassan, 2008). In this section, we focus on this third mechanism of contract enforcement and argue that weaker enforcement of contracts by formal institutions makes it easier for borrowers to get away with nonrepayment of formal loans.

Let us consider a utility maximizing household with a two-period Utility function, $U(C_1; C_2)$. For simplicity, we assume that the household does not have any initial monetary endowment and borrows a loan to finance consumption and investment. However, the household owns land that it can provide as collateral.

In period 1, the household decides to divide the borrowed resources between consumption and investment. Suppose the household consumes x resources in period 1 and uses y resources for investment, such that

$$x + y = 1;$$

we assume that the household will have a positive consumption in period 1 and thus under no circumstances can $x = 0$. Moreover, success, p, is a function of the amount of resources the household invests in period 1, that is, y. For simplicity let us assume that success is a linear function of y; specifically, $p = y$.

If the project is a success, the household gets a net return, π and if it fails, a penalty amount, D, is confiscated by the lender. Hence, the expected consumption in period 2 is

$$E(C_2) = (1 - x)\pi + x(-E(D))$$

that is, the expected consumption of the household in period 2 is equal to the net return the household earns, in the case of success that occurs with a probability $(1 - x)$ plus the amount it has to pay back in the case of failure that occurs with a probability of x.

19.3.1 *Probability of enforcement*

The expected value of D, which is the penalty amount confiscated by the lender in case of default, is a function of the return on investment (produce) and land. This means that in the case of default, the creditor confiscates the entire produce of the household, and if the produce is not enough to meet the amount due, then the collateral (land) is confiscated. The probability of confiscating the land depends on the probability of enforcement of contract:

$$E[D] = Produce + \theta\, Land,$$

where θ is the probability of enforcement.

Thus, the second-period utility function is

$$U = \alpha(C_1) + \beta E(C_2),$$

where β is the household's discount factor for consumption in period 2.

19.3.1.1 Optimal choice of consumption

To find an analytical solution to the problem we assume a Constant Returns to Scale Utility function of the form:

$$\mathrm{U} = [x^{\sigma} + \left(\beta\left[(1-x)\pi - x(Produce + \theta\,Land)\right]^{\sigma}\right]^{\frac{1}{\sigma}}. \tag{19.1}$$

The farmer chooses x consumption in period 1, and effectively invests y in period 1, to maximize lifetime utility. Hence, the first-order condition for optimization yields

$$\frac{dU}{dx} = 0. \tag{19.2}$$

This implies

$$\frac{1}{\sigma}[x^{\sigma} + \left(\beta\left[(1-x)\pi - x(P + \theta\,L)\right]^{\sigma}\right]^{\frac{1-\sigma}{\sigma}} \left[\sigma x^{\sigma-1} + \beta\sigma\left[(1-x)\pi - x(P+\theta L)\right]^{\sigma-1}\right][-\pi - P - \theta L] = 0 \tag{19.3}$$

$$=> \sigma x^{\sigma-1} + \beta\sigma\left[\left[1 - x\pi - x(P+\theta L)\right]^{\sigma-1}\right][-\pi - P - \theta L] = 0 \tag{19.4}$$

$$=> \sigma x^{\sigma-1} = \beta\sigma\left[(1-x)\pi - x(P+\theta L)\right]^{\sigma-1}][\pi + P + \theta L] \tag{19.5}$$

$$=> \frac{x^{\sigma-1}}{\left[(1-x)\pi - x(P+\theta L)\right]^{\sigma-1}} = \beta\left[\pi + P + \theta L\right] \tag{19.6}$$

$$=> \frac{x}{(1-x)\pi - x(P+\theta L)} = \beta\left[\pi + P + \theta L\right]^{\frac{1}{\sigma-1}} \tag{19.7}$$

$$=> x = \mu(1-x)\pi - \mu x(P+\theta L) \tag{19.8}$$

where

$$\mu = \beta\left[\pi + P + \theta L\right]^{\frac{1}{\sigma-1}} \tag{19.9}$$

$$x = \frac{\pi\mu}{1 + \pi\mu + \mu(P+\theta L)} \tag{19.10}$$

$$=> x = \frac{\pi}{\frac{1}{\mu} + (\pi + P + \theta L)} \tag{19.11}$$

19.3.1.2 Comparative statics

We are interested in looking at the change in consumption in the first period when the probability of enforcement changes. In other words, we are interested in $\partial x/\partial\theta$. Taking the partial derivative of x with respect to θ, we obtain

$$\frac{\partial x}{\partial \theta} = -\frac{\pi(L + \dfrac{\left[L\beta\left(\theta L + P + \pi\right)\right]^{\frac{1}{\sigma-1}-1} \beta'\left(\theta L + P + \pi\right)}{1-\sigma}}{\left(\theta L + \beta\left(\theta L + P + \pi\right)^{\frac{-1}{\sigma-1}} + P + \pi\right)^2} \tag{19.12}$$

Since the denominator is a square term, it is positive. Now we consider the numerator.

$$-\pi\left(L + \frac{\left[L\beta\left(\theta L + P + \pi\right)\right]^{\frac{1}{\sigma-1}-1} \beta'\left(\theta L + P + \pi\right)}{1-\sigma}\right)$$

Decreasing marginal rate of substitution implies that $\sigma < 1$ *or* $\sigma = 0$. In our theoretical set up we have assumed that the household has some positive consumption in period 1, which implies that $x \neq 0$. Because of this we can safely assume that in our set-up, $\sigma \neq 1$, because $\sigma = 1$ implies perfect substitutability, which results in corner solutions, with one of the solutions being $x = 0$.

$$\text{Now, } \sigma < 1 \qquad \Rightarrow 1-\sigma > 0.$$

Moreover, each term in the numerator is positive; hence, the whole term is positive. Thus,

$$\frac{\partial x}{\partial \theta} < 0.$$

This implies that as θ increases, x decreases; that is, with the increase in the probability of enforcement, consumption in period 1 decreases. θ determines the level of expected penalty imposed on the borrowers in case of default. Thus, the preceding result suggests that the higher the expected penalty, the lower the consumption in the first period and thus higher the investment will be.

19.3.2 High versus low penalty

In the earlier section, we study the consumption and investment decisions of borrowing households when they are faced with different types of penalties in the event of a default. We consider two specific cases. In the first case, the lender only confiscates the produce of the household in case of a default. Even if the lending contract uses land as collateral, the lender writes off the remaining debt and only confiscates the produce. We define this to be the case of a low penalty. Therefore, a household that is unable to repay its debt in the second period only fears the confiscation of produce. In such a situation, the household knows that if it defaults in period 2, then the creditor will confiscate all the produce. Thus, the household will not have an incentive to produce more in period 2. Instead, it will consume more in period 1 and invest just enough to finance the second-period consumption.

In the second case, the lender confiscates not only the second-period production but also the collateral as per the lending contract, in the event of a default by the household. We define this to be the case of a high penalty. The households are now faced with a higher penalty and fear confiscation of both produce and land. The loss of land affects not only the second-period utility but is likely to affect utility for all future periods since the household will not be able to produce for future consumption. Since land is an illiquid asset, the household will invest more in period 1 to be able to return the debt, out of the fear of losing the land.

19.4 Conclusion

Loan waivers typically create expectations of low penalty amongst households. Under a loan waiver scheme, households who have been unable to repay their debt and have collateral attached to the debt have their loans waived and collateral freed. In other words, loan waivers prevent contract enforcement. With a precedence of loan waivers in agricultural markets, households expect formal institutions to intervene in credit contract enforcements and not seize collateral in case of default. On the other hand, there are no such expectations from informal institutions since they do not come under the purview of government programs.

If households believe that the probability of enforcement is less when the contract is written with formal sources, then they will have a higher tendency to default on the loan by indulging in unproductive consumption. On the other hand, if households believe that the probability of contract enforcement and hence of losing their collateral is high, in the case of loans taken from informal sources, then they are likely to invest and produce enough to repay their loans. To the extent that repeated loan waiver programs affect people's belief about the probability of enforcement, they are likely to generate different behavioral responses from people in their treatment of formal versus informal loans.

References

Bakshi, R and Karim Hassan. "From 'No Collateral No Loan' to 'No Collateral No Default': The Economics of Group Lending Microfinance." *The Social Sciences* 3, no. 8 (2008): 611–625.

Bolton, Patrick, and Howard Rosenthal. "Political Intervention in Debt Contracts." *Journal of Political Economy* 110, no. 5 (2002): 1103–1134.

Ghatak, Maitreesh. "Screening by the Company You Keep: Joint Liability Lending and the Peer Selection Effect." *The Economic Journal* 110, no. 465 (2000): 601–631.

Giné, Xavier, and Martin Kanz. "The Economic Effects of a Borrower Bailout: Evidence from an Emerging Market." World Bank Policy Research Working Paper no. 7109 (2014).

Giné, Xavier, and Dean Karlan. "Peer Monitoring and Enforcement: Long Term Evidence from Microcredit Lending Groups with and without Group Liability." Yale University Department of Economics, Working Paper (January 2008).

Kanz, Martin. *What Does Debt Relief Do for Development? Evidence from India's Bailout Program for Highly-Indebted Rural Households. Evidence from India's Bailout Program for Highly-Indebted Rural Households*. World Bank Policy Research Working Paper, 6258 (November 1, 2012).

Mian, Atif, Amir Sufi, and Francesco Trebbi. "Resolving Debt Overhang: Political Constraints in the Aftermath of Financial Crises." Technical report, National Bureau of Economic Research (2012).

Pandey, B. K. *Rural Development: Towards Sustainability (3 Vols.)*. Gyan Publishing House (2005).

Rajan, Raghuram, and Andrew Winton. "Covenants and Collateral as Incentives to Monitor." *The Journal of Finance* 50, no. 4 (1995): 1113–1146.

Stiglitz, J. E. "Peer Monitoring and Credit Markets." *The World Bank Economic Review*, 4, no. 3 (1990) 351–366.

Wenner, Mark D. "Group Credit: A Means to Improve Information Transfer and Loan Repayment Performance." *The Journal of Development Studies* 32, no. 2 (1995): 263–281.

20 Using path analysis to build a sustainable transport service quality model

B.R. Naveen and Anjula Gurtoo

20.1 Introduction

Transport is an important part of infrastructure, and a developed transport network encourages speedy and satisfactory movement of men and material (Namboodiri, 2007).The demand for transport that provides business accessibility and safe mobility, with minimum negative impacts on natural environment, is growing (Hubschneider, 2011). According to the United Nations (Brundtland, 1987), sustainability is the present generations' ability to fulfill their needs judiciously and allowing generations of the future to meet their own needs without compromise. By improving the operational performance and efficiency of services, the goal of sustainability can be reached in the service sector (Wolfson et al., 2014).

The literature highlights sustainable service as fulfilling customers' needs and that can be maintained for a long duration without affectingthe natural and social environments of the customer. Sustainability should be the core value of the underlying relationship between service providers and customers to deliver sustainable service. In India's comprehensive surface transport network, intercity bus transportation holds an important place because of its potential in connecting cities, smaller communities, rural areas and less populated regions (Fravel, 2003). The attainment of a steady transportation exercise by the social, economic and environmental at different distance-based range of activities is termed as sustainable transportation (Tao and Hung, 2003). Sustainability in transport also gets defined as meeting the present transport and maneuverability demands and allowing generations of the future to meet these needs without compromise (Black, 1996). There are not enough studies on modeling service quality of bus transport with respect to passenger perception (Das and Pandit, 2013).

In this study, we use path analysis to explore the cause–effect relationship model of service quality with respect to the three pillars of sustainability in relation with overall satisfaction of intercity bus transport.The objectives of this study area are (a) the application of path analysis methodology to build a measurement model of a sustainable transport service modeland (b) to build a structural relationship model of a sustainable transport service model. Most studies that explore the relationship in sustainability models are based on analysis of variance, correlation or regression analysis. There are limited scientific studies, however, on the use of path analysis techniques for sustainable development and specifically in the field of sustainable transport. Path analysis, however, has an advantage. The methodology makes it possible to evaluate the relative impact of each explanatory variable on the resulting variables. This helps the intercity transport organizations to devise a step-wise strategy for delivering service quality.

20.1.1 Literature review—Service quality and sustainability

Service quality refers to the judgment regarding the superiority of a service and the assessment of outcomes of the service received by the customer (Parasuraman et al., 1988; Ponrahono et al., 2016). Customer satisfaction is positively influenced by service quality in the context of public transport (Khurshid et al., 2012). Cronin and Taylor (1992) introduced SERVPERF model with the argument that service quality should be measured as an attitude and reinforced the perception-based measurement of service quality. Successful retention of sustainable service development and delivery indicates that the services will, over time, be tailored to the evolving requirements of customers as well as suppliers. Kandampully (1997) point towards the imperativeness of adopt strategies capable of ensuring sustainability over a period.

Finding an appropriate equilibrium between present and future transport needs in terms of environmental, social and economic qualities is defined as sustainable transport (OECD, 1996). Sustainable transportation has been defined as "meeting present transport and mobility needs without compromising the ability of future generations to meet these needs" (Black, 1996), reflective of the Brundtland Report. According to the OECD (2002), sustainable transportation system is one which meets transport needs without disturbing the ecosystem or public health with the judicious use of renewable and nonrenewable resources.

Sustainability, therefore, is the attainment of uninterrupted transportation operations aided by the social, economic and environmental objectives at different distance-based scales of operation (Tao and Hung, 2003). Social, economic and environmental elements should be integrated in service and its long-term effects should be studied (Wolfson et al., 2014). Social, economic and environmental elements have a significant impact on transportation, and it is an important factor in sustainability (Litman, 2008). As per the Centre for Sustainable Transportation (Gilbert and Wong, 2003), a sustainable transportation system is one that (a) allows meeting the fundamental access requirements of people and societies in a way that is safe and compatible with human and ecosystem health and equity within and between generations; (b) is affordable, works effectively, provides transportation choices and promotes a stimulating economy; and (c) minimizes the waste and emissions within the absorption capability of planet, reduces the usage of non-renewable resources and minimizes the usage of renewable resources to the point of sustainable yield, reuse and recycle their parts and reduce land use and noise production.

20.1.2 Social sustainability

Sustainable service creation demands the scope of service delivery be extended to include social elements beyond its economic elements (Gilbert and Wong, 2003). Social indicators should represent the impact on the quality of life of society and individuals, such as accessibility, reliability, cost-quality assurance, empathy, tangibles and responsiveness like safety, equity, and security (Haghshenas and Vaziri, 2012; Verma et al., 2015). Accessibility is defined as nearness of living population to a transit station or the density of the network in an area (Haghshenas and Vaziri, 2012). Reliability is the ability to deliver guaranteed services accurately, dependably, consistently according to the promised schedule and in a timely manner without making a mistake each time (Parasuraman et al., 2002). According to Freitas (2013), the aspects such as customer handling ability, politeness, courteous information dissemination and issue redressal are important for assurance. According to Parasuraman et al. (1991), tangibles are those entities associated with the service delivery such as the appearance of personnel and physical facilities, equipment, physical and communication materials. According to Leong et al. (2015), the willingness of offering individual service to each customer by the service organization is termed as empathy. According to Parasuraman et al. (1991), employees' desire to help the customers by providing adequate safe and secure services is termed as a responsiveness dimension of service quality.

20.1.3 Economic sustainability

Money value is the best possible blend of lifetime price and quality to satisfy the user's requirements and can be evaluated using economic effectiveness and efficiency criteria (Jackson, 2012). According to Jackson (2012), economic efficiency is defined as limiting the input for given output or maximizing the output for given input along with maintaining the quality. Effectiveness is described as obtaining the expected results from an activity successfully. Zeithaml (1988) describes the perceived value as the customer's usefulness evaluation of the products based on the perception of the gains received for the price. He also claims that perceived value can be enhanced by improving perceived advantages or lowering perceived expenses.

The reduction in the value of the ticket fare is not enough for economic sustainability and passenger satisfaction (Freitas, 2013). Customer relationship is a function of the costs and benefits that accrue from the relationship. The balance between compromises and benefits, both monetary and nonmonetary, is based on the customer's assessment of a service (Mouwen, 2015).

20.1.4 Environmental sustainability

Road transport being the largest end-use energy consumer is one of the quickly increasing causes of greenhouse gas emissions and air pollutants (Uherek et al., 2010). Therefore, during the next decades, road transport will continue to be a main industry in mitigating climate change and improving air quality (Uherek et al., 2010). Sustainable transport systems are those that minimize fossil fuel consumption, emissions and the usage of natural land while providing people with easy access (Verma et al., 2015). According to Kamaruddin et al. (2012) a sustainable transport system closely related to the relationship between environment and satisfaction (Bachok et al., 2014). Accidents, air pollution, damage to infrastructure, noise, negative external costs and social costs are some of the negative impacts of transportation (Verhoef, 1994).

However, passenger miles per gallon on an intercity bus is two times more than the fuel efficiency of intercity rail and four times higher than domestic air carriers (Woldeamanuel, 2012). Environmentally sustainable transport objectives can therefore be achieved if a significant improvement in technological growth, rigorous behavioral adaptations and shifts in economic and spatial constructions are presumed (Steg, 2003). Research studies indicate that by promoting the use of technology problems like air and noise pollution caused by traffic congestion, parking issues and energy consumptions can be reduced (Chapman, 2007; Black and Black, 2009).

20.1.5 Technology

Technology to increase service efficiency and profitability and the change from the service mode to the self-service or e-service mode is also a chance to evaluate and imbue the services' resources and effort maps with sustainability (Wolfson et al., 2014). Service sustainability can be developed with technology, benefiting both clients and service providers (Wolfson et al., 2014). Service providers use information technology to minimize costs and provide their customers with value-added services (Zhu et al., 2002) like information provision related to the route (date and time of travel, bus service provider and distance covered), access to online data and onboard entertainment. Individuals prefer technological solutions over changes in behavior because latter is considered as a restriction on freedom to move (Poortinga et al., 2003). Two of the satisfactory variables for travelers are accurate data and the great graphics–text equilibrium of a website (Cheng, 2011).

20.2 Methodology

This study employs the service quality of intercity bus transport data to explain the application of Partial Least Square Structural Equation Modelling (PLS-SEM) technique. This study is based on primary data which is collected to capture the perceptions of service quality of intercity bus transport from the passengers traveling between urban cities/towns/villages like Bangalore–Mysore, Bangalore–Tumkur, Bangalore–Mangalore, Bangalore–Hubli. Data were collected through the survey method using a structured questionnaire. The survey was conducted between August 2016 and March 2017 at the intercity bus stations mainly in the Bangalore region through convenience sampling technique. The intercity bus passenger data based on service quality and satisfaction were not available from the service providers, and the characteristics of the population were not known; hence, a convenience sampling method was employed. According to KSRTC key statistics (2015), on an average, 26.90 lakh passengers travel in Karnataka every day. Taking 26.90 lakh per day as the sample population, with 95% confidence level and 4% margin of error, the sample size in this study is 605. Statistical techniques used for data analysis involvePLS-SEM, mediation analysis and moderation analysis. The variables and respective items are listed in Appendix 20A.

20.2.1 Path analysis (PLS–SEM)

The purpose of path analysis is to provide magnitude estimates and significance of hypothesized causal links between sets of variables. The method emphasizes the heuristic use of a visual diagram that is technically defined as a path diagram. Path analysis uses standardized coefficients of partial regression as coefficients of impact. The path is said to be significant at the 0.05 level (i.e., its estimated path parameter is significant) if the Critical Ration (CR) is > 1.96.

Compared to correlation analysis, the advantage of path analysis is that it makes it possible to evaluate the relative impact of each explanatory variable on the resulting variables. This is achieved by clarifying the fundamental assumptions of causal relationships first and then clarifying the indirect effect of the explanatory variable. The path analysis method involves assuming there are linear, additive asymmetric associations between that collection of variables which can be evaluated at least on a quasi-interval scale. Each dependent variable in the path diagram is regarded as determined by the preceding variables.

Goodness-of-fit (GoF) tests determine whether to accept or reject the model being tested. These overall fit trials do not show that there are important specific paths within the model. If the model is approved, the path coefficients in the model will be further interpreted (in poor-fit models, "significant" path coefficients are not meaningful). Chi-square, normed fit index (NFI) and root mean square residual (SRMR) are the goodness of fit measures and these concepts are explained in detail in further sections.

20.2.2 PLS-SEM model assessment

Partial Least Squares (PLS) variance-based Structural Equation Modelling (SEM) is commonly recognized in the domains of human resource management, business management studies, including information systems management, operations management, organizational behavior and marketing management (Hair et al., 2012).In this study, PLS-SEM is opted for, for the following reasons: (1) It is relevant for theoretical construction research (Sarsted, 2008; Vinzi et al., 2010). (2) It is regarded as suitable for the examination of complicated causal relationship models (Lowry and Gaskin, 2014). (3) It is a nonparametric approach, with fewer limitations on the data distribution and sample size (Vinzi et al., 2010). The PLS-SEM method is used to evaluate the measurement model (outer model) and structural model (inner model). In simple words, PLS-SEM technique can be used when the sample size is small, theory on applications is inadequate and predictive accuracy is important. Smart-PLS (v.3.2.7) is used to perform PLS-SEM to achieve the previously mentioned objectives.

SEM consists of two models, namely, the measurement model and the structural model. In the measurement model, the relationship of a latent variable (unobserved variable) with a set of measured variables (observed variables) is tested. The structural model, also called the path model, tests the causal relationships among unobserved variables and/or observed variables. In the measurement model, if arrows point from latent variable to the observed variables, then such a model is called a reflective measurement model. If arrows point from observed variables to latent variables, then such a model is called a normative measurement model. The reflective measurement model is commonly used in management sciences and psychological research. In this study, reflective measurement model is used. The steps involved in PLS-SEM analysis are as follows.

20.2.2.1 Step 1: Measurement model assessment

First, the measurement model should be constructed and evaluated for its correctness. The observed variables should be connected to the respective proposed unobserved variables and model constructed, and this is called as outer model. The measurement model is evaluated using internal consistency, construct validity and discriminant validity. In PLS-SEM, the measurement model evaluation involves composite reliability (CR) to assess internal consistency, individual indicator reliability and average variance extracted (AVE) to assess convergent validity (Hair et al., 2016). These concepts are explained in detail in the following sections.

20.2.2.2 Step 1a: Evaluating Variance Inflation Factor (VIF)

Using the variance inflation factor (VIF), the multicollinearity of the independent variables can be evaluated and high multicollinearity indicates that some indicators may be redundant. A better way to verify multicollinearity is to calculate the variance inflation factor (VIF) and multicollinearity detection methods are frequently used. As a rule of thumb, VIF value exceeding 5 or 10 indicates the multicollinearity issue (James et al., 2013).

20.2.2.3 Step 1b: Evaluating reliability for internal consistency

The results consistency across same factor items is evaluated using this type of reliability test (Hair et al., 2013). It describes whether the items in their results that represent a variable are the same (Hair et al., 2006). Reliability for internal consistency is obtained through the use of composite reliability (CR). The value of CR should be greater than 0.7 for acceptance.

20.2.2.4 Step 1c: Evaluating convergent validity

This describes the degree to which a measure correlates favorably with substitutes of the same variable (Hair et al., 2013) and AVE is calculated to determine convergent validity. The value of AVE should be greater than 0.5 for acceptance.

20.2.2.5 Step 1d: Evaluating Discriminant Validity

It is the degree to which a variable is really different from other variables, in terms of how much it correlates with other variables and degree to which indicators represent only a single variable (Hair et al., 2013). Fornell and Larcker's (1981) criterion and cross-loading scores are used to determine discriminant validity. The square root of AVE for all latent variables should be higher than the inter-construct correlations to confirm the discriminant validity of latent variables.

20.2.2.6 Step 1e: Evaluating indicator reliability

Indicator reliability describes the extent of variation in an item explained by a variable (Hair et al., 2013). Higher outer loading on a variable demonstrates a relationship with the associated measure (Hair et al., 2013). Items with the loading greater than 0.70 should be retained, and those with less than 0.40 should be omitted and their impact on the variable's AVE and CR should be analyzed (Hair et al., 2013). The given item should be dropped if the variable's AVE and CR value exceed above the threshold value; otherwise, the item should be retained. The construct's convergent validity is adequate if AVE is less than 0.5 but composite reliability is greater than 0.6 (Fornell and Larcker, 1981). By dropping these items, the CR and AVE can be increased above the threshold values of 0.70 and 0.40, respectively (Fornell and Larcker, 1981; Hair et al., 2013, p. 100). From the results of construct validity presented in Table 20.1, the construct values are above the threshold level and fulfill the criteria of validity.

Table 20.1 Construct Validity

Latent Variables	*Cronbach's Alpha*	*Rho A*	*Composite Reliability*	*Average Variance Extracted (AVE)*
Economic	0.839	0.866	0.892	0.675
Empathy	0.721	0.738	0.843	0.642
Environmental	0.839	0.850	0.892	0.676
External Tangibles	0.720	0.761	0.833	0.625
IR	0.846	0.851	0.907	0.764
Luggage Assurance	0.737	0.743	0.851	0.655
Responsiveness	0.771	0.794	0.851	0.589
STR	0.698	0.724	0.867	0.766
Satisfaction	0.606	0.640	0.790	0.560
Tangibles	0.739	0.744	0.851	0.656
Technology	1.000	1.000	1.000	1.000
Women Friendliness	0.860	0.997	0.932	0.872

Note: IR = Information Reliability; STR = Service Time Reliability.

20.2.2.7 Step 2: Structural model assessment

After evaluating the measurement model by determining the reliability and validity of the latent variables, the next step is to evaluate the structural model. The structural model is evaluated to assess the relationship between endogenous and exogenous variables. In SEM terminology, the independent variable or construct is termed as exogenous variable and the dependent variable is termed an endogenous variable. In PLS-SEM, structural model evaluation involves path coefficients for assessing the significance and relevance of structural model relationships, coefficient of determination (R^2) value for assessing the predictive accuracy of the model, predictive relevance (Q^2) value for assessing the predictive relevance of the model and effect size (F^2) for assessing the significant effect of the exogenous variable on an endogenous variable (Hair et al., 2013).

20.2.2.8 Step 2a: Evaluating path-coefficients

This study presents the path coefficient for the direct relationship between social variables of service quality and satisfaction constructs, with the economic and environmental constructs as mediator variables and technology as the moderator variable. For the 605 data points and 500 samples in this study, nonparametric bootstrapping procedure proposed by Vinzi et al. (2010) is used. In the PLS-SEM technique, bootstrapping is employed to build the structural model and explore the causal relationships between constructs. The primary objective of bootstrapping is to calculate the standard error of coefficient estimates to examine the statistical significance of the coefficient (Vinzi et al., 2010).

20.2.2.9 Step 2b: Assessing effect size (f^2)

Effect size (f^2) is assessed as the changein the R^2 value when a given exogenous variable is kept out of the model; the f^2 size effect is the measure to assess the changeover in R^2 value. The effect size is calculated using the following formula:

Formula:

$$f^2 effectsize = \frac{\left(R^2(incl) - R^2(excl)\right)}{\left(1 - R^2(incl)\right)},$$

where R^2 included and R^2 excluded are the R^2 values of endogenous latent variables when a model includes or excludes the selected exogenous variable (Hair et al., 2013) (Table 20.2).

The effect size (F^2) shows the impact on a particular endogenous variable by a particular latent predictor as shown in Table 20.2. In this study, all the exogenous variables have small effect size (f^2) in explaining the satisfaction, economic and environmental constructs. From the structural relationship results presented in Table 20.2, dimensions namely empathy, external tangibles, information reliability, luggage assurance, responsiveness, service time reliability, bus tangibles and technology significantly impacts the overall satisfaction of the service. Factors, namely, external tangibles, responsiveness significantly impacts economic dimension and bus tangibles, significantly impact the environmental dimension.

20.2.2.10 Step 2c: Assessing coefficient of determination (R^2)

The structural model is assessed using the value of R^2 (Coefficient of determination). This coefficient tests the model's predictive accuracy and is computed as the squared correlation between the predictive and real values of a given endogenous construct. The R^2 scores reflect the combined impacts of the exogenous factors on the endogenous latent factors and the amount of variation in the endogenous constructs described by all the exogenous constructs associated with them (Hair et al., 2013). In our study, the endogenous factors are economic, environmental and satisfaction having adjusted R^2 values 0.150, 0.042 and 0.422, respectively. This indicates the predictive relevance of the structural model established in this study.

20.2.2.11 Step 2d: Assessing predictive relevance (Q^2)

The predictive relevance (Q^2) value of the model is determined by employing blindfolding technique. The predictive relevance for each of the endogenous variables in the model can be cross-validated using blindfolding (Stone-Geisser Q^2 value) (Geisser, 1975). By conducting the blindfolding method with an omission range of 8, all endogenous factors generated

Table 20.2 Results of Structural Relationships

Path	*Original Sample*	*Standard Deviation*	*f^2 Effect Size*	*t Statistics*	*p Values*
Economic -> Satisfaction	0.044	0.039	0.003	1.143	0.254
Empathy -> Economic	0.060	0.048	0.003	1.258	0.209
Empathy -> Environmental	0.052	0.057	0.002	0.915	0.361
Empathy -> Satisfaction	0.249	0.051	0.062	4.869	0.000**
Environmental -> Satisfaction	-0.067	0.040	0.007	1.657	0.098
External Tangibles -> Economic	0.100	0.047	0.010	2.105	0.036*
External Tangibles -> Environmental	-0.040	0.048	0.001	0.846	0.398
External Tangibles -> Satisfaction	-0.086	0.035	0.010	2.440	0.015*
IR -> Economic	0.068	0.055	0.004	1.242	0.215
IR -> Environmental	-0.050	0.050	0.002	1.012	0.312
IR -> Satisfaction	0.116	0.043	0.017	2.667	0.008**
Luggage Assurance -> Economic	0.075	0.051	0.005	1.457	0.146
Luggage Assurance -> Environmental	-0.054	0.056	0.002	0.951	0.342
Luggage Assurance -> Satisfaction	0.181	0.042	0.039	4.343	0.000**
Responsiveness -> Economic	0.228	0.056	0.046	4.060	0.000**
Responsiveness -> Environmental	0.075	0.065	0.004	1.149	0.251
Responsiveness -> Satisfaction	0.147	0.044	0.026	3.341	0.001**
STR -> Economic	0.069	0.043	0.005	1.583	0.114
STR -> Environmental	0.075	0.048	0.005	1.559	0.120
STR -> Satisfaction	0.089	0.039	0.011	2.317	0.021*
T_Eco -> Satisfaction	-0.002	0.038	0.000	0.042	0.966
T_Empathy -> Satisfaction	0.008	0.047	0.000	0.172	0.864
T_Env -> Satisfaction	-0.020	0.037	0.001	0.533	0.594
T_Etan -> Satisfaction	0.083	0.036	0.010	2.315	0.021*
T_IR -> Satisfaction	-0.094	0.042	0.012	2.247	0.025*
T_LA -> Satisfaction	-0.038	0.039	0.002	0.955	0.340
T_Resp -> Satisfaction	-0.017	0.039	0.000	0.433	0.665

(Continued)

Table 20.2 (Continued)

Path	*Original Sample*	*Standard Deviation*	*f^2 Effect Size*	*t Statistics*	*p Values*
T_STR -> Satisfaction	−0.038	0.037	0.002	1.049	0.295
T_Tan -> Satisfaction	−0.069	0.049	0.005	1.426	0.154
T_WF -> Satisfaction	0.024	0.038	0.001	0.630	0.529
Tangibles -> Economic	−0.012	0.050	0.000	0.234	0.815
Tangibles -> Environmental	−0.220	0.056	0.035	3.897	0.000**
Tangibles -> Satisfaction	0.147	0.051	0.024	2.907	0.004**
Technology -> Satisfaction	0.102	0.032	0.017	3.227	0.001**
Women Friendliness -> Economic	0.037	0.041	0.002	0.889	0.375
Women Friendliness -> Environmental	0.014	0.047	0.000	0.309	0.757
Women Friendliness -> Satisfaction	0.037	0.039	0.002	0.948	0.344

Small: $0.0 < F^2$ effect size < 0.15; Medium: $0.15 < F^2$ effect size < 0.35; Large: F^2 effect size > 0.35.
$^{**}p < 0.01$, $^{*}p < 0.05$.

cross-validate redundancy Q^2 values (Hair et al., 2013). In this study, economic, environmental and satisfaction have Q^2 values of 0.097, 0.030 and 0.212, respectively. This indicates the small and medium effect sizes, respectively. Because all the Q^2 values are greater than zero (>0) which defines the predictive significance of the PLS structural model. From the results presented in Table 20.3 it is evident that the model larger variance for satisfaction endogenous construct (42.2%) whereas economic and environmental have 15% and 4.2%, respectively. Based on Q^2 values, the satisfaction construct has medium effect size; the economic and environmental constructs have small effect sizes.

20.2.2.12 Step 3: Mediation analysis

By including a mediator, which is a third explanatory variable, the cause–effect relationship between exogenous and endogenous variable can be explored in mediation analysis (Hair et al., 2013). The bootstrapping strategy in PLS-SEM is appropriate for mediation analysis because it can be applied to small sample sizes and bootstrapping does not assume the sampling distribution of statistics (Hair et al., 2013). To perform the PLS-SEM mediation analysis, first the direct effect (i.e. p.13) on the endogenous

Table 20.3 Model Fit Summary and Results of R^2 and Q^2

Endogenous Latent Variable	*R-Squared*	*R-Squared Adjusted*	Q^2	*Effect Size*[a]
Economic	0.161	0.150	0.097	Small
Environmental	0.055	0.042	0.030	Small
Satisfaction	0.442	0.422	0.212	Medium
Model Fit Summary				
Fit Summary		Saturated Model	Estimated Model	
SRMR		0.064	0.064	
d_ULS		2.547	2.601	
d_G		0.853	0.856	
Chi-Square		3,219.920	3,227.367	
NFI		0.640	0.639	

Note: IR = Information Reliability; STR = Service Time Reliability.Note: Q2 values of 0 and below indicates a lack of predictive relevance (Hair et al., 2016)

[a] *Small:* 0.0 < Q2 effect size < 0.15; Medium: 0.15 < Q2 effect size < 0.35; Large: Q2 effect size > 0.35.

variable by the exogenous variable should be evaluated and this should be significant without having mediator (Zhao et al., 2010). If the direct path is significant, then the mediator variable is included as the next step in the PLS path model and the indirect path significance is evaluated (i.e. p12 * p23). For this situation, each individual path significance, p12 and p23, is needed. After operating the bootstrapping operation, the indirect path can be evaluated and if the indirect impact is discovered to be significant, the mediator will absorb some of the direct path (Table 20.4).

To assess the extent of absorbed direct path, variation accounted for (VAF) is calculated as

$$VAF = (p12 \times p23) / (p13 + p12 \times p23)$$

The mediation effect has following conditions based on the VAF value (Hair et al., 2013, p.224):

$0 < VAF < 0.20$ indicates No Mediation.
$0.20 < VAF < 0.80$ indicates Partial Mediation.
$VAF > 0.80$ indicates Full Mediation.

This study attempts to explain the concept of mediation using intercity transport data and mediation analysis is conducted to assess the extent of indirect impact of mediating factors, namely, economic and environmental, on the relationship between exogenous variables, namely, service-time reliability, information reliability, luggage assurance, tangibles, external tangibles, empathy and responsiveness, and the endogenous variable, namely, satisfaction. From the results, it is evident that economic and environmental constructs do not mediate the service quality variables.

Table 20.4 Mediation Analysis

Factors	*P12*	*P23*	*P13 (Direct Effect)*	*Indirect Effect*	*VAF*	*Mediation*
Economic						
Empathy	0.060	0.044	0.249	-0.001	-0.003248053	No
External Tangibles	0.100	0.044	-0.086	0.007	-0.089887446	No
IR	0.068	0.044	0.116	0.006	0.051982179	No
Luggage Assurance	0.075	0.044	0.181	0.007	0.036757995	No
Responsiveness	0.228	0.044	0.147	0.005	0.033815805	No
STR	0.069	0.044	0.089	-0.002	-0.022078653	No
Tangibles	0.012	0.044	0.147	0.014	0.087754617	No
Women Friendliness	0.037	0.044	0.037	0.001	0.017768725	No
Environmental						
Empathy	0.052	-0.067	0.249	-0.001	-0.003248053	No
External Tangibles	-0.040	-0.067	-0.086	0.007	-0.089887446	No
IR	-0.050	-0.067	0.116	0.006	0.051982179	No
Luggage Assurance	-0.054	-0.067	0.181	0.007	0.036757995	No
Responsiveness	0.075	-0.067	0.147	0.005	0.033815805	No
STR	0.075	-0.067	0.089	-0.002	-0.022078653	No
Tangibles	-0.220	-0.067	0.147	0.014	0.087754617	No
Women Friendliness	0.014	-0.067	0.037	0.001	0.017768725	No

Note: Mediating Variable: Technology; Endogenous Variable: overall satisfaction. IR = Information Reliability; STR = Service Time Reliability

20.2.2.13 Step 4: Moderation analysis

The relationship between the constructs is not constant and depends on the third variable values. Such situation is termed a moderation, and the third variable is called a moderator. In the model, the strength and/or the direction of the relationship between the constructs is changed by the moderator variable. In simple words, the moderator variable influences the strength of the relationship between the constructs/variables. This study attempts to explain the concept of moderation using intercity transport data. In this study, the moderating effect of the technology variable between the constructs is determined. From the results presented in Table 20.2, T_Etan and T_IR have significant impact on satisfaction. Technology moderates the influence of external tangibles and information reliability on satisfaction of the service. This indicates that the strength of the relationship between external tangibles and overall satisfaction, information reliability and satisfaction is influenced by the technology.

20.2.2.14 Step 5: Model fit assessment

In PLS-SEM, GoF is regarded an overall model fit metric, and Table 20.3 presents the model fit summary. Henseler et al. (2014) introduced standardized root mean square residual (SRMR) to assess GoF of a path model. For an approximate model fit, the SRMR threshold value recommended for an estimated model is less than 0.08. As per the findings, SRMR value of the model is 0.064 which is less than 0.08 which indicates model fitness.

20.2.2.15 Step 5a: Assessing chi-square values

It is the most popular fit test, also called discrepancy, written by all computer programs. If there is a good-fit model, then significance should not be there in the chi-square valuation, whereas a significant chi-square indicates a lack of a good-fit model. Chi-square is a metric of "badness of fit" which means significance finding implies that the covariance structure of the specified model is significantly distinct from the covariance matrix noted. If chi-square value is less than 0.05, the model is dismissed. Chi-square assessment is used in our study to verify the model fit.

20.2.2.16 Step 5b: Assessing normed fit index values

Bentler and Bonett (1980) suggested NFI as one of the first fit measures in the SEM literature. It calculates the proposed model's chi-square value and compares it to a significant benchmark. Since the suggested model's chi-square valuation does not in itself provide adequate information to decide model fit, the null model's chi-square value is utilized by NFI as a yardstick. The 1 minus the chi-square value of the suggested model divided by the chi-square values of the null model is termed NFI. The values of NFI range from 0 to 1. The nearer the NFI value to 1, the more aligned it is. Usually, NFI scores above 0.9 are acceptable.

20.2.2.17 Step 5c: Assessing standardized root mean square

The SRMR is a measure of the covariance residuals' mean absolute value. The SRMR relies on the transformation into correlation matrices of the predicted covariance matrix and the sample covariance matrix. The difference between the correlation matrix suggested by the model, and the observed correlation is termed as SRMR. Thus, the median extent of the differences between recorded and anticipated correlations can be assessed as a perfect measure of model fit criteria. A value less than 0.10 or 0.08 is regarded as a good fit (Hu and Bentler, 1999). SRMR is the suitable criterion to prevent model mismatching in PLS-SEM (Henseler et al., 2014).The NFI and SRMR scores are used to verify the model fit in our research and the results of this evaluation are shown in Table 20.3.

20.3 Conclusion

Path analysis with PLS-SEM is the nonparametric approach for structural equation modeling. This approach is gaining recognition in exploring the cause–effect relationship between the latent variables. This method is one of the effective approaches which can be applied in sustainable development research to explore the cause–effect relationship between its constructs. Mediation analysis can be applied to explore how the latent construct or mediating variable influences the relationship between exogenous and endogenous constructs. Similarly, moderation analysis can be applied to explore whether a latent construct influences the relationship between endogenous and exogenous constructs. In this study, the case of sustainable service quality in transport is considered to explore the cause–effect relationship between social constructs such as service quality variables and overall satisfaction with economic and environmental constructs as mediators and technology variable as moderator. Findings reveal that social construct greater influence on overall satisfaction than economic and environmental constructs. Furthermore, it indicates that there is no mediation effect of economic and environmental constructs between social construct (service quality dimensions) and overall satisfaction. It is also evident that technology has a moderation effect on tangibles and information reliability with respect to overall satisfaction. This methodology is one of the new approaches which can be applied to other fields of sustainable development to gain more insights.

References

Bachok, Syahriah, Mariana Mohamed Osman, and Zakiah Ponrahono. "Passenger's Aspiration towards Sustainable Public Transportation System: Kerian District, Perak, Malaysia." *Procedia-Social and Behavioral Sciences* 153 (2014): 553–565.

Bentler, Peter M., and Douglas G. Bonett. "Significance Tests and Goodness of Fit in the Analysis of Covariance Structures." *Psychological Bulletin* 88, no. 3 (1980): 588.

Black, Deborah, and John Black. "A Review of the Urban Development and Transport Impacts on Public Health with Particular Reference to Australia: Trans-Disciplinary Research Teams and Some Research Gaps." *International Journal of Environmental Research and Public Health* 6, no. 5 (2009): 1557–1596.

Black, William R. "Sustainable Transportation: A US Perspective." *Journal of Transport Geography* 4, no. 3 (1996): 151–159.

Brundtland, Gro Harlem. *Report of the World Commission on Environment and Development: "Our Common Future"*. United Nations (1987). Retrieved from https://www.are.admin.ch/are/en/home/sustainable-development/international-cooperation/2030agenda/un-_-milestones-in-sustainable-development/1987--brundtland-report.html

Chapman, Lee. "Transport and Climate Change: A Review." *Journal of Transport Geography* 15, no. 5 (2007): 354–367.

Cheng, Yung-Hsiang. "Evaluating Web Site Service Quality in Public Transport: Evidence from Taiwan High Speed Rail." *Transportation Research Part C: Emerging Technologies* 19, no. 6 (2011): 957–974.

CroninJr, J., and Steven A. Taylor. "Measuring Service Quality: A Reexamination and Extension." *Journal of Marketing* 56, no. 3 (1992): 55–68.

Das, Shreya, and Debapratim Pandit. "Importance of User Perception in Evaluating Level of Service for Bus Transit for a Developing Country like India: A Review." *Transport Reviews* 33, no. 4 (2013): 402–420.

Fornell, Claes, and David F. Larcker. "Evaluating Structural Equation Models with Unobservable Variables and Measurement Error." *Journal of Marketing Research* 18, no. 1 (1981): 39–50.

Fravel, Frederic D. "Intercity Bus Links: Moving into New Territory." *TR News* 225 (2003) 24–29.

Freitas, André Luís Policani. "Assessing the Quality of Intercity Road Transportation of Passengers: An Exploratory Study in Brazil." *Transportation Research Part A: Policy and Practice* 49 (2013): 379–392.

Geisser, Seymour. "The Predictive Sample Reuse Method with Applications." *Journal of the American Statistical Association* 70, no. 350 (1975): 320–328.

Gilbert, David, and Robin K. C. Wong. "Passenger Expectations and Airline Services: A Hong Kong Based Study." *Tourism Management* 24, no. 5 (2003): 519–532.

Haghshenas, Hossein, and Manouchehr Vaziri. "Urban Sustainable Transportation Indicators for Global Comparison." *Ecological Indicators* 15, no. 1 (2012): 115–121.

HairJr, Joseph F., G. Tomas M. Hult, Christian Ringle, and Marko Sarstedt. *A Primer on Partial Least Squares Structural Equation Modeling (PLS-SEM)*. Thousand Oaks, CA: Sage Publications (2016).

Hair, Joe F., Marko Sarstedt, Christian M. Ringle, and Jeannette A. Mena. "An Assessment of the Use of Partial Least Squares Structural Equation Modeling in Marketing Research." *Journal of the Academy of Marketing Science* 40, no. 3 (2012): 414–433.

Hair, Joseph F., Christian M. Ringle, and Marko Sarstedt. "Partial Least Squares Structural Equation Modeling: Rigorous Applications, Better Results and Higher Acceptance." *Long Range Planning* 46, no. 1–2 (2013): 1–12.

Hair, Joseph F., William C. Black, Barry J. Babin, Rolph E. Anderson, and Ronald L. Tatham. "*Multivariate Data Analysis* (Vol. 6). Pearson Prentice Hall Upper Saddle River (2006).

Henseler, Jörg, Theo K. Dijkstra, Marko Sarstedt, Christian M. Ringle, Adamantios Diamantopoulos, Detmar W. Straub, David J. KetchenJr, Joseph F. Hair, G. Tomas M. Hult, and Roger J. Calantone. "Common Beliefs and Reality about PLS: Comments on Rönkkö and Evermann (2013)." *Organizational Research Methods* 17, no. 2 (2014): 182–209.

Hu, Li-tze, and Peter M. Bentler. "Cutoff Criteria for Fit Indexes in Covariance Structure Analysis: Conventional Criteria Versus New Alternatives." *Structural Equation Modeling: A Multidisciplinary Journal* 6, no. 1 (1999): 1–55.

Hubschneider, Ing. "Hans." *Transport Master Planning in the Middle East*, May, 2011. Retrieved from PTV AG at http://www.gabf.ghorfa.de/fileadmin/inhalte/wirtschaftsforum/2011/Sessions/Session 203.

Jackson, Penny. *Value for Money and International Development: Deconstructing Myths to Promote a More Constructive Discussion*. Independent Development Evaluation: African Development Bank (2012).

James, Gareth, Daniela Witten, Trevor Hastie, and Robert Tibshirani. *An Introduction to Statistical Learning*. Vol. 112. New York: Springer (2013).

Kamaruddin, Rohana, Ismah Osman, and Che Anizaliana Che Pei. "Public Transport Services in Klang Valley: Customer Expectations and its Relationship Using SEM." *Procedia-Social and Behavioral Sciences* 36 (2012): 431–438.

Kandampully, Jay. "Firms Should Give Loyalty Before They Can Expect It from Customers." *Managing Service Quality: An International Journal* 7, no. 2 (1997): 92–94.

Karnataka State Road Transport Corporation (KSRTC): Key Statistics (2015). Retrieved in 2015 https://ksrtc.in/pages/key-statistics.html

Khurshid, Rida, Hummayoun Naeem, Sana Ejaz, Faiza Mukhtar, and Taha Batool. "Service Quality and Customer Satisfaction in Public Transport Sector of Pakistan: An Empirical Study." *International Journal of Economics and Management Sciences* 1, no. 9 (2012): 24–30.

Leong, Lai-Ying, Teck-Soon Hew, Voon-Hsien Lee, and Keng-Boon Ooi. "An SEM–Artificial–Neural–Network Analysis of the Relationships between SERVPERF, Customer Satisfaction and Loyalty among Low-Cost and Full-Service Airline." *Expert Systems with Applications* 42, no. 19 (2015): 6620–6634.

Litman, T. "Sustainable Transportation Indicators: A Recommended Program to Define a Standard Set of Indicators for Sustainable Transportation Planning." Sustainable Transportation Indicators Subcommittee of the Transportation Research Board (ADD40 [1]) (2008).

Lowry, Paul Benjamin, and James Gaskin. "Partial Least Squares (Pls) Structural Equation Modeling (SEM) for Building and Testing Behavioral Causal Theory: When to Choose It and How to Use It." *IEEE Transactions on Professional Communication* 57, no. 2 (2014): 123–146.

Mouwen, Arnoud. "Drivers of Customer Satisfaction with Public Transport Services." *Transportation Research Part A: Policy and Practice* 78 (2015): 1–20.

Namboodiri, U. "*State Policies Affecting Competition: Passenger Road Transportation Sector*." April 2007. Retrieved fromhttp://www.cci.gov.in/sites/default/files/2statepolicesaffecting_20080508111218.pdf

OECD. OECD Proceedings: Towards Sustainable Transportation. Conference Organized by the OECD, Hosted by the Government of Canada, Vancouver, British Columbia, March 24–27 (1996). Retrievedfrom:/www.oecd.org/dataoecd/28/54/2396815.pdf

OECD. *Towards Sustainable Transportation*. Paris: OECD Publications (2002).

Parasuraman, A., L. Berry, and V. Zeithaml. "Refinement and Reassessment of the SERVQUAL scale." *Journal of Retailing* 67, no. 4 (2002): 114.

Parasuraman, Ananthanarayanan, Valarie A. Zeithaml, and Leonard L. Berry. "Servqual: A Multiple-Item Scale for Measuring Consumer Perc." *Journal of Retailing* 64, no. 1 (1988): 12.

Ponrahono, Zakiah, Syahriah Bachok, Mansor Ibrahim, and Mariana Mohamed Osman. "Assessing Passengers' Satisfaction Level on Bus Services in Selected Urban and Rural Centres of Peninsular Malaysia." *Procedia-Social and Behavioral Sciences* 222 (2016): 837–844.

Poortinga, Wouter, Linda Steg, Charles Vlek, and Gerwin Wiersma. "Household Preferences for Energy-Saving Measures: A Conjoint Analysis." *Journal of Economic Psychology* 24, no. 1 (2003): 49–64.

Sarstedt, Marko. "A Review of Recent Approaches for Capturing Heterogeneity in Partial Least Squares Path Modelling." *Journal of Modelling in Management* 3, no. 2 (2008): 140–161.

Steg, Linda. "Factors Influencing the Acceptability and Effectiveness of Transport Pricing," Schade, J. and Schlag, B. (Ed.) *Acceptability of Transport Pricing Strategies*, Emerald Group Publishing Limited (2003): pp. 187–202.

Tao, Chi-Chung, and Chia-Chi Hung. "A Comparative Approach of the Quantitative Models for Sustainable Transportation." *Journal of the Eastern Asia Society for Transportation Studies* 5 (2003): 3329–3344.

Uherek, Elmar, Tomas Halenka, Jens Borken-Kleefeld, Yves Balkanski, Terje Berntsen, Carlos Borrego, Michael Gauss, et al. "Transport Impacts on Atmosphere and Climate: Land Transport." *Atmospheric Environment* 44, no. 37 (2010): 4772–4816.

Verhoef, Erik. "External Effects and Social Costs of Road Transport." *Transportation Research Part A: Policy and Practice* 28, no. 4 (1994): 273–287.

Verma, Ashish, T. M. Rahul, and Malvika Dixit. "Sustainability Impact Assessment of Transportation Policies—A Case Study for Bangalore City." *Case Studies on Transport Policy* 3, no. 3 (2015): 321–330.

Vinzi V.E., Trinchera L., Amato S. (2010) PLS Path Modeling: From Foundations to Recent Developments and Open Issues for Model Assessment and Improvement. In: Esposito Vinzi V., Chin W., Henseler J., Wang H. (eds) *Handbook of Partial Least Squares*. Springer Handbooks of Computational Statistics. Springer, Berlin, Heidelberg. (2010): 47–82.

Woldeamanuel, Mintesnot. "Evaluating the Competitiveness of Intercity Buses in Terms of Sustainability Indicators." *Journal of Public Transportation* 15, no. 3 (2012): 5.

Wolfson, Adi, Dorith Tavor, and Shlomo Mark. "CleanServs: Clean Services for a More Sustainable World." *Sustainability Accounting, Management and Policy Journal* 5, no. 4 (2014): 405–424.

Zeithaml, Valarie A. "Consumer Perceptions of Price, Quality, and Value: A Means-End Model and Synthesis of Evidence." *Journal of Marketing* 52, no. 3 (1988): 2–22.

Zhao, Xinshu, John G. LynchJr, and Qimei Chen. "Reconsidering Baron and Kenny: Myths and Truths about Mediation Analysis." *Journal of Consumer Research* 37, no. 2 (2010): 197–206.

Zhu, Faye X., Walter Wymer, and Injazz Chen. "IT-Based Services and Service Quality in Consumer Banking." *International Journal of Service Industry Management* 13, no. 1 (2002): 69–90.

Appendix 20A

Table 20.A1 Variables of Complete Data Model

Name of Variables	*Item Number*	*Items*
Service Time Reliability (STR)	Q11	On time arrival
	Q12	On time departure
Information Reliability (IR)	Q99	Information on arrival and departure
	Q100	Information announcement in bus
	Q101	Information sufficiency
Luggage Assurance	Q35	Luggage place
	Q36	Carrying luggage
	Q38	Luggage safety
Bus Tangibles	Q14	Clean bus
	Q15	Good condition seats
	Q16	Comfortable seats
External Tangibles	Q22	Clean toilets at bus stops
	Q23	Clean drinking water at bus stops
	Q24	Eateries at bus stops
Empathy	Q29	Driver courteousness
	Q30	Conductor courteousness
	Q33	Eateries
Responsiveness	Q53	Conductor individual attention
	Q54	Other staff individual attention
	Q55	Getting reserved seats for special passengers
	Q56	Getting reserved seats for general passengers
Women Friendliness	Q144	Women friendly
	Q145	Safety and Security

(Continued)

Table 20.A1 (Continued)

Name of Variables	*Item Number*	*Items*
Economic	Q57	Bus fare satisfaction
	Q58	Service satisfaction for price paid
	Q59	Service to price paid
	Q60	Service to cost
Environmental	Q62	Air pollution
	Q63	Noise pollution
	Q64	Abnormal vibration
	Q65	Disturbance due to vibration
Overall Satisfaction	Q28	Bus as a safe mode
	Q112	Service consistency
	Q116	Using bus service in future

For Product Safety Concerns and Information please contact our EU representative GPSR@taylorandfrancis.com
Taylor & Francis Verlag GmbH, Kaufingerstraße 24, 80331 München, Germany

www.ingramcontent.com/pod-product-compliance
Ingram Content Group UK Ltd.
Pitfield, Milton Keynes, MK11 3LW, UK
UKHW020156200926
475702UK00016B/2733
* 9 7 8 0 3 6 7 6 9 2 3 1 5 *